CRC Handbook
of
Comparative Opioid
and
Related Neuropeptide
Mechanisms

Volume I

Editor

George B. Stefano, Ph.D.
Chairperson
ADAMHA-MARC Project Director
Associate Professor
Department of Biological Sciences
State University of New York/
College at Old Westbury
Old Westbury, New York

CRC Press, Inc.
Boca Raton, Florida

Library of Congress Cataloging-in-Publication Data
Main entry under title:

Handbook of comparative opioid and related neuropeptide
  mechanisms.

  Bibliography: p.
  Includes index.
  1. Neuropeptides.   2. Opioids.   3. Physiology,
Comparative.   I. Stefano, George B., 1945-   .
QP552.N39H36   1986   591'.0188   85-16225
ISBN 0-8493-3257-5 (v. 1)

Direct all inquiries to CRC Press, Inc., 2000 Corporate Blvd., N.W., Boca Raton, Florida, 33431.

© 1986 by CRC Press, Inc.
International Standard Book Number 0-8493-3257-5 (Volume I)
International Standard Book Number 0-8493-3258-3 (Volume II)

Library of Congress Card Number 0-85-16225
Printed in the United States

SYMPOSIUM:

COMPARATIVE ASPECTS OF OPIOID AND RELATED MECHANISMS

May 28-31, 1984
Campus Center Recital Hall
State University of New York /College at Old Westbury

ORGANIZER:

Dr. George B. Stefano   Department of Biological Sciences, SUNY/College at Old Westbury, New York

ORGANIZATIONAL COMMITTEE:

Dr. Michael J. Greenberg   Director, C. V. Whitney Laboratory, St. Augustine, Florida

Dr. Michael K. Leung   Department of Chemistry, SUNY/College at Old Westbury, New York

Dr. Maynard H. Makman   Departments of Molecular Pharmacology and Biochemistry, Albert Einstein College of Medicine, New York

Dr. Berta Scharrer   Departments of Anatomy and Neuroscience, Albert Einstein College of Medicine, New York

SPONSORS:

State University of New York/College at Old Westbury
State University of New York Research Foundation
National Science Foundation
Alcohol, Drug Abuse, Mental Health Agency
The Upjohn Company
Hoffmann La Roche, Inc.
New England Nuclear
E. I. DuPont De Nemours & Company
Schering-Plough Corporation
Morrell Instrument Company, Inc.
Zeiss, Inc.
Waters Associates

PREFACE

The program of this symposium has focused on various approaches to the elucidation of the general significance, diverse functions, and evolutionary history of neuropeptides. By providing a forum for the coordination of current information on invertebrate as well as vertebrate systems, the symposium highlighted the heuristic value of the comparative approach. Existing data in both invertebrate and vertebrate nervous systems indicated that there exists high stability in short peptide sequences (i.e., enkephalins, Phe-Met-Arg-Phe-NH_2, etc.) over long evolutionary periods. In addition, neurohormonal and neurotransmitter/neuromodulator functions appear to be quite stable, including the interaction between opioid and dopamine, opioid and substance P, etc. The great versatility in the expression of these sequences in different neuronal systems may be due to varying posttranslational processing patterns. It is quite evident that the biosynthesis of opioid and related neuropeptides is governed by "old" genes. Primary emphasis of the symposium was on discussion of these timely topics of growing importance.

THE EDITOR

George B. Stefano, Ph.D. is an Associate Professor of Cell Biology, Chairman of the Biological Sciences Program and the Project Director of the Alcohol, Drug Abuse Mental Health Administration — Minority Access to Research Career Grant at the State University of New York, College at Old Westbury.

Dr. Stefano graduated from Wagner College with a B.Sc. degree in Biology in 1967. He received a M.S. degree in Cellular Physiology from Fordham University in 1969. He received a Certificate in Mammalian Radiation from Oak Ridge Associate Universities in 1971. He received a Ph.D. in Cell Biology from Fordham University in 1973. From 1973 to 1979, Dr. Stefano was an Assistant Professor at The City University of New York, New York City Community College. During this time he also was a research associate in the Biological Sciences Program at Fordham University. In addition, he has worked with several investigators at Einstein College of Medicine since 1979. From 1980 to 1983, Dr. Stefano served in the capacity of Research Coordinator of the Department of Anesthesiology at St. Joseph's Hospital and Medical Center in Paterson, New Jersey. From 1979 to 1982, Dr. Stefano was a Principal Investigator on a NIH-MBRS Grant at Medgar Evers College of the City University of New York. He is the author of more than 70 publications, including original research papers in referred journals and review articles as well as chapters in books. His research efforts have been supported by grants from the National Institutes of Health MBRS as well as ADAMHA-MARC. Dr. Stefano has committed himself to involving undergraduate minority students in research activities in the hope of interesting these students in research careers. He also has received several National Academy of Science Grants involving Research Visits to Eastern Europe. Previously, he has served as an invited speaker at international symposia as well as a member of several organizing committees for such meetings. He is the recipient of an Alumni Achievement Award from Wagner College.

Dr. Stefano's other activities include founding a small neuroscience foundation The East Coast Neuroscience Foundation, Inc., in Northport, New York. He is also co-founder of the international journal "Cellular and Molecular Neurobiology" and serves as a member of the editorial board. He also has served in the capacity of reviewer for NSF and ADAMHA Grant agencies. Dr. Stefano has recently been invited as a major speaker at the Gordon Research Conference on the Biology of Aging. Dr. Stefano is interested in the application of neurochemical and neuropharmacological techniques to study opioid mechanisms in "simple" invertebrate nervous systems. Since these mechanisms as well as these neuropeptides are present in both vertebrate and invertebrate neuronal tissues, it is of interest to study how their metabolism and interaction change in older organisms. It is also of interest to note that these studies provide valuable information regarding the origins and evolvement of neuropeptides in general.

CONTRIBUTORS

Hans-Urs Affolter
Laboratory of Cell Biology
National Institute of Mental Health
Bethesda, Maryland

Heinz Bodenmüller
ZMBH
Universität Heidelberg
Heidelburg, West Germany

H. H. Boer
Department of Zoology
Vrije Universiteit
Amsterdam, The Netherlands

Danièle Brossard
Collaborateur Technique
Biologie Cellulaire et Générale
Université de Champagne Ardenne
Reims, France

Antonio Carolei, M. D.
Assistant Professor
Scienze Neurologiche
University of Rome
Rome, Italy

G. A. Cottrell, Ph.D., D.Sc.
Professor
Department of Physiology and
  Pharmacology
University of St. Andrews
Fife, United Kingdom

Noel Wyn Davies
Department of Physiology and
  Pharmacology
University of St. Andrews
Fife, United Kingdom

Hanne Duve, D.Sc.
School of Biological Sciences
Queen Mary College
University of London
London, England

R. H. M. Ebberink
Department of Zoology
Vrije Universiteit
Amsterdam, The Netherlands

Lee E. Eiden, Ph.D.
Research Scientist
Laboratory of Cell Biology
National Institute of Mental Health
Bethesda, Maryland

Thomas Flanagan, Ph.D.
Neurobiology Group
Cold Spring Harbor Laboratory
Cold Spring Harbor, New York

W. P. M. Geraerts
Department of Zoology
Vrije Universiteit
Amsterdam, The Netherlands

Pierre Giraud
Laboratoire de Médecine Expérimentale
Centre National de la Recherche
  Scientifique
Marseille, France

Michael J. Greenberg, Ph.D.
Scientific Director
C. V. Whitney Laboratory
University of Florida
St. Augustine, Florida

C. J. P. Grimmelikhuijzen, Ph.D.
Zoological Institute
University of Heidelberg
Heidelberg, West Germany

Carola Haas
Department of Neurochemistry
Max-Planck-Institut für Psychiatrie
Martinsried, West Germany

Bente Langvad Hansen, Ph.D.
Institute of Medical Microbiology
University of Copenhagen
Copenhagen, Denmark

Georg Nørgaard Hansen, D.Sc.
Institute of Cell Biology and Anatomy
University of Copenhagen
Copenhagen, Denmark

László Hiripi, Ph.D.
Biological Research Institute
Tihany, Hungary

Maurice Hirst, Ph.D.
Professor of Pharmacology
Department of Pharmacology and
  Toxicology
University of Western Ontario
London, Ontario, Canada

Adair J. Hotchkiss
Clinical Hematology Branch
NIADDK
National Institute of Health
Bethesda, Maryland

Peter P. Jaros, Ph.D.
Department of Zoophysiology
University Oldenburg
Oldenburg, West Germany

Kenneth V. Kaloustian, Ph.D.
Associate Professor of Biology
Chairman, Department of Medical Lab
  Sciences
Quinnipiac College
Hamden, Connecticut

Martin Kavaliers
Department of Psychology
University of Alberta
Edmonton, Alberta, Canada

Daniel L. Kilpatrick, Ph.D.
Roche Research Center
Roche Institute of Molecular Biology
Nutley, New Jersey

Richard M. Kream, Ph.D.
Assistant Professor
Departments of Anesthesiology and
  Biochemistry and Pharmacology
Tufts University School of Medicine
Boston, Massachusetts

Scott M. Lambert
C. V. Whitney Laboratory and
Department of Pharmacology and
  Therapeutics
College of Medicine
University of Florida
Gainesville, Florida

Herman K. Lehman, Ph.D.
C. V. Whitney Laboratory
University of Florida
St. Augustine, Florida

Michael K. Leung, Ph.D.
Associate Professor
Department of Chemistry
State University of New York/College at
  Old Westbury
Old Westbury, New York

**Professor Maynard H. Makman,
  M.D., Ph.D.**
Departments of Biochemistry and
  Molecular Pharmacology
Albert Einstein College of Medicine
Bronx, New York

Vito Margotta
Associate Professor
Biologia Animale e Dell'Uomo
University of Rome
Rome, Italy

Rainer Martin, Ph.D.
Professor
Electron Microscopy
Universität Ulm
Ulm, West Germany

Jeffrey McKelvy, Ph.D.
Professor
Department of Neurobiology and
 Behavior
State University of New York/Stony
 Brook
Stony Brook, New York

Guido Palladini, M. D.
Associate Professor
Scienze Neurologiche
University of Rome
Rome, Italy

David A. Price, Ph.D.
Assistant Research Scientist
C. V. Whitney Laboratory
University of Florida
St. Augustine, Florida

Christian Rémy
Professor
Laboratoire de Neuroendocrinologie
Université de Bordeaux I
Talence, France

Peter J. Rzasa, M.S.
Supervisor, Radioassay Lab
Department of Nuclear Medicine
Hospital of St. Raphael
New Haven, Connecticut

H. Chica Schaller, Ph.D.
ZMBH
Universität Heidelberg
Heidelberg, West Germany

Berta Scharrer, Ph.D.
Departments of Anatomy and
 Neuroscience
Albert Einstein College of Medicine
Bronx, New York

L. P. C. Schot
Department of Zoology
Vrije Universiteit
Amsterdam, The Netherlands

Alan George Scott, M.Sc.
Analytical Scientist
School of Biological Sciences
Queen Mary College
London, England

D. G. Smyth, Ph.D.
Doctor
Laboratory of Peptide Chemistry
National Institute for Medical Research
London, England

George B. Stefano, Ph.D.
Chairperson
ADAMHA-MARC Project Director
Associate Professor
Department of Biological Sciences
State University of New York/
 College at Old Westbury
Old Westbury, New York

Alan Thorpe, Ph.D.
Reader in Zoology
School of Biological Sciences
Queen Mary College
University of London
London, England

Giorgio Venturini
Associate Professor
Biologia Animale e Dell'Uomo
University of Rome
Rome, Italy

Karl-Heinz Voigt, M.D.
Institute for Clinical Pharmacology
Free University Berlin
Berlin, West Germany

Jeffrey D. White
Postdoctoral Research Associate
Department of Neurobiology and
  Behavior
State University of New York/Stony
  Brook
Stony Brook, New York

Birgit Zipser, Ph.D.
Visiting Scientist
Department of Adult Psychiatry
National Institute of Mental Health
Bethesda, Maryland and
Associate Research Scientist
Department of Zoology
University of Maryland
College Park, Maryland

TABLE OF CONTENTS

Biosynthesis (Posttranslational Processing)

NEUROREGULATION OF ADRENAL PROENKEPHALIN GENE EXPRESSION

D. L. Kilpatrick

SUMMARY

Adrenal enkephalin-containing (EC) peptides are known to increase 10- to 15-fold in the rat following surgical denervation of the gland. In this report it is shown that the increase is preceded by a lag of several hours. The major species of newly appearing EC peptides appears to be the intact precursor, proenkephalin. Processing of proenkephalin to smaller EC peptides in the denervated glands is slow and limited. It has been further demonstrated that the steady-state levels of preproenkephalin mRNA are increased severalfold following denervation. Together these data demonstrate that the observed increase in rat adrenal EC peptides occurs entirely by a pretranslational mechanism. An interesting observation was the dissociation between the effects of denervation on EC peptides and catecholamines indicating that the effects of denervation on these two classes of hormones are discoordinate.

INTRODUCTION

Over the past several years our laboratory has been involved in elucidating the biosynthetic pathway of the opioid peptides, Met-enkephalin and Leu-enkephalin. By using a combination of protein microsequencing and recombinant DNA techniques, it has been shown that in bovine adrenal medulla, these pentapeptides are derived from a multienkephalin precursor, proenkephalin. Proenkephalin not only gives rise to Met- and Leu-enkephalin, but to a spectrum of different-sized enkephalin-containing (EC) peptides. While some of the larger EC peptides may only serve as precursors to enkephalins, others possess significant opioid activity and may themselves be important physiologically. Although the structure of proenkephalin and the EC peptides derived from it have been well characterized, little is known about the mechanisms involved in the biosynthesis and processing of this gene product. In most animals the adrenal medulla is rich in EC peptides that are derived from proenkephalin. Several years ago Schultzberg et al.[1] noted that immunohistochemical staining for enkephalin was highly positive in adrenals of all the animal species they tested except for the rat. They noted, however, that when the adrenals were denervated by sectioning the splanchnic nerve, there was an increase in the number of chromaffin cells that stained positively for enkephalins. The rat adrenal is also unusual in that the composition of proenkephalin-derived products is normally quite different from those found in adrenals of other species.[2] Free enkephalins have been shown to be present in only trace amounts along with comparable amounts of EC peptides in the range of 20 kdaltons. Intermediate-sized EC peptides, which are prominent in adrenals of other species, could not be detected in rat adrenals. In a previous study, we examined the effects of denervation on the size distribution of rat adrenal EC peptides.[2] The first change observed after denervation was an increase in the largest EC peptides (~20 kdaltons). By 48 hr this had increased to over 10 times control values, with little change in free enkephalins. After 96 hr there was an increase in free enkephalins and intermediate-sized EC peptides. Those findings indicated that denervation markedly influences the metabolism of adrenal proenkephalin. The present studies were undertaken to investigate this phenomenon further in an attempt to elucidate the mechanisms involved. Since enkephalins and their congeners are localized in adrenal chromaffin granules along with catecholamines,[1] the effects of denervation of both classes of compounds were compared.

MATERIALS AND METHODS

Sprague-Dawley rats (males, 150 to 200 g), with their left adrenal glands denervated by transection of the splanchnic nerve, were obtained from Taconic Farms (Germantown, N.Y.). At specific time intervals, rats were killed by decapitation and the glands were removed immediately and frozen in liquid nitrogen. In experiments involving intervals shorter than 24 hr all the rats were killed at Taconic Farms at the specific times and the glands were removed and shipped to us in dry ice. Pooled glands were homogenized in 1 M acetic acid/ 0.1% 2-mercaptoethanol, containing 1 μg/mℓ each of pepstatin, leupeptin, and phenylmethylsulfonyl fluoride. Glands were homogenized in 2 mℓ of the solvent in a glass/glass homogenizer or a Polytron homogenizer. The homogenates were centrifuged for 30 min at 800 g and the supernatants were centrifuged for 16 hr at 100,000 g. Total EC peptides were estimated on aliquots of the extracts by radioimmunoassay for Met-enkephalin following treatment with trypsin and carboxypeptidase B to release the pentapeptide from larger EC peptides. EC peptides were separated using a Sephadex G-75 column (1.2 × 95 cm) that was developed with 1 M acetic acid/0.1% 2-mercaptoethanol at 20 mℓ/hr, collecting 2- to 4-mℓ fractions. Each fraction was assayed for Met-enkephalin following treatment with trypsin and carboxypeptidase B. Met-enkephalin-Arg^6-Phe^7 was assayed by radioimmunoassay with a specific antiserum perpared in this laboratory.[3] There was no significant crossreactivity (<0.02%) with Met-enkephalin, Leu-enkephalin, or the octapeptide, Met-enkephalin-Arg^6-Gly^7-Leu^8. Peptide B, which contains the heptapeptide at its carboxyl terminus, crossreacts 120%. Crude proenkephalin preparations crossreact to a small extent.

The heptapeptide Met-enkephalin-Arg^6-Phe^7 was cleaved out from the high-molecular-weight EC peptide(s) by treatment with Lys-C endoproteinase (Boehringer-Mannheim Biochemicals). Release of the octapeptide [Met]enkephalin-Arg^6-Gly^7-Leu^8 (as the Arg$^\circ$ octapeptide) required an additional incubation step with carboxypeptidase B (Boehringer-Mannheim Biochemicals).

For the determination of the heptapeptide Met-enkephalin-Arg^6-Phe^7 in column fractions, aliquots were evaporated to dryness and redissolved in 100 $\mu\ell$ of 0.2 M N-ethylmorpholine acetate buffer, pH 7.4. To each aliquot, 20 $\mu\ell$ of Lys-C endoproteinase (1.0 mg/mℓ in the same buffer) was added and the samples incubated for 16 hr at 37°C. The samples were then heated to 100°C for 30 min to destroy residual enzymatic activity. Arg$^\circ$ heptapeptide released was assayed by specific radioimmunoassay for the heptapeptide.

For catecholamine assay, three innervated and three denervated glands at each time point were homogenized in 5% trichloroacetic acid/100 mM HCl. Samples were centrifuged for 10 min at 1700 g and catecholamines were determined[4] on 20$\mu\ell$ aliquots taken from the supernatants. Protein was assayed as described before.[5]

For studies of preproenkephalin mRNA, intact and denervated glands at each time point were pooled separately and poly(A)$^+$ RNA was isolated using the guanidinium thiocyanate/ CsCl gradient procedure[6] followed by oligo-dT column chromatography.[7] Northern analysis of poly(A)$^+$ RNA fractions was carried out according to Thomas[8] using Baby Blots (Bethesda Research Laboratories, Bethesda, Md.) Preproenkephalin mRNA was quantified by RNA blotting.[9] Rat brain poly(A)$^+$ RNA was used as a standard. Quantitation of autoradiograms was carried out using a soft laser scanning densitometer (LKB Instruments, Inc., Rockville, Md.) and a Minigrator® (Spectra-Physics, Piscataway, N.J.). Densities were converted to picograms of hybridized preproenkephalin cDNA by scintillation counting of rat brain poly(A)$^+$ RNA dots.

Two different ^{32}P-labeled proenkephalin DNA probes were used in these experiments. A synthetic 30-nucleotide-long oligodeoxyribonucleotide which is complementary to a portion of both the human and bovine preproenkephalin mRNA sequences was labeled at its 5′-end using polynucleotide kinase (P.L. Biochemicals, Milwaukee, Wis.) and γ-^{32}P-ATP (Amer-

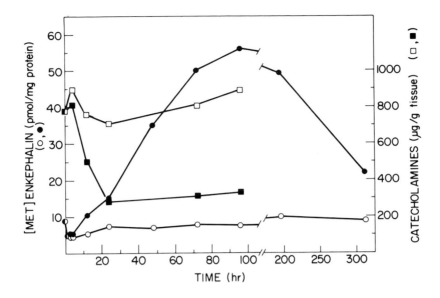

FIGURE 1.   Changes in Met-EC peptides and catecholamines in rat adrenal glands following denervation. Aliquots of the tissue extracts of innervated (open circles) and denervated (closed circles) glands (10 to 100 $\mu\ell$) were lyophilized, redissolved in 100 $\mu\ell$ of 0.2 *M* *N*-ethyl-morpholine acetate buffer, pH 8.0, and treated with trypsin and carboxypeptidase B. Met-enkephalin released was assayed by a specific radioimmunoassay. Aliquots of tissue extracts in 5% TCA per 100 m*M* HCl (20 $\mu\ell$) of the innervated (open boxes) and denervated (closed boxes) glands were assayed for catecholamines (see Section III). (From Fleminger, G., Lahm, H. -W., and Udenfriend, S., *Proc. Natl. Acad. Sci. U.S.A.*, 81, 3587, 1984. With permission.)

sham; >5000 Ci/mmol). Human preproenkephalin cDNA (919 bp) was isolated from the plasmid pHPE-9 (a gift from Drs. M. Comb and E. Hebert). The cDNA was labeled by nick translation using $5'$-[α-^{32}P]dCTP (Amersham; 3000 Ci/mmol) to a specific activity of 1 to 4 $\times$ 10^8 cpm/μg.

Total RNA and poly(A)$^+$ RNA were measured by UV absorption (1 A_{260}/mℓ = 40 μg). DNA was measured in guanidinium thiocyanate extracts using the diaminobenzoic acid fluorescence assay.[10]

RESULTS

Effects of Denervation of EC Peptides and Catecholamines

The levels of both Met-EC peptides and catecholamines in denervated and innervated rat adrenal glands (Sprague-Dawley) as a function of time following denervation (30 min to 2 months) are shown in Figure 1. Thirty minutes after surgery the levels of EC peptides in both denervated and innervated glands decreased by 50%, from 9.3 to about 5.0 pmol/mg of extractable protein. Similar effects on EC peptides were observed in sham-operated rats (data not shown). These early changes are apparently a result of the surgical procedure. After an initial lag period of about 4 to 10 hr, EC peptides in the denervated glands started to rise and attained levels 10- to 15-fold higher than the contralateral glands several days after denervation. There was a slow return to control values such that 13 days after denervation, the glands still contained about three times as much EC peptides as did their contralateral innervated controls. Following the initial fall there was little further change in EC peptides in glands from controls or sham-operated animals. The effects on catecholamines were quite different. Following denervation, adrenal catecholamines decreased over the first 24 hr and plateaued at about 30 to 40% of control values.

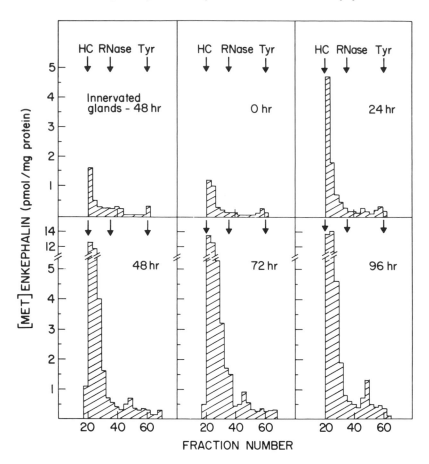

FIGURE 2.   Size distribution of Met-EC peptides in tissue extracts at various time intervals after denervation on a Sephadex G-75 column. Aliquots (20 to 200 μℓ from each fraction were lyophilized, treated with trypsin and carboxypeptidase B, and assayed for Met-enkephalin as described in Figure 1. All the panels, except the first two, show Met-EC peptides in denervated rat adrenal glands. The column was calibrated with markers: HC, cyt C, and Tyr represent elution positions of hemocyanin, cytochrome C, and tyrosine, respectively. (From Fleminger, G., Lahm, H. -W., and Udenfriend, S., *Proc. Natl. Acad. Sci. U.S.A.*, 81, 3587, 1984. With permission.)

Nature of the EC Peptides in Denervated Glands

The composition of EC peptides at various times after denervation was determined by chromatographing the extracts on a Sephadex G-75 column and assaying eluted fractions for free Met-enkephalin produced by treatment with trypsin and carboxypeptidase B. As shown in Figure 2, the bulk of the released Met-enkephalin was at all times associated with an EC peptide(s) that eluted at the void volume of the column.

Some processing of the high-molecular-weight material was observed with time. A peak of Met-EC peptide(s) in the region of $M_r = 3000$ to 6000 was observed which continually increased during the experiment (Figure 3, upper panels). When the same fractions were assayed for Met-enkephalin-Arg[6]-Phe[7] immunoreactivity (without treatment with trypsin and carboxypeptidase B), a peak was observed at the same elution volume (Figure 3, lower panels). From its elution volume and its crossreactivity with the highly specific antiserum to the heptapeptide, it seemed likely that this material was peptide B, the 3.6 kdaltons carboxyl terminal sequence of proenkephalin.[12] Peptide B, which was originally isolated from bovine adrenal medulla, contains the heptapeptide sequence and has been shown to

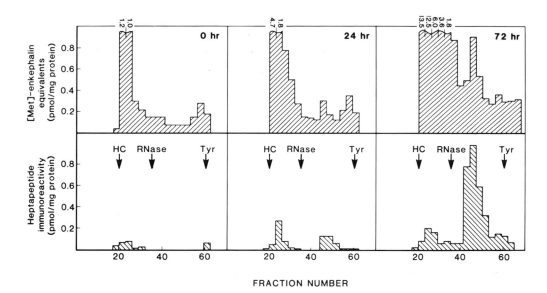

FIGURE 3. Processing of proenkephalin in denervated rat adrenal glands. Aliquots (200 μℓ) of the fractions of a Sephadex G-75 column (see Figure 2) were lyophilized and assayed with specific antibodies to the heptapeptide Met-enkephalin-Arg[6]-Phe[7] (lower panels). The upper panels are an enlargement of the corresponding panels in Figure 2. Numbers represent the amounts of Met-enkephalin that are higher than 2 pmol/mg protein. (From Fleminger, G., Lahm, H. -W., and Udenfriend, S., *Proc. Natl. Acad. Sci. U.S.A.*, 81, 3587, 1984. With permission.)

crossreact with the heptapeptide antiserum at least as well as the heptapeptide itself (data not shown). The very gradual increase in "peptide B" after denervation indicates that the processing of proenkephalin in these glands is slow and is limited to the carboxyl end of the proenkephalin molecule at these times. Since "peptide B" was not found in extracts of innervated glands or at shorter time intervals after denervation, it may represent an intermediate which under steady-state conditions in vivo is further processed to the heptapeptide and perhaps free enkephalin or is excrete from the cells. There were concomitant increases in the fraction containing free enkephalins and small EC peptides (Figure 2). This fraction increased more slowly; even after 13 days it represented only 10% of the total EC peptides.

Identification of Proenkephalin in Denervated Glands

The slow rate of processing to low-molecular-weight EC peptides suggested that the majority of the high-molecular-weight, enkephalin-containing fraction represented proenkephalin itself. Since the largest adrenal EC protein to be previously identified was the 18.2 kdaltons protein from bovine adrenal medulla,[13] it was of interest to further characterize the EC protein(s) from denervated rat adrenals.

One hundred and fifty glands from seventy-five bilaterally denervated rats were removed 48 hr after surgery. The glands were extracted and the extract (3.7 nmol of EC peptides) was applied to a Sephadex G-75 column. The bulk of the EC proteins was eluted close to the void volume and contained material larger than 20 kdaltons (not shown). An aliquot of the high-molecular-weight material eluting from the G-75 column was hydrolyzed with trypsin and carboxypeptidase B and the hydrolysate was applied to a calibrated RP-18 column; immunoassay of the fractions obtained from this column for Met- and Leu-enkephalin showed a ratio of 5.5:1. Proenkephalin contains six Met-enkephalin residues and one Leu-enkephalin.[14]

The remaining pooled high molecular fractions from the G-75 column were applied to an RP-8 HPLC column. A single peak of EC material was eluted at a concentration of 32 to 34% 1-propanol (not shown). By contrast, the 18.2 kdaltons EC protein from bovine adrenal

Table 1
**EC PEPTIDES GENERATED FROM PROENKEPHALIN
AFTER TREATMENT WITH LYS-C ENDOPROTEINASE
AND CARBOXYPEPTIDASE B**

Peptide	Expected number of equivalents[a]	Found[b]
Met-enkephalin	1	0.71
Arg° Met-enkephalin	2	3.1
Arg° Leu-enkephalin	1	0.92
Arg° Met-enkephalin-Arg^6-Gly^7-Leu^8	1	0.71
Arg° Met-enkephalin-Arg^6-Phe^7	1	0.90
Peptide E derivative[c]	1	0.95

[a]   Numbers are based on the sequences of bovine and rat proenkephalin.
[b]   Each number represents the average of two independent experiments. One equivalent of proenkephalin is defined as the amount of Met-enkephalin equivalents, divided by 6, determined by a radioimmunoassay after treatment with trypsin and carboxypeptidase B.
[c]   Arg°-Des(Arg° Leu-enkephalin) peptide E (proenkephalin $_{181-200}$).

medulla is known[13] to elute from a similar column at a concentration of 18 to 20% propanol. Assay of Met- and Leu-enkephalin within the peak of the EC protein gave an almost constant ratio of Met- to Leu-enkephalin of approximately 6 for each fraction.

Based on Met-enkephalin and protein assays, we were able to characterize the putative proenkephalin by its content of EC peptides. Of the six Met-enkephalin sequences in proenkephalin, four are bracketed by sequences of dibasic amino acids. The other two, those contained in the octapeptide Met-enkephalin-Arg^6-Gly^7-Leu^8 and in the heptapeptide Met-enkephalin-Arg^6-Phe^7, are followed by a single arginine residue. Treatment with trypsin and carboxypeptidase B, which is commonly used to release free Met-enkephalin from larger EC peptides, cleaves the hepta- and octapeptides. To gain more information on the nature of the high-molecular-weight EC peptide, we treated it with Lys-C endoproteinase which cleaves peptide bonds specifically on the carboxyl terminal side of lysine residues. When an aliquot of the active peak from the RP-8 column was incubated with this enzyme, a quantitative, time-dependent release of heptapeptide immunoreactivity was observed. After 16 hr of incubation, 132 fmol of heptapeptide immunoreactivity was released from a sample containing a total of 880 fmol of Met-enkephalin which gives a ratio of 6.7 of Met-enkephalin to heptapeptide. The expected ratio for intact proenkephalin is 6:1. Complete digestion of bovine or human proenkephalin with Lys-C endoproteinase and carboxypeptidase B should lead to the mixture of EC peptides shown in Table 1. When the putative proenkephalin that had been partially purified was subjected to hydrolysis by Lys-C endoproteinase followed by hydrolysis by carboxypeptidase B, it yielded the pattern of EC peptides on an RP-18 column shown in Figure 4 and Table 1. The peaks of Arg° hepta- and Arg° octapeptide were identified by their crossreactivity with the specific hepta- and octapeptide antisera, respectively, and by the release of stoichiometric amounts of Met-enkephalin after treatment of each peak with trypsin and carboxypeptidase B. Arg° Met- and Arg° Leu-enkephalin were identified by their crossreactivity with Met- and Leu-enkephalin antisera, respectively, and by the generation of almost stoichiometric amounts of free Met- and Leu-enkephalin after incubation with trypsin, as determined by HPLC and immunoreactivity. A small peak eluting close to the elution position of Met-enkephalin-Arg^6 was identified as oxidized Arg° Met-enkephalin. Treatment of this peak with 0.5% 2-mercaptoethanol for 30 min at 100°C resulted in the elution of this material as Arg° Met-enkephalin when rerun on the same RP-18 column. A peak of Met-enkephalin immunoreactivity was also eluted from the column with 20%

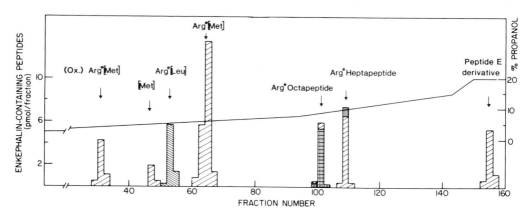

FIGURE 4.   Mapping of EC peptides derived from "proenkephalin". An aliquot (1.0 mℓ) equivalent to 8 pmol of proenkephalin of the pooled fractions from RP-8 chromatography (Figure 3) was evaporated to dryness and redissolved in 200 μℓ of 0.2 *M* N-ethylmorpholine acetate buffer, pH 7.4. Lys-C endoproteinase (40 μℓ, 1 mg/mℓ) was added and the solution was incubated at 37°C for 16 hr. The solution was then heated to 100°C and after cooling purified carboxypeptidase B (40 μg/mℓ) was added. The solution was again incubated at 37°C for 4 hr. 2-Mercaptoethanol was then added (0.5% final concentration) and the solution was heated to 100°C for 30 min. After cooling ^{125}I-Met-enkephalin was added and the solution was applied to an RP-18 column (0.46 × 10 cm) and developed with a gradient of 1-propanol. Each fraction (1 mℓ) was dried down and redissolved in 200 μℓ of 0.2*M* N-ethylmorpholine buffer, pH 8.0. Met- and Leu-enkephalin immunoreactivities were assayed after treatment with trypsin and carboxypeptidase B. Octapeptide and heptapeptide were assayed directly without any further enzymatic treatment. "Peptide E derivative" refers to the peptide Arg°-Des(Arg° Leu-enkephalin) peptide E. (From Fleminger, G., Howells, R. D., Kilpatrick, D. L., and Udenfriend, S., *Proc. Natl. Acad. Sci. U.S.A.,* 81, 7885, 1984.)

propanol. This material, which contained no Leu-enkephalin, heptapeptide, or octapeptide immunoreactivity, was assumed to be Arg°-Des(Arg° Leu-enkephalin) peptide E (proenkephalin$_{181-200}$). Because of the presence of an Arg-Arg sequence following the Met-enkephalin sequence in this peptide, it is not cleaved further by Lys-C endoproteinase. Altogether these data demonstrate that the major EC protein that accumulates in denervated rat adrenal glands is the authentic mature precursor, proenkephalin.

Effect of Denervation on Rat Adrenal Preproenkephalin mRNA

The time course for the increase in EC peptides as well as the identification of proenkephalin as the major EC protein that accumulates in rat adrenals suggested that denervation induces an increase in rat adrenal proenkephalin biosynthesis. However, in the absence of turnover studies, one cannot entirely rule out other explanations for this increase, such as a selective decrease in the rate of opioid peptide secretion or degradation. To determine whether the changes in rat adrenal EC peptides following splanchnic nerve sectioning reflected an increase in preproenkephalin mRNA, adrenal poly(A)$^+$ RNA fractions from rats that had been unilaterally denervated for 24 hr were submitted to Northern analysis (Figure 5). Using a synthetic ^{32}P-labeled oligodeoxyribonucleotide probe, no preproenkephalin mRNA was detectable in the poly(A)$^+$ RNA fraction from innervated adrenal glands. This is undoubtedly due to the relatively low abundance of preproenkephalin mRNA normally present in rat adrenal glands as well as to the use of a nonhomologous DNA probe. However, in poly(A)$^+$ RNA fractions from denervated rat adrenal glands, preproenkephalin mRNA was markedly increased and therefore readily detectable (Figure 5). The time course for the denervation-induced increase in preproenkephalin mRNA was examined using an RNA blotting procedure. As early as 12 hr after denervation, an increase in preproenkephalin mRNA was already evident (Table 2). Preproenkephalin mRNA continued to increase with time, reaching maximal values 24 to 48 hr postdenervation and declining thereafter. The

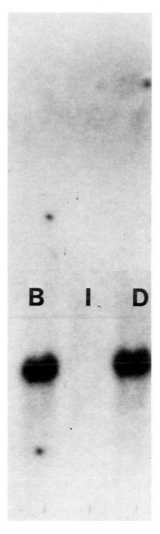

FIGURE 5.   Effect of splanchnicectomy on rat adrenal preproenkephalin mRNA. Innervated and denervated adrenal glands (0.3 to 0.4 g) were obtained from rats that had been unilaterally denervated for 24 hr. Poly(A)⁺ RNA was prepared as described in Section III and the entire sample from each group of glands was subjected to electrophoresis on 1% agarose gels containing 10 mM methyl mercury hydroxide [30 μg from innervated adrenals (I) and 6 μg from denervated adrenals (D)]. As a standard, 6 μg of rat brain poly(A)⁺ RNA (B) was included. After transfer onto nitrocellulose, the filters were baked and then prehybridized for 4 to 12 hr at 30°C in 0.7 M NaCl/0.07 M Na citrate/0.5 M Na phosphate (pH 7)/0.02% BSA/0.02% Ficoll®/0.02% polyvinylpyrrolidone/0.1 mg/mℓ salmon sperm DNA and 50% formamide. Hybridization with ³²P-labeled synthetic oligodeoxynucleotide (2 × 10⁷ cpm) was carried out at 30°C for 12 to 18 hr in prehybridization buffer containing 10% dextran sulfate. The filter was washed 4 times in 0.3 M NaCl/0.03M Na citrate/0.1% SDS at room temperature for 10 min each, followed by 4 washes at 50°C in 15 mM NaCl/1.5 mM Na citrate/0.1% SDS for 30 min. An overnight autoradiogram was then taken at −70°C. Similar results were obtained in five different experiments. (From Kilpatrick, D. L., Howells, R. D., Fleminger, G., and Udenfriend, S., *Proc. Natl. Acad. Sci. U.S.A.*, 81, 7221, 1984.)

Table 2
EFFECT OF DENERVATION ON RAT ADRENAL
NUCLEIC ACIDS AND EC PEPTIDES

	Total RNA (μg)	Poly (A)⁺ RNA (μg)	Proenkephalin mRNA (pg equivalents[a])	EC peptides (pmol/mg protein)
0 hr				
Inn	56	4.9	0.0	5
Den	53	4.9	0.1	5
4 hr				
Inn	56	5.5	0.3	5
Den	51	4.8	0.2	6
12 hr				
Inn	59	6.3	0.3	6
Den	28	2.6	1.4	10
24 hr				
Inn	62	5.8	0.0	7
Den	21	1.9	1.7	15
48 hr				
Inn	68	7.4	0.2	6
Den	29	2.6	2.8	35
72 hr				
Inn	66	7.6	0.2	7
Den	38	2.6	1.7	49
96 hr				
Inn	66	7.2	0.3	6
Den	46	3.4	0.6	55

Note: Data are not corrected for recovery. RNA values are expressed per adrenal gland.

[a] Values refer to picograms of nick-translated human proenkephalin cDNA hybridized per adrenal gland; 1 μg of rat brain poly(A)⁺ RNA standard bound 1.8 pg of cDNA.

appearance of the specific mRNA preceded that of the gene product, proenkephalin, and started to decline while proenkephalin was still increasing (Table 2).

The nonhomologous (human) cDNA probe used in these experiments did not permit accurate measurement of the small amounts of preproenkephalin mRNA in innervated rat adrenals. The observed values for innervated glands (Table 2) thus may only represent upper limits. However, the increase in preproenkephalin mRNA on denervation was so great that experimental variations in the control innervated glands are not a factor. It is worth noting that the content of preproenkephalin mRNA expressed per individual adrenal gland as well as per microgram of rat adrenal poly(A)⁺ RNA increased following splanchnicectomy (Table 2). The effect of denervation on preproenkephalin mRNA is therefore selective and not due to a general increase in the population of adrenal poly(A)⁺ RNA. In fact, the amounts of total adrenal RNA as well as poly(A)⁺ RNA actually decreased by as much as 75% after denervation (Table 2). The decreases in RNA content followed a time course essentially identical to that seen for the increase in preproenkephalin mRNA. The extent to which this reflects changes in medullary vs. cortical RNA content has not been determined. The DNA content of rat adrenal glands remained unchanged by denervation, however (data not shown). The observed decreases in RNA content therefore represent changes in RNA metabolism and are not indicative of cell death.

DISCUSSION

The studies described here clearly demonstrate that splanchnic nerve sectioning causes a marked increase in the steady-state levels of rat adrenal preproenkephalin mRNA. The magnitude of this increase (greater than tenfold) is comparable to that observed for adrenal proenkephalin, indicating that the effects of denervation result from a pretranslational mechanism. Whether this involves an increase in the rate of transcription of the preproenkephalin gene or is the result of changes in one or more posttranscriptional events (e.g., mRNA stabilization or processing) remains to be determined.

Another unanswered question concerns the cellular mechanisms which are responsible for this response. One possibility is that denervation causes an induction of preproenkephalin gene expression by stimulating the release of acetylcholine from splanchnic nerve endings. Tyrosine hydroxylase has been shown to undergo transsynaptic induction in response to increased splanchnic nerve activity.[17] However, the fact that catecholamines do not increase following denervation of rat adrenal glands (Figure 1) suggests that a transsynaptic mechanism does not explain our findings. An alternative explanation is that splanchnic innervation of the rat adrenal medullar exerts an inhibitory trophic influence on preproenkephalin gene expression and that denervation leads to derepression. The expression of the skeletal muscle proteins apolipoprotein A_1[18] and tropomyosin and troponin[19] also appear to be under innervation-dependent repression that is reversible by denervation. Analogous mechanisms may operate in rat adrenal chromaffin cells to regulate preproenkephalin gene expression. Based on such a model, the transient nature of the increase in preproenkephalin mRNA following denervation is somewhat unexpected. However, this could conceivably represent tissue reinnervation in these experiments. Innervation-dependent repression may also explain the unusually small amounts of EC peptides in the rat adrenal gland as compared to other species. The factor(s) responsible for this putative repression remains to be determined. Recent studies[20] suggest that acetylcholine may play such a role in the rat adrenal gland, but other repressive trophic factors must still be considered. The large decreases in RNA content following splanchnic nerve sectioning indicate that neural input also has a dramatic effect on rat adrenal RNA metabolism. Similar decreases in both total and poly(A)$^+$ RNA populations have been observed in skeletal muscle after denervation.[21]

ACKNOWLEDGMENTS

I would like to acknowledge the contributions of Drs. G. Fleminger, R. Howells, H. - W. Lahm, and S. Udenfriend to this work.

REFERENCES

1. **Schultzberg, M., Lundberg, J. M., Hokfelt, T., Terenius, L., Brandt, J., Elde, R. P., and Goldstein, M.,** Enkephalin-like immunoreactivity in gland cells and nerve terminals of the adrenal medulla, *Neuroscience*, 3, 1169, 1978.
2. **Lewis, R. V., Stern, A. S., Kilpatrick, D. L., Gerber, L. D., Rossier, J., Stein, S., and Udenfriend, S.,** Marked increases in large enkephalin-containing polypeptides in the rat adrenal gland following denervation, *J. Neuroscience*, 1, 80, 1981.
3. **Kilpatrick, D. L., Howells, R. D., Lahm, H. -W., and Udenfriend, S.,** Evidence for a proenkephalin-like precursor in amphibian brain, *Proc. Natl. Acad. Sci. U.S.A.*, 80, 5772, 1983.
4. **Merrills, R. J.,** A semiautomatic method of determination of catecholamines, *Anal. Biochem.*, 6, 272, 1963.
5. **Bohlen, P., Stein, S., Dairman, W., and Udenfriend, S.,** Fluorometric assay of proteins in the nanogram range, *Arch. Biochem. Biophys.*, 155, 213, 1973.
6. **Chirgwin, J. M., Przybyla, A. E., MacDonald, R. J., and Rutter, W. J.,** Isolation of biologically active ribonucleic acid from sources enriched in ribonuclease, *Biochemistry*, 18, 5294, 1979.
7. **Norgard, M. V., Tocci, M. J., and Monahan, J. J.,** On the cloning of eukaryotic total poly(A)-RNA populations in *Escherichia coli*, *J. Biol. Chem.*, 225, 7665, 1980.

8. **Thomas, P. S.,** Hybridization of denatured RNA and small DNA fragments transferred to nitrocellulose, *Proc. Natl. Acad. Sci. U.S.A.,* 77, 5201, 1980.

9. **White, B. A. and Bancroft, F. S.,** Cytoplasmic dot hybridization, *J. Biol. Chem.,* 267, 8569, 1982.

10. **Vytasek, R.,** A sensitive fluorometric assay for the determination of DNA, *Anal. Biochem.,* 120, 243, 1982.

11. **Fleminger, G., Lahm, H. -W., and Udenfriend, S.,** Changes in rat adrenal catecholamines and proenkephalin metabolism after denervation, *Proc. Natl. Acad. Sci. U.S.A.,* 81, 3587, 1984.

12. **Stern, A. S., Jones, B. N., Shively, J. E., Stein, S., and Udenfriend, S.,** Two adrenal opioid polypeptides: proposed intermediates in the processing of proenkephalin, *Proc. Natl. Acad. Sci. U. S. A.,* 78, 1962, 1981.

13. **Kilpatrick, D. L., Jones, B. N., Lewis, R. V., Stern, A. S., Kojima, K., Shively, J. E., and Udenfriend, S.,** An 18,200-dalton adrenal protein that contains four [Met]enkephalin sequences, *Proc. Natl. Acad. Sci. U.S.A.,* 79, 3057, 1982.

14. **Udenfriend, S. and Kilpatrick, D. L.,** Biochemistry of the enkephalins and enkephalin-containing peptides, *Arch. Biochem. Biophys.,* 221, 309, 1983.

15. **Fleminger, G., Howells, R. D., Kilpatrick, D. L., and Udenfriend, S.,** Intact proenkephalin is the major enkephalin-containing peptide produced in rat adrenal glands following denervation, *Proc. Natl. Acad. Sci. U.S.A.,* 81, 7885, 1984.

16. **Kilpatrick, D. L., Howells, R. D., Fleminger, G., and Udenfriend, S.,** Denervation of rat adrenal glands markedly increases preproenkephalin mRNA, *Proc. Natl. Acad. Sci. U.S.A.,* 81, 7221, 1984.

17. **Guidoitti, A. and Costa, E.,** Trans-synaptic regulation of tyrosine 3-mono-oxygenase biosynthesis in rat adrenal medulla, *Biochem. Pharmacol.,* 26, 817, 1977.

18. **Shackelford, J. E. and Lebherz, H. G.,** Regulation of Apolipoprotein A₁ synthesis in avian muscles, *J. Biol. Chem.,* 258, 14829, 1983.

19. **Matsuda, R., Spector, D., and Strohman, R. C.,** Denervated skeletal muscle displays discoordinate regulation for the synthesis of several myofibrillar proteins, *Proc. Natl. Acad. Sci. U.S.A.,* 81, 1122, 1984.

20. **Bohn, M. C., Kessler, J. A., Golight, L., and Black, I. B.,** Appearance of enkephalin-immunoreactivity in rat adrenal medulla following treatment with nicotinic antagonists or reserpine, *Cell Tissue Res.,* 231, 469, 1983.

21. **Metafora, S., Felsani, A., Cotrufo, R., Tajana, G. F., DiIorio, G., DelRio, A., DePrisco, P. P., and Esposito, V.,** Neural control of gene expression in the skeletal muscle fibre: the nature of the lesion in the muscular protein-synthesizing machinery following denervation, *Proc. R. Soc. London Ser. B,* 209, 239, 1980.

BIOSYNTHESIS AND PROCESSING OF OPIOID PEPTIDES IN THE CENTRAL NERVOUS SYSTEM

Jeffrey D. White and Jeffrey F. McKelvy

SUMMARY

There is now a wealth of information with respect to the location of enkephalins within the central nervous system, the relative amounts of the enkephalins in the central and peripheral nervous system, and the types of enkephalin receptors on opioid receptive cells. However, there is relatively little known about the precise biosynthesis of the enkephalins, and virtually nothing is known regarding the biosynthesis and processing of the enkephalin precursor in a defined projection system within the central nervous system. We present here data that demonstrate the feasibility of studying the in vivo biosynthesis, posttranslational processing, and transport of Met-enkephalin-containing peptides in the central nervous system of individual freely moving rats. ^{35}S-methionine was infused above the paraventricular nucleus of the hypothalamus through stereotaxically implanted cannulae. Upon completion of the labeling period, the posterior pituitary and median eminence were dissected and the radiolabeled peptides extracted. In the presence of unlabeled carrier peptides, ^{35}S-labeled Met-enkephalin-containing peptides were purified to constant specific activity using sequential HPLC purifications coupled with chemical modifications. This technique allowed us to examine enkephalin biosynthesis in normal and lactating female rats and to determine that the production of the enkephalins is stimulated during lactation.

INTRODUCTION

Since their isolation and identification in 1975,[1] the enkephalins have been among the best-studied neuropeptides within the nervous system. The subsequent isolation and characterization of β-endorphin in proopiomelanocortin[2] and of the dynorphin peptides[3,4] have led to the realization that the enkephalins are part of an opioid peptide family. The availability of commercially synthesized enkephalins has fostered the development of radioimmunoassays as well as the immunohistochemical mapping of the location of the enkephalins throughout the central and peripheral nervous system.[5,6] The availability of synthetic peptides has also allowed the initiation of investigation into the nature of opiate receptors within the brain. An ever-increasing number of studies have demonstrated the existence of multiple opiate receptor subtypes within the brain whose specificity is dependent upon the amino acid sequence of the ligand (see Reference 7 for a review). The immunoassay techniques have also demonstrated that the opioids are widely but not uniformly distributed throughout the nervous system. Their anatomical localization correlates very well with the localization of the opiate receptors.[8]

The recent advances in the techniques of molecular biology have permitted the cDNA cloning of mRNAs which specifically direct the synthesis of separate precursors for each of the opiate peptides, i.e., proopiomelanocortin,[9] proenkephalin[10-12] and prodynorphin, or proenkephalin B.[13] These cDNA sequences have demonstrated that β-endorphin, Met-enkephalin, and the dynorphins are all derived from different precursor proteins. However, Leu-enkephalin can be derived from either the polyenkephalin precursor or the prodynorphin precursor.[10-13] These cDNA sequences have also shown that the enkephalins are flanked in the precursor by pairs of basic amino acids. Thus, their liberation from the precursor could be by a tryptic-like cleavage followed by a carboxypeptidase cleavage.[14] The recent isolation and characterization of enzymes from the brain that have these cleavage specificities appear to support this hypothesis.[15]

In addition to the enkephalin pentapeptides, several other enkephalin-containing peptides have been isolated and sequenced, primarily from adrenal tissue.[16-20] Figure 1 shows the protein structure for human proenkephalin predicted from its cDNA sequence. The precursor codes for six copies of Met-enkephalin, two of which are extended beyond the pentapeptide sequence prior to the paired basic residues, one copy of Leu-enkephalin and an N-terminal sequence which is devoid of opiate sequences. Also shown in this figure are the locations of the larger enkephalin-containing sequences referred to above. Each of the larger peptides that has been isolated and sequenced to date has been found exactly within the predicted amino acid sequence. The position of the recently isolated octapeptide metorphamide,[21] also known as adrenorphin,[22] is also indicated.

As is evident from the discussion above, a great deal has been learned about the biochemistry of the enkephalins; however, virtually nothing is known regarding the precise steps in enkephalin biosynthesis or regarding the regulation of enkephalin biosynthesis and processing. Rossier et al.[24] have shown that cultured adrenal chromaffin cells incorporated ^{35}S-methionine into a protein that, upon subsequent enzymatic treatment, liberated authentic Met-enkephalin. Other authors have been able to demonstrate synthesis of Met-enkephalin and Leu-enkephalin in vitro by guinea pig striatal slice preparations.[25] More recently, La Gamma et al.[26] have used radioimmunoassay techniques to examine the increased synthesis of Leu-enkephalin in explant cultures of rat adrenal medullae. However, in no instance has the biosynthesis and transport of Met-enkephalin and its analogs been studied in a defined neuronal projection system within the central nervous system. It is to this problem that the studies described below have been addressed. The discussion above clearly demonstrates that a complete understanding of the role of the enkephalins in any projection system is fundamentally grounded in an understanding of the nature of the enkephalin-containing peptides present in and released by that system.

MATERIALS AND METHODS

For these studies, we have used our previously published methods for studying neuropeptide biosynthesis in vivo.[27] Sprague-Dawley rats were stereotaxically cannulated with 26 gauge cannula above the PVN. After an overnight recovery period, 500 µCi of ^{35}S-methionine (>1000 Ci/mmol) was infused in artificial extracellular fluid using an Alzet osmotic mini-pump delivery system at a flow rate of 1 µℓ/hr; infusion was for 2 hr. After the appropriate chase period, during which time artificial extracellular fluid was pumped through the cannulae, the animals were killed by decapitation and the tissue was harvested. Normal females were in diestrus during the label infusion and lactating females were at 9 to 12 days of lactation. The median eminence and posterior pituitary were dissected from fresh tissue and the PVN was punched from frozen coronal brain sections to verify proper cannulae placement. Tissue was homogenized in the buffer of Bennett et al.[28] and carrier peptides were added at this time. The sulfoxidation reaction was in 100 µℓ of 2 *M* acetic acid with 1% hydrogen peroxide at room temperature for 60 min; sulfone formation was performed in 50% freshly prepared performic acid in a total volume of 100 µℓ for 60 min on ice. All analyses were performed on a Hewlett-Packard® model 1082B high performance liquid chromatograph. Reverse-phase columns were from Brownlee. The gradient profiles are described in the figure legends.

RESULTS

Figure 2 is a flow diagram summarizing the labeling and purification protocol we have used in these experiments. We chose the PVN projection system for several reasons: (1) recent data have suggested that the enkephalins are present in oxytocin-secreting neurons;[29,30]

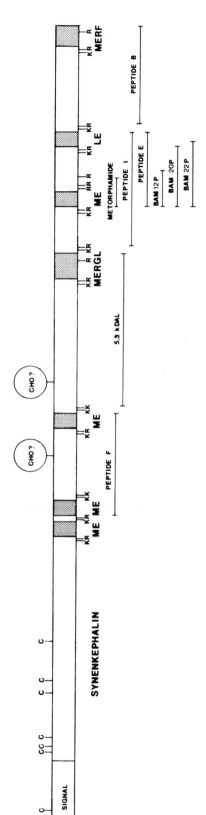

FIGURE 1.   The structure of the human proenkephalin precursor. Shown above is the predicted structure of the precursor protein to the enkephalins as deduced from the nucleotide sequence of the cDNA.[10] Also shown are the locations of Met-enkephalin (ME), Leu-enkephalin (LE), Met-Arg-Gly-Leu-enkephalin (MERGL), and Met-Arg-Phe-enkephalin (MERF). The seven cysteine residues in the N-terminal region of the protein are indicated (C) as are the two potential glycosylation sites (CHO). The putative cleavage signals of arginine (R) and lysine (K) are indicated. Also shown are the large enkephalin-containing peptides that have been isolated from bovine adrenal tissue[16-20] as well as the amidated octapeptide metorphamide[21,22] and synenkephalin.[23]

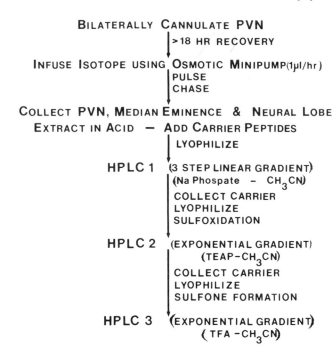

BILATERALLY CANNULATE PVN
>18 HR RECOVERY

INFUSE ISOTOPE USING OSMOTIC MINIPUMP(1μl/hr)
PULSE
CHASE

COLLECT PVN, MEDIAN EMINENCE & NEURAL LOBE
EXTRACT IN ACID — ADD CARRIER PEPTIDES
LYOPHILIZE

HPLC 1 (3 STEP LINEAR GRADIENT)
(Na Phospate – CH₃CN)
COLLECT CARRIER
LYOPHILIZE
SULFOXIDATION

HPLC 2 (EXPONENTIAL GRADIENT)
(TEAP–CH₃CN)
COLLECT CARRIER
LYOPHILIZE
SULFONE FORMATION

HPLC 3 (EXPONENTIAL GRADIENT)
(TFA –CH₃CN)

FIGURE 2. A flow diagram of the method for studying neuropeptide biosynthesis in vivo. All HPLC purification steps use elution of the peptides from reverse-phase columns. CH₃CN = acetonitrile; TEAP = triethylammonium phosphate;[27] TFA = trifluoroacetic acid; PVN = paraventricular nucleus of the hypothalamus. Sulfoxide and sulfone formation are described in Section III.

(2) we have previously been successful in labeling oxytocin and vasopressin in this system[31] (3) both the median eminence and the posterior pituitary are regions that are relatively pure nerve terminals and are devoid of cell bodies, which would tend to simplify the number of possible explanations for any given set of results; and (4) this system can be physiologically stimulated. For these experiments, we have chosen to initially examine only the small Met-enkephalin peptides that have been isolated. There are currently no data available regarding the precise sequence of the nonopiate sequences in the rat proenkephalin precursor, hence synthetic peptides to these regions are unavailable. However, the study of the biosynthesis of these smaller peptides does allow us to draw conclusions regarding the processing of the enkephalin precursor, and any identification of either the heptapeptide or of the octapeptides is definitive evidence of biosynthesis of the enkephalin precursor by cells in the PVN.

Figure 3 shows the chromatographic profile from the tissue extract of a median eminence sample. This animal had been infused with 500 μCi of ³⁵S-methionine for 2 hr, then an additional 4 hr was allowed for synthesis, processing, and transport of the enkephalin peptides from the PVN to the median eminence. As the figure clearly demonstrates, each of the enkephalin peptides was well separated from all others and from both oxytocin and vasopressin. As is apparent from the figure, clear peaks of radioactivity comigrating with the peptides of interest were not evident at this initial stage of purification. Thus, the second and third steps of HPLC purification were needed. The recovery of peptide carrier at this first step was always quite good (>90% for each of the peptides).

Each of the carrier peptides was collected separately in no more than 2 mℓ from this first purification step, lyophilized then resuspended in the presence of hydrogen peroxide to generate the methionine sulfoxide form of the peptide. This chemical modification of the peptides results in a change in their chromatographic mobility and thus should accomplish

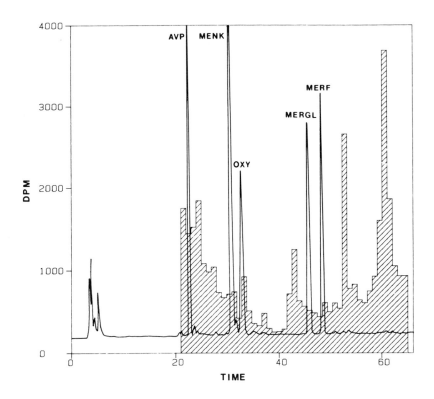

FIGURE 3. HPLC and radioactivity profiles from a median eminence tissue sample. After lyophilization, the sample extract was resuspended in 0.1% TFA and applied to the first HPLC purification system. Buffer A was 0.1 M NaH$_2$PO$_4$ + 0.1 M H$_3$PO$_4$; buffer B was CH$_3$CN. The gradient was 5 to 17% B over 20 min, a 12-min isocratic step, 17 to 27% B over the following 20 min then to 37% B by the end of the gradient. The flow rate was 1.0 mℓ/min. One-minute fractions were collected and 100 $\mu\ell$ was removed for radioactivity determination. The radioactivity plotted is the total ^{35}S-radioactivity present in each fraction. The ultraviolet absorbance, monitored at 210 nm, is also plotted. AVP = arginine vasopressin; OXY = oxytocin; MENK = Met^5-enkephalin; MERGL = Met^5-Arg^6-Gly^7-Leu^8-enkephalin; MERF = Met^5-Arg^6-Phe^7-enkephalin.

an additional purification step, since contaminating peptides which comigrated with the peptide of interest in the initial HPLC step would be predicted to undergo a different change in chromatographic mobility. Coupled with this chemical modification, the second HPLC purification step used an ion-pairing buffer system and used an exponential gradient shape, both of which should facilitate the separation of closely related peptides. Again, after this step, each of the carrier peptide peaks was collected and lyophilized.

Following resuspension, each of the carrier peptides was treated with performic acid to form the methionine sulfone form of the peptide. The rationale for this step is the same as for the second purification step. Figure 4 shows the chromatographic profile for each of the three Met-enkephalin peptides. In this final HPLC step, the buffer system was 0.1% trifluoracetic acid and the gradient was again exponentially shaped. The figure demonstrates that a clear peak of radioactivity comigrates with each of the carrier peptides. In this figure, all the radioactivity values have been corrected for the recovery of the standard to permit direct comparison of the peptides. In all cases, the peptides had been purified to constant specific activity, i.e., there was a constant ratio of radioactivity to integrated carrier peak area between steps 2 and 3. The recovery of the initially added carrier peptide standard averaged 53% for all three peptides through the final step.

These data represent the first demonstration of the biosynthesis of the polyenkephalin

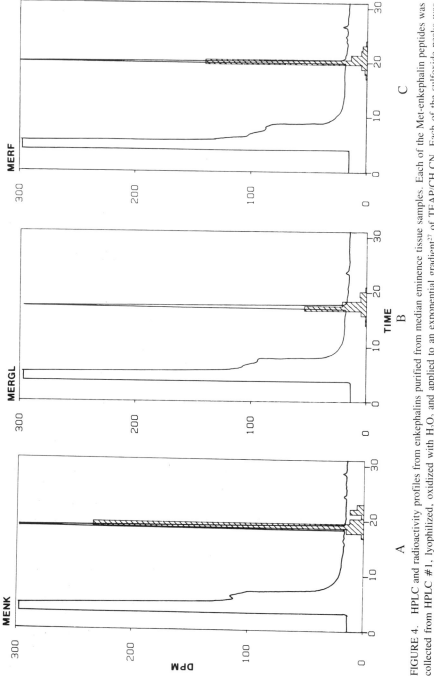

FIGURE 4.   HPLC and radioactivity profiles from enkephalins purified from median eminence tissue samples. Each of the Met-enkephalin peptides was collected from HPLC #1, lyophilized, oxidized with H_2O_2, and applied to an exponential gradient[27] of TEAP/CH_3CN. Each of the sulfoxide peaks was collected, lyophilized, and converted to the methionine-sulfone form with performic acid, as described in Section III. The samples were then applied to the final HPLC step, shown above. Buffer A is 0.1% TFA and buffer B is 0.1% TFA in CH_3CN. In all cases the gradient is from 5 to 23% B over 40 min, after 5 min at initial conditions. Note the peak of radioactivity that comigrates with the carrier peptide peak in all cases. The radioactivity plotted has been corrected for the recovery of the carrier peptide to permit direct comparison. (A) Met[5]-sulfone enkephalin. (B) Met[5]-sulfone MERGL. (C) Met[5]-sulfone MERF.

Table 1
THE RATIOS OF ³⁵S-RADIOACTIVITY IN EACH OF THE MET-ENKEPHALIN PEPTIDES PURIFIED FROM THE MEDIAN EMINENCE TISSUE EXTRACTS

	Pulse	Chase	MENK:MERGL:MERF
Normal	2	4	3.1:0.3:1
Lactating	2	2	3.3:0.5:1
	2	4	2.8:0.4:1
	2	6	3.4:0.4:1

Note: The values above represent the mean of the ratios of radioactivity, after correction for recovery of the initially added standard, for the three peptides after the final HPLC purification step. In all cases, the value for MERF has been arbitrarily set equal to one. There were two to four animals in each group.

precursor in a defined set of cell bodies and the transport of mature Met-enkephalin-containing peptides to a defined terminal field. We have not shown the data from peptides purified from the posterior pituitary. Generally, the absolute amount of radioactivity recovered from the posterior pituitary samples was severalfold less than that from the median eminence samples. Thus, identification of the presence of the peptides was always possible; however, absolute and accurate quantitation of the radioactivity was not.

As alluded to above, one primary reason for choosing this projection system was that it was amenable to physiological stimulation. It has long been known that the suckling of pups during lactation specifically initiates the release of oxytocin from the posterior pituitary of the mother to cause the milk let-down reflex in the mammillary epithelia.[32] It has been shown that the oxytocin neurons in the PVN fire in a burst preceding this release, whereas the vasopressin neurons are silent to this stimulus.[32] It is this specificity of response that makes the system particularly appealing. Thus, in these experiments the biosynthesis and processing of the enkephalin precursor were compared between normal (one day postestrus) and lactating (9 to 12 days of lactation) females. It should be noted that the cannulation procedure did not appear to affect the normal females with respect to their progression through the estrous cycle nor were the lactating mothers adversely affected. The data shown in Figures 3 and 4 were from a lactating female.

In these experiments we initially chose to examine the processing of the polyenkephalin precursor by comparing the ratio of radioactivity present in the three Met-enkephalin-containing peptides described above. This method facilitates the comparison of data between animals, because we do not need to rely on reproducibility of absolute amounts of radioactivity between animals. Furthermore, these ratios are at least partially a direct reflection of the processing of the polyenkephalin precursor since all three peptides are solely derived from it.

The data in Table 1 summarize our findings from normal and lactating animals. The data from the animals exposed to a 2-hr pulse of radiolabel suggest that there is little, if any, difference in the relative ratios of the three peptides between the two physiological states. The finding that the ratio remained constant in the lactating animals at each of the chase times examined suggested that this ratio was established within the 2-hr chase period. It is noteworthy that the absolute amount of incorporation into the peptides was greater in all of the lactating animals than in any of the normal animals, suggesting that during lactation enkephalin biosynthesis was stimulated.

DISCUSSION

The results presented above demonstrate that the investigation of the biosynthesis and processing of the proenkephalin precursor is possible within a defined projection system within the CNS in single animals. These studies are the first to demonstrate this. Using our previously described methods for studying neuropeptide biosynthesis in vivo,[27] we have purified MENK, MERGL, and MERF to constant specific activity using sequential HPLC and chemical modification techniques. In preliminary experiments (data not shown), we have also identified radiolabeled peptide that comigrates with synthetic metorphamide through these three purification steps. These studies have not relied on immunobinding techniques for any of the purification steps and, therefore, these methods should be suitable for application in any neurochemistry laboratory.

There are several relevant conclusions to be drawn from the results presented above. As has been shown for other peptides,[33,34] enkephalin appears to be transported from the PVN to the median eminence and posterior pituitary via the fast component of axonal transport, because radiolabeled peptides were purified from both samples after a total elapsed time of 4 hr. Furthermore, the absence of a significant change in the radioactivity ratios of the three enkephalin peptides during any of the labeling times examined suggests that posttranslational processing of the enkephalin precursor occurs at the level of the cell body and/or during axonal transport.

The recovery of radiolabeled enkephalins from the median eminence suggests that they may serve in a neuroendocrine role. This suggestion is strengthened by the observation that enkephalin biosynthesis appeared to be increased during lactation. There is also a great deal of experimental evidence to suggest that the enkephalins are functionally important in this system. Several studies have shown that the opiates can modulate the release of oxytocin or vasopressin.[35] Furthermore, there is evidence for a role of both the opiates[36] and for oxytocin[37] in the control of prolactin secretion. Thus, the enkephalin and oxytocin neurons in the PVN may be sensitive to either changing steroid concentrations or to direct neuronal stimulation or both and may be regulated in concert during lactation.

The presence of metorphamide in the terminals of the median eminence and the finding that the ratio of radioactivity of MENK:MERGL:MERF was not 4:1:1 both suggest that the processing of the enkephalin precursor in this projection system is not simply at paired basic residues. The cleavage of metorphamide from the precursor involves cleavage at a single arginine residue followed by C-terminal amidation. Thus, the processing of the enkephalin precursor must be complex in this system. With the elucidation of the precise structure for the rat enkephalin precursor, the presence of the larger enkephalin-containing peptides can be examined for further insight into the regulation of the processing of the enkephalin precursor in this system.

The data presented in Table 1 indicate that the processing of the enkephalin precursor appeared to be invariant between lactating and normal animals. These data cannot yet determine that there is no regulation of processing of the precursor during lactation. However, they do suggest that any regulation is at the level of the larger peptides, e.g., in the portion of the precursor coding for MERGL and metorphamide. It should be noted that BAM 12P has been measured in the pituitary and hypothalamus by immunological techniques.[38]

The finding that the level of incorporation of radioactivity into enkephalin-containing peptides was lower in the posterior pituitary than in the median eminence was unexpected, as judged from the levels of enkephalin in the two regions. There are several simple explanations for this finding including (1) a difference in the turnover rate of the peptides between the posterior pituitary and the median eminence; (2) a difference in the processing of the enkephalin precursor between the neurons projecting to the pituitary vs. those that project to the median eminence; (3) the relative amount of enkephalin peptides is less in those neurons projecting to the posterior pituitary than in those terminating in the median eminence; and (4) a larger proportion of enkephalin-synthesizing neurons in the PVN project to the median eminence than to the posterior pituitary. Further biochemical studies and anatomical pathway tracing techniques will be necessary to rule out any of these explanations.

Future experiments to be conducted in this system will include the determination of the peptides that are released from the median eminence in vitro. These experiments will allow us to determine if the peptide ratio observed in the terminals is a direct reflection of the peptides that are released upon stimulation. Measurement of mRNA levels in the PVN from normal animals and from lactating animals will allow us to determine if the increased synthesis of the enkephalins during lactation is reflected in an increase in mRNA content. Additionally, we can examine the possible regulation of this system under acute stimulation by the direct injection of steroid.

Finally, experiments such as those described here must be conducted in other projection systems. Projection-specific processing has been described in the POMC system.[39] It will be essential for a complete understanding of the physiological role of the enkephalins to determine the nature of the enkephalin peptides present in specific terminal fields in the central nervous system. The elucidation of the mechanisms of neuronal communication will come about only after an understanding of the nature of protein interactions between neurons.

ACKNOWLEDGMENTS

We thank Dr. J. E. Krause for many helpful discussions and J. Demian for photographic assistance. This work was supported by NSF BNS-7684506 to Jeffrey F. McKelvy.

REFERENCES

1. **Hughes, J., Smith, T. W., Kosterlitz, H. W., Fothergill, L. A., Morgan, B. A., and Morris, H. R.,** Identification of two related pentapeptides from the brain with potent opiate agonist activity, *Nature (London)*, 258, 577, 1974.
2. **Li, C. H. and Chung, D.,** Isolation and structure of an untriakontapeptide with opiate activity from camel pituitary glands, *Proc. Natl. Acad. Sci. U.S.A.*, 73, 1145, 1976.
3. **Goldstein, A., Tachibana, S., Lowney, L. I., Hunkapillar, M., and Hood, L.,** Dynorphin-(1-13), an extraordinary potent opioid peptide, *Proc. Natl. Acad. Sci. U.S.A.*, 76, 6666, 1979.
4. **Weber, E., Evans, C. J., and Barchas, J. D.,** Predominance of the amino-terminal octapeptide fragment of dynorphin in rat brain regions, *Nature (London)*, 299, 77, 1982.
5. **Miller, R. J., Chang, K. -J., Cooper, B., and Cuatrecasas, P.,** Radioimmunoassay and characterization of enkephalins in rat tissues, *J. Biol. Chem.*, 253, 531, 1978.
6. **Williams, R. G. and Dockray, G. J.,** Distribution of enkephalin related peptides in rat brain: immuno-histochemical studies using antisera to Met-enkephalin and Met-enkephalin Arg⁶Phe⁷, *Neuroscience*, 9, 563, 1983.
7. **Snyder, S. H.,** Drug and neurotransmitter receptors in the brain, *Science*, 224, 22, 1984.
8. **Atweh, S. F. and Kuhar, M. J.,** Distribution and physiological significance of opioid receptors in the brain, *Br. Med. Bull.*, 39, 47, 1983.
9. **Nakanishi, S., Inoue, A., Kita, T., Nakamura, M., Chang, A. C. Y., Cohen, S. N., and Numa, S.,** Nucleotide sequence of cloned cDNA for bovine corticotropin-β-lipotropin precursor, *Nature (London)*, 278, 423, 1979.
10. **Comb, M., Seeeburg, P. H., Adelman, J., Eiden, L., and Herbert, E.,** Primary structure of the human Met- and Leu-enkephalin precursor and its mRNA, *Nature (London)*, 295, 663, 1982.
11. **Noda, M., Furutani, Y., Takahishi, H., Toyosato, M., Hirosa, T., Inayama, S., Nakanishi, S., and Numa, S.,** Cloning and sequence analysis of cDNA for bovine adrenal proenkephalin, *Nature (London)*, 295, 202, 1982.
12. **Gubler, U., Seeburg, P., Hoffman, B. J., Gage, L. P., and Udenfriend, S.,** Molecular cloning establishes proenkephalin as precursor of enkephalin-containing peptides, *Nature (London)*, 295, 206, 1982.

13. **Kakidani, H., Furutani, Y., Takahashi, H., Noda, M., Morimoto, Y., Hirose, T., Asai, M., Inayama, S., Nakanishi, S., and Numa, S.,** Cloning and sequence analysis of cDNA for porcine β-neo-endorphin/dynorphin precursor, *Nature (London)*, 298, 245, 1982.

14. **Fleminger, G., Ezra, E., Kilpatrick, D. L., and Udenfriend, S.,** Processing of enkephalin-containing peptides in isolated bovine adrenal chromaffin granules, *Proc. Natl. Acad. Sci. U.S.A.*, 80, 6418, 1983.

15. **Fricker, L. D. and Snyder, S. H.,** Purification and characterization of enkephalin convertase, an enkephalin-synthesizing carboxypeptidase, *J. Biol. Chem.*, 258, 10950, 1983.

16. **Jones, B. N., Stern, A. S., Lewis, R. V., Kimura, S., Stein, S., Udenfriend, S., and Shively, J. E.,** Structure of two adrenal polypeptides containing multiple enkephalin sequences, *Arch. Biochem. Biophys.*, 204, 392, 1980.

17. **Mizuno, K., Minamino, N., Kangawa, K., and Matsuo, H.,** A new family of "Big" Met-enkephalins from bovine adrenal medulla: purification and structure of docosa- (BAM 22P) and eicosapeptide (BAM 20P) with very potent opiate activity, *Biochem. Biophys. Res. Commun.*, , 97, 1283, 1980.

18. **Kilpatrick, D. L., Taniguchi, T., Jones, B. N., Stern, A. S., Shively, J. E., Hullihan, J., Kimura, S., Stein, S., and Udenfriend, S.,** A highly potent 3200-dalton adrenal opioid peptide that contains both a [Met]- and [Leu]enkephalin sequence, *Proc. Natl. Acad. Sci. U.S.A.*, 78, 3265, 1981.

19. **Stern, A. S., Jones, B. N., Shively, J. E., Stein, S., and Udenfriend, S.,** Two adrenal opioid polypeptides: proposed intermediates in the processing of proenkephalin, *Proc. Natl. Acad. Sci. U.S.A.*, 78, 1962, 1981.

20. **Jones, B. N., Shively, J. E., Kilpatrick, D. L., Kojima, K., and Udenfriend, S.,** Enkephalin biosynthetic pathway: a 5300-dalton adrenal polypeptide that terminates at its COOH end with the sequence [Met]enkephalin-Arg-Gly-Leu-COOH, *Proc. Natl. Acad. Sci. U.S.A.*, 79, 1313, 1982.

21. **Weber, E., Esch, F. S., Bohlen, P., Paterson, S., Corbett, A. D., McKnight, A. T., Kosterlitz, H. W., Barchas, J. D., and Evans, C. J.,** Metorphamide: isolation, structure, and biologic activity of an amidated opioid octapeptide from bovine brain, *Proc. Natl. Acad. Sci. U.S.A.*, 80, 7362, 1983.

22. **Matsuo, H., Miyata, A., and Mizuno, K.,** Novel C-terminally amidated opioid peptide in human phaeochromocytoma tumour, *Nature (London)*, 305, 721, 1983.

23. **Liston, D. R., Vanderhaeghen, J. J., and Rossier, J.,** Presence in brain of synenkephalin, a proenkephalin-immunoreactive protein which does not contain enkephalin, *Nature (London)*, 302, 62, 1983.

24. **Rossier, J., Trifaro, J. M., Lewis, R. V., Lee, R. W. H., Stern, A., Kimura, S., Stein, S., and Udenfriend, S.,** Studies with [^{35}S]methionine indicate that the 22,000-dalton [Met]enkephalin-containing protein in chromaffin cells is a precursor of [Met]enkephalin, *Proc. Natl. Acad. Sci. U.S.A.*, 77, 6889, 1980.

25. **McKnight, A. T., Hughes, J., and Kosterlitz, H. W.,** Synthesis of enkephalins by guinea-pig striatum *in vitro*, *Proc. R. Soc. Lond. B.*, 205, 199, 1979.

26. **La Gamma, E. F., Krause, J. E., Adler, J. E., White J. D., McKelvy, J. F., and Black, I. B.,** Regulation of proenkephalin mRNA and Leu-enkephalin in explanted rat adrenal medullae, *Soc. Neurosci. Abstr.*, 10, 15, 1984.

27. **McKelvy, J. F., Krause, J. E., and White, J. D.,** Methods for the study of the biosynthesis of neuroendocrine peptides *in vivo* and *in vitro*, in *Methods in Enzymology: Neuroendocrine Peptides*, Vol. 103, Conn, P. M., Ed., Academic Press, New York, 1983, 511.

28. **Bennett, H. P. J., Browne, C. A., and Solomon, S.,** Characterization of eight forms of corticotropin-like intermediate lobe peptide from the rat intermediate pituitary, *J. Biol. Chem.*, 257, 10096, 1982.

29. **Vanderhaeghen, J. J., Lotstra, F., Liston, D. R., and Rossier, J.,** Proenkephalin, [Met]enkephalin, and oxytocin immunoreactivities are colocalized in bovine hypothalamic magnocellular neurons, *Proc. Natl. Acad. Sci. U.S.A.*, 80, 5139, 1983.

30. **Martin, R., Geis, R., Holl, R., Schafer, M., and Voigt, K. H.,** Co-existence of unrelated peptides in oxytocin and vasopressin terminals of rat neurohypophyses: immunoreactive methionine^5-enkephalin, leucine^5-enkephalin and cholecystokinin-like substances, *Neuroscience*, 8, 213, 1983.

31. **White, J. D., Krause, J. E., and McKelvy, J. F.,** *In vivo* biosynthesis and transport of oxytocin, vasopressin and neurophysins to posterior pituitary and nucleus of the solitary tract, *J. Neurosci.*, 4, 1262, 1984.

32. **Poulain, D. A. and Wakerley, J. B.,** Electrophysiology of hypothalamic magnocellular neurons secreting oxytocin and vasopressin, *Neuroscience*, 7, 773, 1982.

33. **Pickering, B. T., Jones, C. W., Burford, G. D., McPherson, M., Swann, R. W., Heap, P. F., and Morris, J. F.,** The role of the neurophysin proteins: suggestions from the study fo their transport and turnover, *Ann. N. Y. Acad. Sci.*, 248, 15, 1975.

34. **Krause, J. E. and White, J. D.,** Substance P biosynthesis and axonal transport in the rat striatonogral system: evidence for rapid precursor processing and fast axonal transport from pulse-chase experiments, submitted.

35. **White, J. D., Stewart, K. D., Krause, J. E., and McKelvy, J. F.,** The biochemistry of peptide-secreting neurons, *Physiol. Rev.*, 65, 553, 1985.

36. **Ragavan, V. V. and Frantz, A. G.,** Opioid regulation of prolactin secretion: evidence for a specific role of β-endorphin, *Endocrinology,* 109, 1769, 1981.

37. **Lumpkin, M. D., Samson, W. K., and McCann, S. M.,** Hypothalamic and pituitary sites of action of oxytocin to alter prolactin secretion in the rat, *Endocrinology,* 112, 1711, 1983.

38. **Baird, A., Ling, N., Bohlen, P., Benoit, R., Klepper R., and Guillemin, R.,** Molecular forms of the putative enkephalin precursor BAM-12P in bovine adrenal, pituitary and hypothalamus, *Proc. Natl. Acad. Sci. U.S.A.,* 79, 2023, 1982.

39. **Herbert, E. and Uhler, M.,** Biosynthesis of polyprotein precursors to regulatory peptides, *Cell,* 30, 1, 1982.

REGULATION OF ENKEPHALIN GENE EXPRESSION, PROHORMONE PROCESSING AND SECRETION IN BOVINE CHROMAFFIN CELLS

Lee E. Eiden, Pierre Giraud, Adair J. Hotchkiss, and Hans-Urs Affolter

SUMMARY

Enkephalin peptides are co-released with catecholamines from bovine chromaffin cells in primary culture by acetylcholine and nicotinic cholinergic agonists. Nicotine in 5 to 10 μM concentrations causes a rapid ($<$15 min) release of 10 to 20% of total cellular methionine-enkephalin pentapeptide. Nicotine also elicits a rapid increase in intracellular messenger RNA coding for proenkephalin (mRNA^{enk}) which is maximal (170 to 250% of control levels) by 8 to 10 hr of exposure to nicotine. This is followed by a gradual increase in enkephalin peptide levels over a 24- to 72-hr period. Forskolin and cholera toxin, which, like nicotine, elevate intracellular cAMP in chromaffin cells, also mimic the effect of nicotine to elevate mRNA^{enk}, but do not cause acute enkephalin or catecholamine release. Reserpine, which blocks catecholamine reuptake into storage vesicles among other potential effects, causes a 100 to 200% increase in Met-enkephalin pentapeptide levels within 24 hr, but this is accompanied by a gradual decrease, rather than an increase in mRNA^{enk}. Gel permeation chromatography of chromaffin cell extracts followed by radioimmunoassay of column fractions before and after treatment with trypsin/carboxypeptidase B reveals that forskolin and nicotine cause an increase in both Met-enkephalin pentapeptide and higher molecular weight enkephalin-containing peptides, while reserpine treatment results in a shift of enkephalin immunoreactivity from high-molecular-weight to low-molecular-weight ($<$800 daltons) forms, suggesting that reserpine acts to enhance enkephalin precursor processing.

Forskolin and reserpine represent two classes of pharmacological agents that can regulate enkephalin biosynthesis at pre- and posttranslational steps, respectively. The similarity of forskolin and nicotine action on enkephalin biosynthesis and mRNA^{enk} suggests that nicotinic receptor stimulation may act in vivo to couple enkephalin secretion and biosynthesis via an increase in enkephalin gene transcription, which may be mediated by cAMP.

INTRODUCTION

Catecholamine and enkephalin secretion from the adrenal medulla in vivo are under the control of the sympathetic nervous system. Acetylcholine released from the splanchnic nerve stimulates nicotinic cholinergic receptors on the chromaffin cells of the medulla, which results in calcium influx and exocytotic release of catecholamines, enkephalins, chromagranins, and other secretory products into the general circulation.[1-4] Increased sympathetic outflow thus causes a depletion of medullary stores of these secreted hormones. A critical issue in understanding how a particular endocrine hormone phenotype is maintained by neuroendocrine cells, such as chromaffin cells, is understanding how the cell maintains a constant supply of peptide hormone available for release, in the face of fluctuating demand for the exported peptide. Potential mechanisms of regulation include (1) alteration in the rate of peptide hormone structural gene transcription during various states of secretory activity; (2) increased mRNA translation upon increased secretion, coupled to a constant rate of gene transcription, and (3) increased activity of processing enzymes whose substrate is the hormone precursor molecule, leading to a mobilization from prohormone stores of mature peptide available for release. The mechanisms of biosynthetic regulation of secretory peptides such as enkephalin can only be adequately addressed by concomitant measurement of the mature peptide hormone, its prohormone precursor(s), and the messenger RNA coding for the polypeptide in question. This report summarizes recent data, obtained using enkephalin biosynthesis in cultured bovine chromaffin cells as a model system, which may provide a better understanding of the intracellular dynamics of endocrine hormone homeostasis.

MATERIALS AND METHODS

Chromaffin Cell Culture

Chromaffin cells were obtained from fresh bovine adrenal glands by retrograde venous perfusion with a buffered solution of 0.1% collagenase, followed by mincing and further digestion with collagenase and purification by differential centrifugation as described in Reference 5. Cells were plated at a density of 300,000 to 1,000,000 cells/mℓ/16-mm diameter well (Costar® 24-well plates) in the Basal Medium of Eagle with Earle's Salts containing 25 mM sodium HEPES, 292 μg/mℓ glutamine, 100 units/mℓ penicillin, 100 μg/mℓ streptomycin, 10 μg/mℓ cytosine arabinofuranoside, and supplemented with 5% fetal bovine serum (Hyclone), in a humidified 95% air/5% CO_2 atmosphere. After 72 to 96 hr of culture, cells were treated with various drugs added in fresh medium for 24 to 96 hr. In experiments designed to measure release from the cells, medium was removed by aspiration and cells were washed once with standard release medium (SRM —118 mM NaCl, 4.6 mM KCl, 10 mM Na HEPES, 2.2 mM CaCl$_2$, 1.2 mM MgSO$_4$, 0.1% BSA, pH 7.3). Following incubation in 400 $\mu\ell$ SRM for 15 min at 37°C, test substances were added to each well in a volume of 100 $\mu\ell$, and incubation continued for an additional 15 min.

Met-Enkephalin, Catecholamine, and cAMP Measurements

All experiments were terminated by removal of medium (for storage at -20°C prior to direct assay) and extraction of cells with 0.1 N HCl or 0.5 M acetic acid (for storage at -20°C prior to clarification by centrifugation, vacuum evaporation, and reconstitution in assay buffers). Met-enkephalin was measured using a radioimmunoassay (RIA) highly selective for Met-enkephalin pentapeptide (antiserum RB-4 was kindly provided by Dr. Steve Sabol, NHLBI, NIH).[6,7] Total catecholamines were determined fluorimetrically[7,8] and cyclic AMP was determined using a commercially available RIA (New England Nuclear Corp.).

Chromatography

0.1 N HCl extracts of chromaffin cells were dried under vacuum and chromatographed after reconstitution in phosphate buffer on a TSK-2000 SW sieving column. Eluate fractions were assayed for Met-enkephalin by RIA before or after digestion with trypsin and carboxypeptidase B as previously described.[7]

Quantitation of mRNA[enk]

A modification of the procedure of Rowe et al.[9] was used for RNA extraction. Cells were rinsed once with phosphate-buffered saline and incubated at 42°C for 90 min in 10 mM Tris pH 7.5, 5 mM EDTA, 1% SDS, 100 μg/mℓ RNase-free proteinase K (Worthington), 500 $\mu\ell$ per well. Incubates were twice extracted with 1.6 volumes of Tris-saturated phenol-chloroform (1:1, v/v) and total nucleic acids precipitated by centrifugation from ethanol. RNA was denatured in buffer A (20 mM MOPS, 5 mM sodium acetate, 1 mM EDTA, pH 7.0) containing 2.2 M formaldehyde and 50% formamide, and electrophoresed on 1% agarose gels containing 2.2 M formaldehyde in buffer A, with buffer A as the running buffer. Duplicate gels were processed for RNA blot hybridization or total RNA quantitation.[7]

For RNA blot hybridization, nucleic acids were transferred from agarose to Gene Screen (New England Nuclear Corp.) by anodic electroelution in 25 mM NaPO$_4$ pH 6.5 in a Bio-Rad Trans-Blot® apparatus. Filters were baked at 80°C for 90 min *in vacuo* and prehybridized for 2 to 5 hr at 42°C in 50% formamide, 50 mM NaPO$_4$, 0.1% SDS, 0.1% Ficoll®, BSA, and polyvinylpyrrolidine and 5× SSC (750 mM NaCl, 75 mM Na citrate) containing 250

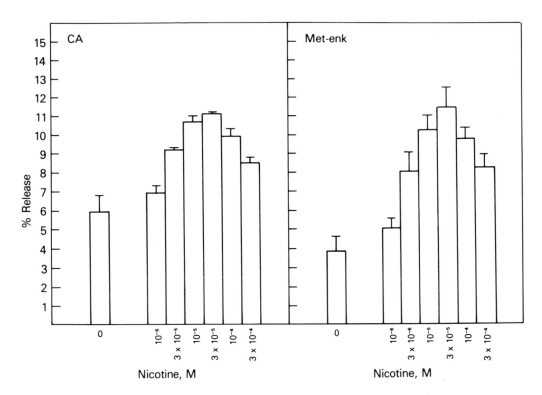

FIGURE 1. The effect of nicotine on catecholamine and Met-enkephalin release from cultured chromaffin cells. Percent release at each dose of nicotine was calculated as the percentage of total cellular immunoreactive enkephalin or total catecholamine (epinephrine plus norepinephrine) released into the medium during a 15-min incubation period. Values are the mean ± S.E.M. of 4 individual determinations.

μg/mℓ each of denatured sonicated salmon sperm DNA and yeast tRNA. Filters were hybridized at 42°C for 18 to 24 hr in the same buffer plus 10% dextran sulfate and ^{32}P-labeled cDNA^{enk} (10 ng/mℓ, 1 to 3 $\times$ 10^8 cpm/μg of the internal Pst I fragment of plasmid pbovenk[10] purified by preparative agarose gel electrophoresis and labeled by nick translation with α-^{32}P-dATP and dCTP). Following a 42°C wash with 0.2 $\times$ SSC/0.1% SDS, filters were autoradiographed and mRNA^{enk} quantitated by densitometric scanning of the latter with an LKB soft laser scanner.

Total RNA was quantitated by staining of duplicate gels in buffer A containing 30 μg/mℓ ethidium bromide followed by a 4-hr wash in buffer A. Negatives of UV-transilluminated photographs of the gel were scanned densitometrically, and total RNA determined by comparison with a dilution series of standard adrenomedullary RNA of known concentration.

RESULTS

Effect of Nicotine on Catecholamine and Enkephalin Release from Chromaffin Cells

Nicotine elicited a dose-dependent release of both Met-enkephalin and catecholamines from cultured bovine chromaffin cells (Figure 1), as earlier reported by Livett and co-workers.[11] Carbachol and acetylcholine, as well as the nicotine agonists lobeline and 1,1-dimethyl-4-phenylpiperazine, also cause a parallel release of Met-enkephalin and catecholamines upon acute exposure to chromaffin cells.[12-15a] The time course of nicotinic stimulation of release is rapid, and reaches a plateau between 5 and 15 min of stimulation (unpublished observations). For both enkephalin and catecholamine release, nicotinic stimulation is max-

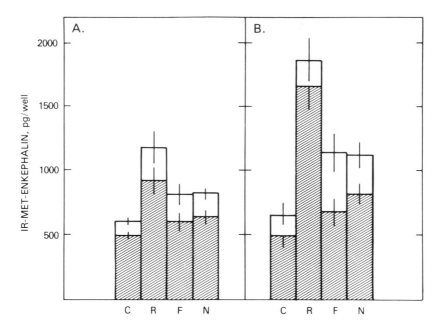

FIGURE 2.   The effect of reserpine, forskolin, and nicotine on Met-enkephalin levels of cultured chromaffin cells. Cells were exposed to 10 μ*M* reserpine, 25 μ*M* forskolin, or 5 μ*M* nicotine for (A) 24 hr or (B) 72 hr. Cell Met-enkephalin is shown by the shaded bars; medium Met-enkephalin by the unshaded portion of each bar; and total (cell + medium) enkephalin by the total height of each bar. Values represent the mean ± S.E.M. of three individual determinations (wells).

imal at around 10^{-5} *M,* and exhibits a characteristic decline at higher doses presumed to reflect a rapid dose-dependent desensitization of the nicotinic receptor (Figure 1). Reserpine and forskolin had no effect on acute peptide or catecholamine release at the maximum doses employed (10 and 50 μ*M*, respectively).

Effect of Reserpine, Forskolin, and Nicotine on Met-Enkephalin Levels in Chromaffin Cells

Reserpine causes a relatively rapid increase in immunoreactive Met-enkephalin, resulting in a doubling of intracellular Met-enkephalin levels, with no effect on enkephalin release into the medium, within 24 hr (Figure 2). By contrast, both forskolin and nicotine produce a less rapid increase in enkephalin biosynthesis, mostly reflected in an increase in Met-enkephalin secreted into the medium (Figure 2), even though forskolin has no effect on Met-enkephalins release (or potentiation of nicotine-induced enkephalin release) upon acute (15 to 60 min) exposure (unpublished observations). The effect of nicotine and forskolin to increase enkephalin pentapeptide biosynthesis is completely abolished by pretreatment of chromaffin cells with the protein synthesis inhibitor cycloheximide (0.5 μg/mℓ), while the increase in enkephalin levels elicited by reserpine is only partially blocked by cycloheximide (Reference 7 and unpublished observations).

Changes in Intracellular mRNA[enk] Elicited by Reserpine, Forskolin, and Nicotine

Both nicotine and forskolin cause a 50 to 300% increase in intracellular mRNA[enk] which is maximal at 24 hr and sustained for up to 48 hr of exposure to these two agents (Figure 3 and Table 1). The time course of mRNA[enk] elevation by forskolin and nicotine is rapid; maximal induction occurs within 12 hr of treatment with either agent (Figure 4), suggesting that increased enkephalin gene transcription, rather than inhibition of mRNA[enk] degradation,

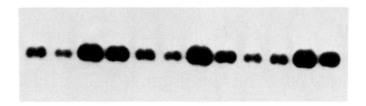

CR FNCR FNCR FN

A

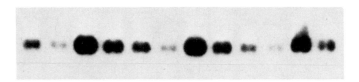

CRFNCRFNCRFN

B

FIGURE 3.   The effect of reserpine, forskolin, and nicotine on mRNA[enk]. RNA harvested from cells treated with 10 μM reserpine (R), 50 μM forskolin (F), 6 μM nicotine (N), or vehicle (0.1% ethanol in culture medium) (C) for (A) 24 or (B) 48 hr was blotted, hybridized with ^{32}P-labeled cDNA[enk] and autoradiographed for 18 to 24 hr. Data corrected for total RNA contained in each sample are summarized in Table 1.

Table 1
EFFECT OF RESERPINE, FORSKOLIN, AND
NICOTINE ON mRNA[enk]

24 hr	mRNA[enk] (% of control)
Control	100 ± 12
10 μM reserpine	88 ± 14
50 μM forskolin	376 ± 22
6 μM nicotine	252 ± 44
48 hr	
Control	100 ± 9.5
10 μM reserpine	35 ± 3.1
50 μM forskolin	384 ± 25
6 μM nicotine	163 ± 5.3

Note: Data are those presented in Figure 3, normalized to mRNA[enk] levels present in untreated (control) cultures at each time point. Mean $\pm$ S.E.M. of the three determinations shown in Figure 3.

is the mechanism of mRNA[enk] induction. Reserpine, on the other hand, causes a gradual decrease in mRNA[enk] over a 48-hr period (Figure 3 and Table 1). This effect is selective for mRNA[enk] present in unstimulated chromaffin cells, since reserpine causes a 70% decrease in mRNA[enk] by itself, but only a 22% decrease in mRNA[enk] newly synthesized following treatment with forskolin.[7]

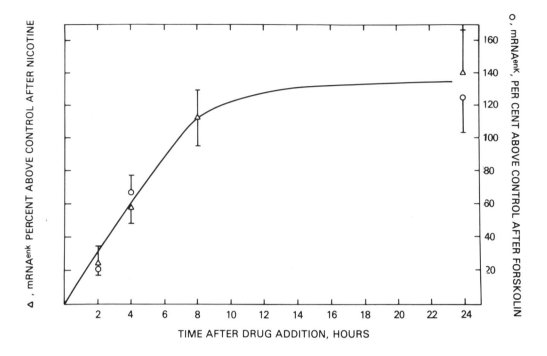

FIGURE 4.   Time course of mRNA[cnk] induction by forskolin and nicotine. Cells were exposed to 25 μM forskolin or 10 μM nicotine for 2, 4, 8, or 24 hr. mRNA[cnk] was quantitated by blot hybridization, corrected for total RNA per sample, and normalized to control (untreated) mRNA[cnk] levels at each time point. Data represent the mean $\pm$ S.E.M. of three to six separate determinations. (Data for nicotine taken from Reference 15.)

Size Distribution of Enkephalin-Containing Polypeptides after Treatment with Forskolin or Reserpine

Extracts of chromaffin cells treated with reserpine or forskolin for 72 hr were subjected to gel permeation chromatography followed by trypsin/carboxypeptidase B digestion of column fractions to liberate immunologically cryptic Met-enkephalin from precursors containing the pentapeptide sequence (Figure 5). Forskolin caused an increase in both high- and low-molecular-weight enkephalin-containing peptides. Reserpine caused no apparent increase in total immunoreactive enkephalin, but rather a relative shift in Met-enkephalin from high- to low-molecular-weight species, consistent with an effect of reserpine to stimulate enkephalin pentapeptide processing from the higher-molecular-weight enkephalin-containing precursor polypeptides, rather than a stimulation of *de novo* enkephalin biosynthesis (Figure 5). Extracts of nicotine-treated cells yielded a chromatographic profile similar to that obtained with forskolin (data not shown), further supporting the hypothesis that nicotine and forskolin stimulated enkephalin biosynthesis by a similar mechanism.

Effect of Nicotine on Intracellular Levels of Cyclic AMP

As previously reported by others, nicotine increases intracellular cAMP in chromaffin cells, causing an approximately 60% increase in cAMP as early as 5 min after addition of nicotine to the culture medium (Table 2). The sustained increase in cAMP levels over a 2- to 4-hr period suggests that there is not a rapid desensitization of this effect of nicotinic stimulation, or alternatively that cAMP metabolism is very slow in cultured chromaffin cells, perhaps due to a low activity of cyclic AMP phosphodiesterase.

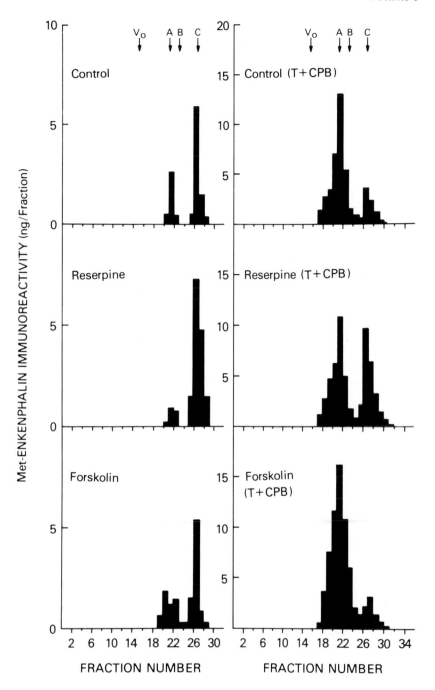

FIGURE 5.   Size distribution of Met-enkephalin pentapeptide-containing polypeptides in reserpine- or forskolin-treated chromaffin cells. Cells were exposed to 5 μM reserpine, 25 μM forskolin, or no drug (control) for 72 hr and immunoreactive Met-enkephalin measured in column fractions obtained by gel permeation chromatography of each cell extract, either directly (left panel) or after digestion with trypsin and carboxypeptidase B (T + CPB; right panel). A, B, and C represent the positions of elution of cytochrome C (12.5 kdaltons aprotinin (6.5 kdaltons) and Met-enkephalin Arg[6]Phe[7] (0.9 kdaltons), respectively. (Data taken from Reference 7.)

Table 2
TIME COURSE OF NICOTINE-INDUCED
ELEVATION IN INTRACELLULAR cAMP

	cAMP (fmol/well)	
Time (min)	Control	5 μM nicotine
5	203 ± 24	340 ± 24
40	198 ± 17	305 ± 22
70	242 ± 29	333 ± 44
130	232 ± 15	425 ± 29
240	273 ± 25	423 ± 42

Note: Values are the mean ± S.E.M. of six determinations (wells).

DISCUSSION

In order to understand how enkephalin biosynthesis might be regulated in an endocrine cell which stores and secretes it, the effect of three prototypical agents on enkephalin biosynthesis and mRNA levels in chromaffin cells was studied. Reserpine is a member of a class of drugs, which also includes tetrabenazine, that cause depletion of intracellular catecholamines by inhibiting their reuptake into secretory granules, thus causing catechol-amine efflux into the cytosol and subsequent metabolism by monoamine oxidase.[16,17] Reserpine probably has other intracellular actions as well, since it is known to increase protein synthesis in mucus-secreting intestinal cells.[18] Forskolin acts directly on adenylate cyclase to stimulate production of cyclic AMP.[19] Elevated intracellular cAMP has been suggested as the messenger which, through stimulation of cyclic AMP-dependent protein kinase, may enhance the transcription of specific structural genes following its translocation to the nucleus.[20-22] Nicotine is a stable analog of acetylcholine, the natural secretagogue which causes enkephalin release from adrenomedullary chromaffin cells, and therefore depletion of secretory vesicle stores of the peptide.

Reserpine appears to increase Met-enkephalin content of chromaffin cells by stimulating the proteolytic processing of enkephalin-containing peptides. Evidence for this mechanism of reserpine action is (1) reserpine causes an increase in intracellular Met-enkephalin pentapeptide without a concomitant increase (in fact a decrease) in mRNA[enk]; (2) the increase in enkephalin is only partially blocked by the protein synthesis inhibitor cycloheximide; and (3) the increase in Met-enkephalin pentapeptide occurs at the expense of higher-molecular-weight enkephalin-containing peptides with no apparent increase in the total cellular content of enkephalin (high- plus low-molecular-weight species). Recently, a carboxypeptidase B-like and a trypsin or aminopeptidase enzymatic activity have been isolated from highly purified adrenomedullary secretory vesicles, which cleave basic amino acids from the C- and N-terminus, respectively, of Met-enkephalin.[23-25] Both of these enzymes are markedly pH dependent.[23,25] If reserpine acts to alter intragranular pH or oxidation-reduction state by inhibiting catecholamine entry (which is linked to proton transfer across the granule membrane),[26] then it may enhance enkephalin precursor processing by altering the enzymatic activity of these putative processing enzymes. Alternatively, reserpine may act indirectly by increasing cytoplasmic norepinephrine, epinephrine, or dopamine concentrations, the latter stimulating enkephalin processing by some as yet unknown mechanism. Storage vesicles are thought to be capable of catecholamine reuptake following partial exocytosis and recycling from the plasma membrane.[27] If this is so, reserpine may be mimicking the physiological process whereby vesicles that have undergone exocytosis and endocytotic recycling achieve repletion of both catecholamine and enkephalin content; the former by uptake of catecholamines from the cytoplasm and the latter by Met-enkephalin mobilization via processing of polypeptide precursors already present in the secretory granule.

Nicotinic stimulation results in enkephalin release, elevation of intracellular cAMP, elevation of mRNA coding for proenkephalin, and an increase in enkephalin pentapeptide and the enkephalin-containing polypeptides which are its biosynthetic precursors. All of these effects, with the exception of acute stimulation of enkephalin release, are mimicked by forskolin, cholera toxin, and isobutylmethylxanthine (References 6 and 7; unpublished observations) which act by stimulation of adenylate cyclase or inhibition of cAMP phosphodiesterase. In addition, the effect of nicotine to induce $mRNA^{enk}$ is blocked by the calcium-channel blocker D-600, while forskolin induction of $mRNA^{enk}$ is not (Reference 15). Interestingly, forskolin, which causes a five- to tenfold higher maximal elevation of intracellular cAMP than nicotine, also causes a two- to threefold greater induction of $mRNA^{enk}$ than does nicotine. If nicotine and forskolin both increase $mRNA^{enk}$ by activation of protein kinase, with subsequent phosphorylation of a nuclear protein which can activate enkephalin gene transcription, the lability of this phosphoprotein may allow a very tight coupling between intracellular levels of cAMP and the rate of enkephalin gene transcription, accounting for the increased efficacy of forskolin relative to nicotine to increase intracellular $mRNA^{enk}$.

Nicotinic stimulation could enhance cAMP production in several ways: by a direct coupling of the nicotinic receptor to adenylate cyclase, by causing the release of a substance (catecholamines, ATP, enkephalin) which activates adenylate cyclase via a membrane receptor, or by increasing intracellular calcium which could potentiate nicotinic activation of adenylate cyclase or activate it directly upon binding to cytosolic calmodulin. Alternatively, a calcium-dependent but cAMP-independent mechanism may be responsible for nicotinic activation of the enkephalin gene.

Future studies employing chromaffin cells to study enkephalin gene regulation should be devoted to the isolation and identification of the protein factor(s) which actually interact with the enkephalin structural gene or flanking DNA sequences to effect transcriptional activation, because it is the factor(s) which finally are responsible for the maintenance of the enkephalin phenotype in chromaffin and other neuroendocrine cells.

ACKNOWLEDGMENTS

We wish to thank Chang-Mei Hsu for performing the peptide radioimmunoassays and Patricia Thurston for the preparation of this manuscript. P. Giraud gratefully acknowledges the financial support of the Fondation de la Recherche Medicale Française.

REFERENCES

1. **Douglas, W. W.,** Stimulus-secretion coupling — the concept and clues from chromaffin and other cells, *Br. J. Pharmacol.,* 34, 451, 1968.
2. **Blaschko, H., Sayers, G., and Smith, A. D., Eds.,** Endocrinology. Adrenal gland, in *Handbook of Physiology, Section 7,* Vol. 6, American Physiological Society, Washington, D. C., 1975.
3. **Kirshner, N. and Kirshner, A. G.,** Chromogranin A, dopamine β-hydroxylase and secretion from the adrenal medulla, *Philos. Trans. R. Soc. London Ser. B.,* 261, 279, 1971.
4. **Carmichael, S. W.,** *The Adrenal Gland,* Eden Press, New York, 1979.
5. **Eiden, L. E., Eskay, R. L., Scott, J., Pollard, H., and Hotchkiss, A. J.,** Primary cultures of bovine chromaffin cells synthesize and secrete vasoactive intestinal polypeptide (VIP), *Life Sci.,* 33, 687, 1983.
6. **Eiden, L. E. and Hotchkiss, A. J.,** Cyclic adenosine monophosphate regulates vasoactive intestinal polypeptide and enkephalin biosynthesis in cultured bovine chromaffin cells, *Neuropeptides,* 4, 1, 1983.

7. **Eiden, L. E., Giraud, P., Affolter, H. -U., Herbert, E., and Hotchkiss, A. J.,** Alternate modes of enkephalin biosynthesis regulation by reserpine and cyclic AMP in cultured chromaffin cells, *Proc. Natl. Acad. Sci. U.S.A.,* 81, 3949, 1984.
8. **Euler, U. S. V. and Flooding, I.,** A fluorimetric micro-method for differential estimation of adrenaline and noradrenaline, *Acta Hystiol. Scand.,* 33 (Suppl. 18), 45, 1955.
9. **Rowe, D. W., Moen, R. C., Davidson, J. M., Byers, P. H., Bornstein, P., and Palmiter, R. D.,** Correlation of procollagen mRNA levels in normal and transformed chick embryo fibroblasts with different rates of procollagen synthesis, *Biochemistry,* 17, 1581, 1978.
10. **Gubler, U., Seeburg, P., Hoffman, B. J., Gage, L. P., and Udenfriend, S.,** Molecular cloning establishes proenkephalin as the precursor of enkephalin-containing peptides, *Nature (London),* 295, 202, 1982.
11. **Livett, B. G., Dean, D. M., Whelan, L. G., Udenfriend, S., and Rossier, J.,** Co-release of enkephalin and catecholamines from cultured adrenal chromaffin cells, *Nature (London),* 289, 317, 1981.
12. **Stine, S. M., Yang, H.-Y. T., and Costa, E.,** Release of enkephalin-like immunoreactive material from isolated bovine chromaffin cells, *Neuropharmacology,* 19, 683, 1980.
13. **Eiden, L. E., Giraud, P., Hotchkiss, A., and Brownstein, M. J.,** Enkephalins and VIP in human pheochromocytomas and bovine adrenal chromaffin cells, in *Regulatory Peptides, From Molecular Biology to Function,* Costa, E. and Trabucchi, M., Eds., Raven Press, New York, 1982, 387.
14. **Eiden, L. E.,** The use of a cloned cDNA probe and peptide radioimmunoassay to examine transcriptional and translational regulation of enkephalin expression in eukaryotic cells, in *Insect Neurochemistry and Neurophysiology,* Borkovec, A. B. and Kelly, T. J., Eds., Plenum Press, New York, 1984, 299.
15. **Eiden, L. E., Giraud, P., Dave, J., Hotchkiss, A. J., and Affolter, H.-U.,** Nicotinic receptor stimulation activates both enkephalin release and biosynthesis in adrenal chromaffin cells, *Nature,* 312, 661, 1984.
15a. **Hotchkiss, A. J. et. al.,** unpublished observations.
16. **Viveros, O. H., Arqueros, L., and Kirshner, N.,** Mechanism of secretion from the adrenal medulla, *Mol. Pharmacol.,* 7, 434, 1971.
17. **Wilson, S. P., Chang, K.-J., and Viveros, O. H.,** Synthesis of enkephalins by adrenal medullary chromaffin cells: reserpine increases incorporation of radiolabeled amino acids, *Proc. Natl. Acad. Sci. U.S.A.,* 77, 4364, 1980.
18. **Misch, D. W. and Kim, W.-K.,** Action of reserpine on mucous secretion in hamster intestine: development of an animal model for autonomic control of exocrine secretion, *J. Cell Biol.,* 95, 403a, 1982.
19. **Seamon, K. D., Padgett, W., and Daly, J. W.,** Forskolin: unique diterpene activator of adenylate cyclase in membranes and in intact cells, *Proc. Natl. Acad. Sci. U.S.A.,* 78, 3363, 1981.
20. **Jungmann, R. A., Lee, S.-G., and DeAngelo, A. B.,** Translocation of cytoplasmic protein kinase and cyclic adenosine monophosphate-binding protein to intracellular acceptor sites, in *Advances in Cyclic Nucleotide Research,* Vo. 5, Drummond, G. I., Greengard, P., and Robison, G. A., Eds., Raven Press, New York, 1975, 281.
21. **Murdoch, G. H., Rosenfeld, M. G., and Evans, R. M.,** Eukaryotic transcriptional regulation and chromatin-associated protein phosphorylation by cyclic AMP, *Science,* 218, 1315, 1982.
22. **Kumakura, K., Guidotti, A., and Costa, E.,** Primary cultures of chromaffin cells: molecular mechanisms for the induction of tyrosine hydroxylase mediated by 8-Br-cyclic AMP, *Mol. Pharmacol.,* 16, 865, 1979.
23. **Hook, V. Y. H., Eiden, L. E., and Brownstein, M. J.,** A carboxypeptidase processing enzyme for enkephalin precursors, *Nature (London),* 295, 341, 1982.
24. **Fricker, L., Supattapone, S., and Snyder, S.,** Enkephalin convertase: a specific enkephalin synthesizing carboxypeptidase in adrenal chromaffin granules, brain and pituitary gland, *Life Sci.,* 31, 1841, 1982.
25. **Hook, V. Y. H. and Eiden, L. E.,** Two peptidases that convert ^{125}I-lys-arg-(met)enkephalin and ^{125}I-(met)enkephalin-arg^6, respectively, to ^{125}I-(met)enkephalin in bovine adrenal medullary chromaffin granules, *FEBS Lett.,* 172, 212, 1984.
26. **Ungar, A. and Phillips, J. H.,** Regulation of the adrenal medulla, *Physiol. Rev.,* 63, 787, 1983.
27. **Holtzmann, E.,** The origin and fate of secretory packages, especially synaptic vesicles, *Neuroscience,* 2, 327, 1977.

FLEXIBILITY IN THE PROCESSING OF β-ENDORPHIN

D. G. Smyth

SUMMARY

β-Endorphin is generated from a multifunctional prohormone which is processed to give both active and inactive forms of the peptide. The proportions of the peptides generated appear to be characteristic of individual tissues. Evidence has now been obtained that the patterns can exhibit a degree of variation and experiments with pars intermedia cells grown in monolayer culture show that the processing reactions are sensitive to environmental stimuli. It is suggested that the regulation of bioactive β-endorphin takes place in part at the level of processing of the prohormone. The inactivation of β-endorphin, which takes place intracellularly, is seen as an important step in the regulation of multiple activities generated from a common prohormone.

INTRODUCTION

The potent analgesic properties of the 31-residue peptide β-endorphin[1-3] were first recognized when it was isolated as one of a series of peptides present in porcine pituitary. Another of the new peptides isolated was a shorter form of β-endorphin, a 27-residue peptide which lacked four C-terminal residues of β-endorphin and which possessed only slight analgesic properties.[3,4] This surprising finding, that a highly potent peptide is biosynthesized in company with a related but essentially inactive form, was compounded when it was found that β-endorphin also occurs in an N-acetylated form which was completely devoid of opiate activity.[5] More recently another naturally occurring form of β-endorphin, which lacked the C-terminal histidine of β-endorphin 1-27, was identified and this peptide too was found also in an N-acetylated form;[6] indeed N-acetyl β-endorphin 1-26 proved to be the major form of β-endorphin in rat pars intermedia.[7]

The existence of six forms of β-endorphin (Figure 1), only one of which possesses potent opiate activity, invites question on the physiological control of the processing reactions involved in the elaboration of these peptides. Do the six forms occur in the same proportions in different tissues? Does a particular tissue or group of cells always have the same pattern of peptides? Examination of the β-endorphin-related peptides in regions of rat pituitary and brain has indicated that each tissue has a characteristic processing pattern,[8] but more recent experiments are showing that the patterns can exhibit a degree of variation. Furthermore, with the aid of cells grown in monolayer culture, evidence has been obtained that the processing patterns change in response to environmental stimuli.

METHODS

The processing patterns exhibited by peptides related to β-endorphin were studied in two regions of rat pituitary and a number of regions of rat brain. The peptides were extracted from the dissected tissues under acidic conditions and resolved by chromatography under conditions that have been shown to prevent degradation by tissue enzymes; gel filtration was performed on columns of Sephadex G75 in 50% acetic acid and ion-exchange chromatography was carried out on SP-Sephadex C25 (pyridinium form) with a linear gradient from 50% acetic acid to 1 M pyridine in 50% acetic acid. The recoveries of the peptides were in all cases greater than 80%. The inclusion of radioactive marker peptides allowed the corresponding endogenous peptides to be located and identified and the amounts of the β-endorphin 1-31 → β-endorphin 1-27 → β-endorphin 1-26

↓   ↓   ↓

N-acetyl   → *N*-acetyl   → *N*-acetyl
β-endorphin 1-31   β-endorphin 1-27   β-endorphin 1-26

FIGURE 1.   Related forms of β-endorphin identified in pig, ox, and rat pituitary and brain. It is notable that only β-endorphin 1-31 possesses potent analgesic properties.

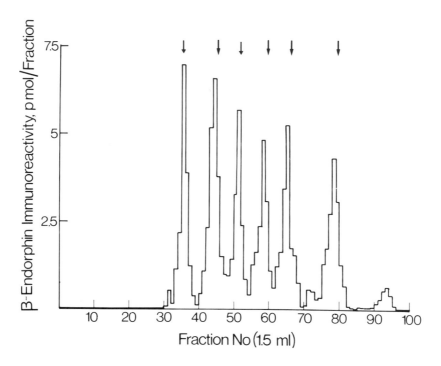

FIGURE 2.   Resolution of β-endorphin-related peptides extracted from rat pars intermedia. The arrows indicate the elution positions (L → R) of acetyl β-endorphin 1-26, acetyl β-endorphin 1-27, β-endorphin 1-26, β-endorphin 1-27, acetyl β-endorphin 1-31, and β-endorphin 1-31, respectively. Note that β-endorphin 1-31 (Peak 6) is a minor component.

extracted peptides were determined by radioimmunoassay using an antiserum that was raised against porcine β-endorphin and which had a similar affinity for each of the six β-endorphin-related peptides.[9]

Rat pars intermedia cells were grown in monolayer culture and incubations were performed for 2 hr at 37°C in the presence and absence of exogenous dopaminergic agents. The details of the experimental procedures are described elsewhere.[10,11]

RESULTS AND DISCUSSION

It was found by gel filtration of peptides extracted from the tissues that the degree of proteolysis of the ACTH-β-endorphin prohormone was much greater in the pars intermedia than in the anterior pituitary. Furthermore, ion-exchange chromatography showed that the major form of β-endorphin in the anterior pituitary was β-endorphin 1-31, whereas the major peptides in the pars intermedia were the *N*-acetylated forms of β-endorphin 1-26 and β-endorphin 1-27 (Figure 2). The results show clearly that β-endorphin exhibits tissue-specific processing.

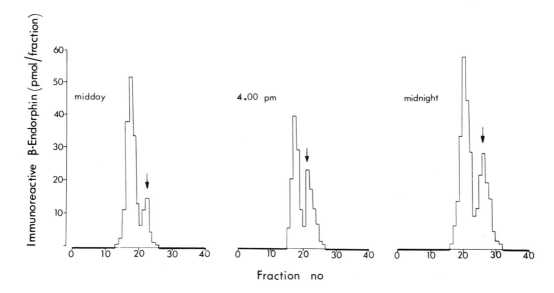

FIGURE 3. Diurnal variation in the processing of β-endorphin in rat anterior pituitary. The patterns show the separation of β-endorphin (indicated by an arrow) from lipotropin, its biosynthetic precursor. The peptides in the extracts of rat anterior pituitary were resolved by gel filtration.

Contrasting patterns of processing have also been seen in specific regions of rat brain.[12] Notably, the potent analgesic peptide, β-endorphin 1-31, is the principal β-endorphin-related peptide in the hypothalamus, while the more processed forms, β-endorphin 1-27 and β-endorphin 1-26, are prevalent in the midbrain and amygdala. In brain, the *N*-acetylated forms of β-endorphin appear to be relatively minor.

The β-endorphin-related peptides are always accompanied by corticotropin (ACTH) or α-melanotropin (α-MSH), which are formed as fragments of the same prohormone.[13] Like the β-endorphin-related peptides, ACTH and α-MSH exhibit tissue-specific processing. In the anterior pituitary ACTH is the major form; in the pars intermedia acetylated forms of α-MSH predominate; and in the hypothalamus the major form is desacetyl α-MSH; its acetylated counterpart, acetyl α-MSH, occurs only in the more distal regions of the brain. Thus it can be seen that in each region of pituitary or brain the formation of biologically active peptides takes place in pairs. In the anterior pituitary, ACTH is accompanied by β-endorphin; in the pars intermedia diacetylated α-MSH is accompanied by unmodified, bioactive β-endorphin; and in the hypothalamus β-endorphin is accompanied by desacetyl α-MSH.

Recent studies have shown that the processing patterns that seemed characteristic of a specific tissue exhibit a degree of variation. In the anterior pituitary, for example, the relative proportions of β-endorphin to lipotropin (its inactive 91-residue precursor peptide) change significantly during diurnal rhythms (Figure 3) and major variations in the proportions of β-endorphin 1-31, 1-27, and 1-26 have been observed in regions of rat brain.[14] The evidence indicates that the processing reactions involved in the elaboration of β-endorphin are dynamic.

To test whether the processing reactions are influenced by physiological or pharmacological stimuli, rat pars intermedia cells were grown in monolayer culture and the effect of dopaminergic agents examined on the pattern of peptides produced. Under control conditions, in the absence of dopaminergic innervation or exogenous dopamine, the processing pattern was found to differ markedly from the pattern that is characteristic of the pars intermedia *in situ;* the major form of β-endorphin in the cultured cells was β-endorphin 1-31, whereas in vivo the acetylated forms of β-endorphin 1-27 and 1-26 predominate. In contrast, when the cells were incubated for 2 hr with 10^{-5} *M* ergocryptine, a dopamine agonist, there was a substantial increase in proteolysis and acetylation with the result that the acetylated forms of β-endorphin 1-26 and β-endorphin 1-27 became the major peptides.[10] The experiments show that the processing of β-endorphin in the pars intermedia cells is sensitive to specific signals, since the pattern of peptides produced by the cells changes in response to the environmental conditions.

The simultaneous formation of peptides with different activities, generated from a common prohormone, allows the possibility that the activities may be coordinated in a specific manner. Mechanisms certainly exist which lead to the potentiation or attenuation of individual peptides. In the case of the ACTH-β-endorphin prohormone, proteolysis and acetylation lead to the activation of α-MSH, but the same reactions lead to the inactivation of β-endorphin. Changes in the degree of proteolysis or acetylation therefore generate α-MSH in company with reduced amounts of bioactive unmodified β-endorphin and the proportions of the two peptides are thereby susceptible to regulation. From the results of the present experiments, it is suggested that with peptides produced from a multifunctional prohormone the control of processing offers a new level at which physiological regulation is likely to take place.

REFERENCES

1. **Feldberg, W. S. and Smyth, D. G.,** The C-fragment of lipotropin, a potent analgesic, *J. Physiol. (London),* 260, 30, 1976.
2. **Loh, H.H., Tseng, L. F., Wei, E., and Li, C. H.,** β-Endorphin is a potent analgesic agent, *Proc. Natl. Acad. Sci. U.S.A.,* 73, 2895, 1976.
3. **Feldberg, W. S. and Smyth, D. G.,** C-fragment of lipotropin an endogenous potent analgesic peptide, *Br. J. Pharmacol.,* 60, 445, 1977.
4. **Deakin, J. F. W., Doströvsky, J. O., and Smyth, D. G.,** Influence of NH_2-terminal acetylation and COOH-terminal proteolysis on the analgesic activity of β-endorphin, *Biochem. J.,* 189, 501, 1980.
5. **Smyth, D. G., Massey, D. E., Zakarian, S., and Finnie, M. D. A.,** Endorphins are stored in biologically active and inactive forms; isolation of α, *N*-acetyl peptides, *Nature (London),* 279, 252, 1979.
6. **Smyth, D. G., Smith, C. C. F., and Zakarian, S.,** Isolation and identification of two new peptides related to β-endorphin, in *Advances in Endogenous and Exogenous Opioids,* Takagi, H., Ed., Elsevier, Amsterdam, 1981, 145.
7. **Zakarian, S. and Smyth, D. G.,** β-Endorphin is processed differently in specific regions of rat pituitary and brain, *Nature (London),* 296, 250, 1982.
8. **Zakarian, S. and Smyth, D. G.,** Distribution of β-endorphin related peptides in rat pituitary and brain, *Biochem. J.,* 202, 561, 1982.
9. **Zakarian, S. and Smyth, D. G.,** Distribution of active and inactive forms of endorphins in the pituitary and brain of the rat, *Proc. Natl. Acad. Sci. U.S.A.,* 76, 5972, 1979.
10. **Smyth, D. G.,** Chromatography of peptides related to β-endorphin, *Anal. Biochem.,* 136, 127, 1984.
11. **Ham, J. and Smyth, D. G.,** Regulation of bioactive β-endorphin processing in rat pars intermedia, *FEBS Lett.,* 175, 407, 1984.
12. **Smyth, D. G. and Zakarian, S.,** β-endorphin in brain, *Brain Res.,* 55, 123, 1982.
13. **Roberts, J. L. and Herbert, E.,** Characterization of a common precursor to corticotropin and β-lipotropin: cell free synthesis of the precursor and identification of corticotropin peptides in the molecule, *Proc. Natl. Acad. Sci. U.S.A.,* 74, 4826, 1977.
14. **Toogood, C. I. A., Ham, J., and Smyth, D. G.,** unpublished results.

ISOLATION-IDENTIFICATION OF OPIOIDS IN INVERTEBRATES

Michael K. Leung and George B. Stefano

SUMMARY

An acid extract of pedal ganglia of the mollusc *Mytilus edulis* was fractionated by high-pressure liquid chromatography with a reverse-phase column. Peak fractions with retention times of those of Met-enkephalin, Leu-enkephalin, and Met-enkephalin-Arg[6]-Phe[7] were subjected to binding assays in tissues. The results showed that these fractions have the same binding activities as authentic enkephalins. Peptides from these fractions were purified by high-pressure liquid chromatography under isocratic conditions. Sequential amino acid analyses showed that these peptides have the same primary structures as Met-enkephalin, Leu-enkephalin, and Met-enkephalin-Arg[6]-Phe[7]. These results with *M. edulis* suggest that invertebrates possess an enkephalinergic system similar to that of higher organisms.

INTRODUCTION

Since the initial identification of Met- and Leu-enkephalin by Hughes et al.[1] about 10 years ago, a tremendous amount of research effort has been concentrated on the study of these two enkephalins and their related peptides. Although many aspects of the opioid system require more intensive investigations, there is enough information available thus far to allow a general understanding of the biological activities as well as the biochemistry of opioids.[2-4] One of the areas that has generated great interest is the processing of opioids from their precursors.

At present, there are three known macromolecular precursors which contain enkephalin sequences; namely, proopiomelanocortin (POMC), proenkephalin, and prodynorphin. Each one of these is capable of being processed to different opioid peptides as well as other neuroactive molecules. Current findings indicate that the processing events of the precursors in mammalian neural tissues are very complex. It appears that the same precursor may be processed through different pathways to different end products, depending upon a number of parameters, such as the regions of the nerve tissues and external nerve stimuli. Thus far, numerous intermediates of various sizes from all these macromolecular precursors have been identified. One interesting aspect that emerged from studies of these intermediates is that a significant number of the neuropeptides in the precursors are bracketed both on the C- and N-terminal by a pair of basic amino acids. However, as a result of the complexity of the mammalian nerve tissues, the study of any isolated event in these tissues is extremely difficult.

There are two principal goals in our study of opioids in the mollusc *Mytilus edulis*. First, it is to establish definitively the existence of an enkephalinergic system in *Mytilus* through the isolation and identification of enkephalins as well as their precursors and intermediates. Second, it is to use *Mytilus* as a simple model for the study of the processing of opioid precursors. The present report will provide results which demonstrate the presence of Met-enkephalin, Leu-enkephalin, and Met-enkephalin-Arg[6]-Phe[7] in *Mytilus* pedal ganglia (Pg).

The selection of *Mytilus* Pg for this study is based upon several physiological and pharmacological studies which provided strong evidence for the presence of an enkephalinergic system in this tissue. Enkephalins have been shown to increase intraganglionic dopamine levels,[5] to depress cyclic AMP (cAMP) levels, and to antagonize dopamine-stimulated adenylate cyclase in *Mytilus*.[6] In addition, the ganglia contain both high and lower affinity opiate binding sites which have thus far been found to resemble mammalian delta receptors.[6-8] These earlier studies also demonstrate a close interrelationship between opioid and dopaminergic mechanisms.[9]

METHODS AND RESULTS

Extraction of Pedal Ganglia

Mytilus Pg were homogenized with a Brinkmann Polytron homogenizer with four 3-sec bursts in an extraction solution (25 pedal ganglia per milliliter) containing 1 M acetic acid, 20 mM HCl, 1 μg each of phenylmethyl sulfonyl fluoride and pepstatin A per milliliter, and 1% 2-mercaptoethanol. The homogenate was clarified by centrifugation at 27,000 × g for 1 hr. The supernatant was then deproteinized by the addition of 50% (wt/vol) trichloroacetic acid to reach a final concentration of 10% (wt/vol) and centrifuged at 27,000 × g for 30 min. The lipids and trichloroacetic acid in the supernatant were removed by extracting three times with equal volumes of diethyl ether. The aqueous portion was lyophilized. All of these steps were carried out at 4°C.

High-Pressure Liquid Chromatography (HPLC)

All steps subsequent to extraction were carried out at room temperature. The lyophilized extract was dissolved in a buffer (one pedal ganglion per microliter) consisting of 10 mM ammonium acetate, pH 4.0. The HPLC system used was a Beckman model 334 liquid chromatograph equipped with a Beckman/Altex 210 sample injector and a Beckman/Altex C-RIA integrator. Before injection, the sample was clarified by centrifugation at 1000 × g for 10 min. An aliquot was then subjected to HPLC on a Brownlee RP-300 reverse-phase column (4.6 × 250 mm). The column was eluted at a flow rate of 2 mℓ/min with a solvent system consisting of the HPLC buffer and a linear gradient of 5 to 25% (vol/vol) 2-propanol in 15 min.[10] For the isolation of crude fractions, the highly polar components of the extract were removed by an initial isocratic elution with 5% (vol/vol) 2-propanol for 19 min before the gradient was started. Collected consecutively were 1-mℓ fractions, starting from 16.5 min after injection. A typical elution pattern of 250 μℓ of acid extract is represented in Figure 1. Salts and highly polar materials in the extract were eluted in the initial isocratic phase. Numerous peaks were observed in the acid extract, and the Met-enkephalin, Leu-enkephalin, and Met-enkephalin-Arg[6]-Phe[7] elution positions were localized by analyzing samples with and without added enkephalin standards. Three peaks with retention times corresponding to the enkephalin standards are identified. Assuming that the peptides in the peaks thus identified were enkephalins, the weights of enkephalins in the acid extract were estimated from the integrated areas under the peaks. The results indicate the presence of weights equivalent of 3 to 4 ng of Met- and Leu-enkephalin each and 7 ng of Met-enkephalin-Arg[6]-Phe[7] per pair of pedal ganglia.

Fractions from each of the enkephalin peak regions were then assayed for biological activities. These fractions corresponded to fractions 11 and 12 for the Met-enkephalin region, fractions 18 to 20 for the Leu-enkephalin region, and fractions 24 to 26 for the Met-enkephalin-Arg[6]-Phe[7]. Preliminary examination of the HPLC fractions by assaying for the suppression of dopamine-stimulated adenylate cyclase activity was as described.[11] The results indicated that the highest enkephalin activities were in fraction 11 of the Met-enkephalin region, fraction 19 of the Leu-enkephalin region, and fraction 25 of Met-enkephalin-Arg[6]-Phe[7] region. These three fractions were then subjected to binding studies in mussel neural tissues.

Enkephalin Binding Analysis with Mussel Neural Tissues

Subtidal *M. edulis* were harvested from Long Island Sound at Northport, N.Y. For binding studies, pedal ganglia from fresh *M. edulis* were processed as described.[7] Immediately prior

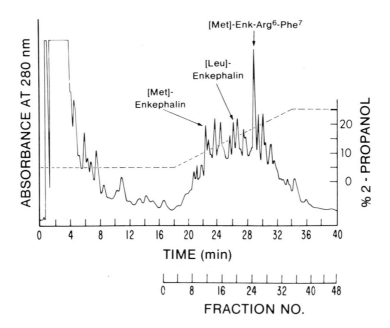

FIGURE 1.   HPLC of mussel ganglia extract. A sample (250 μℓ) was injected into a Brownlee RP-300 reverse-phase column. The column was eluted with 10 mM ammonium acetate buffer, pH 4.0, with 5% 2-propanol at a flow rate of 2 mℓ/min for 19 min. This elution was followed by a linear gradient of 5 to 25% 2-propanol for 15 min, and then elution with buffer with 25% 2-propanol was continued for an additional 6 min. Fractions (1 mℓ) were collected starting from 16.5 min. Met-enkephalin, Leu-enkephalin, and Met-enkephalin-Arg⁶-Phe⁷ regions were identified with known standards. (From *Proc. Natl. Acad. Sci. U.S.A.*, 81, 955, 1984. With permission.)

to the binding experiment, the supernatant was centrifuged at 30,000 × g for 15 min, and the resulting pellet was washed with 50 vol of sucrose/Tris-HCl buffer. The Tris-HCl buffer used was 150 mM KCl in 10 mM Tris-HCl buffer, pH 7.4, supplemented with 0.1% bovine serum albumin. Binding assays were carried out by a modification of the method of Pert and Snyder.[12] Aliquots of membrane suspension (0.2 mℓ, 0.12 mg of membrane protein) were incubated in triplicate at 4°C for 90 min with (D-Ala^2, Met^5) enkephalinamide (DAMA) labeled with ^{3}H in the Tyr^1 residue (New England Nuclear; 15.1 Ci/mmol; 1 Ci = 37 GBq) in the presence of 10 μM dextrorphan or levorphanol in Tris-HCl buffer. Free ligand was separated from membrane-bound labeled ligand by filtration under reduced pressure through Whatman GF/B glass fiber filters which had been soaked (45 min, 4°C) in Tris-HCl buffer containing 0.5% bovine serum albumin. The filters were rapidly washed with 2.5-mℓ portions of the incubation buffer (4°C), containing 2% polyethylene glycol 6000 (Baker). Radioactivity of (^{3}H) DAMA bound to the filters was then measured in a Packard 460 CD liquid scintillation counter. Stereospecific binding is defined as binding in the presence of 10 μM dextrorphan minus binding in the presence of 10 μM levorphanol. Protein concentration was determined in membrane suspensions (without bovine serum albumin) by the method of Lowry et al.[13] Displacement analyses were described by Kream et al.[7]

   The results of binding analyses of fractions 11, 19, and 25 together with known opiates are shown in Table 1. The concentrations of Met-enkephalin, Leu-enkephalin, and Met-enkephalin-Arg⁶-Phe⁷ samples prepared from fractions 11, 19, and 25 were calculated on the basis of the estimated weight. From IC$_{50}$ measurements, it is clear that both fractions 11 and 19, with a value of 1.2 nM, have the same IC$_{50}$ values as authentic Met-enkephalin and Leu-enkephalin and fraction 25, with an IC$_{50}$ value of 38.9 nM has the same value as authentic Met-enkephalin-Arg⁶-Phe⁷ (39.1 nM).

Table 1
**RELATIVE POTENCIES OF OPIATES IN DECREASING
BINDING OF (^{3}H) DAMA (1 nM) T0 MEMBRANE
SUSPENSIONS OF *MYTILIS EDULIS* PEDAL GANGLIA**

Opioid peptide	IC$_{50}$ (nM)	Relative potency
DAMA	1.0 ± 0.15	1.0
(D-Ala^2-, D-Leu^5) enkephalinamide	1.0 ± 0.1	1.0
(Met)enkephalin	1.2 ± 0.5	0.83
(Leu)enkephalin	1.3 ± 0.3	0.77
(Met)enkephalin-Arg^6-Phe^7	39.1 ± 1.1	0.026
(D-Ala^2)-β-endorphin	0.7 ± 0.1	1.4
Dynorphin-(1-13)	0.9 ± 0.15	1.1
Fraction 11	1.2 ± 0.2	0.83
Fraction 19	1.2 ± 0.5	0.83
Fraction 25	38.9 ± 0.9	0.026

Note: Aliquots of pedal ganglion membrane suspension were incubated with nonradioactive opioid compounds at eight concentrations for 10 min at 22°C and then with
(^{3}H) DAMA (1 nM) for 60 min at 4°C. One hundred percent binding is (^{3}H) DAMA
bound in the presence of 10 μM levorphanol. IC$_{50}$ is defined as the concentration
of drug that elicits half-maximal inhibition of specific (^{3}H) DAMA binding. The
mean ± S.E.M. for three experiments is reported for each compound tested. In
column 3 the IC$_{50}$ values are expressed relative to the value of DAMA, which is
defined as 1.0.

HPLC Purification of Crude Fractions

The crude fractions from the Met-enkephalin, Leu-enkephalin, and Met-enkephalin-Arg^6-
Phe^7 regions, which contained the highest biological activities, were lyophilized in separate
vials and redissolved in 100 μℓ of HPLC buffer. Aliquots of these solutions were subjected
to rechromatography under isocratic conditions with the HPLC system described in the
present report. The rechromatography of the materials from the crude fractions of Met-
enkephalin and Leu-enkephalin and Met-enkephalin-Arg^6-Phe^7 was carried out by use of
solvents consisting of HPLC buffer with 5% 2-propanol, 8% 2-propanol, and 12% 2-pro-
panol, respectively. The HPLC chromatograms are shown in Figure 2. Figure 2A represents
the elution pattern of the rechromatographed material from fraction 11. Only one major
peak, which eluted with the same retention time as authentic Met-enkephalin, was observed.
However, a small shoulder was also detected on the leading slope of the peak. This indicated
the presence of a minor contaminant with a retention time similar to that of Met-enkephalin.
Figure 2B shows the elution pattern of the rechromatography of the material from fraction
19. Only one major peak, with the same retention time as authentic Leu-enkephalin, was
observed. Figure 2C represents the elution pattern of the material from fraction 25. Several
major peaks were observed and one of these peaks has the same retention time as the Met-
enkephalin-Arg^6-Phe^7 standard. The peak fractions (2 mℓ) with the same retention time as
authentic Met-enkephalin, Leu-enkephalin, and Met-enkephalin-Arg^6-Phe^7 were collected
and lyophilized.

Amino Acid Sequence Determination

Portions of the purified materials were used for amino acid sequence determination. Amino
acid sequence determination was performed by a commercial laboratory (Sequemat, Water-
town, Mass.). The method used was manual Edman degradation, and the resulting amino
acid derivatives were identified by HPLC.

Amino acid sequence determination showed the sequence of the peptide from Figure 2A
to be Tyr-Gly-Gly-Phe-Met. No additional amino acid was detected beyond the fifth residue.

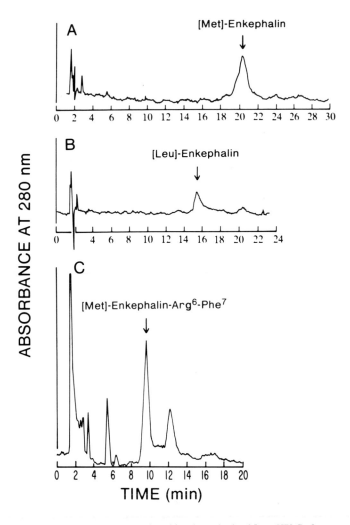

FIGURE 2. Rechromatography of fractions obtained from HPLC of mussel ganglia acid extract (Figure 1). Fractions 11, 19, and 25 were separately lyophilized, redissolved in HPLC buffer, and applied to the column, which was eluted under isocratic conditions. (A) Material from fraction 11 eluted with 5% 2-propanol. (B) Material from fraction 19 was eluted with 8% 2-propanol. (C) Material from fraction 25 was eluted with 12% 2-propanol. (From *Proc. Natl. Acad. Sci. U.S.A.*, 81, 955, 1984. With permission.)

However, amino acid analysis of the last residue showed the presence of glycine, phenylalanine, and leucine together with methionine. The presence of glycine and phenylalanine could be a result of the backwash from the previous residues, while the leucine may have come from the minor contaminant in the shoulder of the leading slope. This contaminant cannot be Leu-enkephalin, because with the solvent system used it is eluted in a totally separated peak after Met-enkephalin.

The amino acid sequence of the peptide from Figure 2B was determined to be Tyr-Gly-Gly-Phe-Leu. In this case, no other amino acid was detected in any of the amino acid fractions.

The results of analysis of the peptide from Figure 2C showed the sequence to be Tyr-Gly-Gly-Phe-Met-Arg-Phe. No additional amino acid was detected beyond the seventh residue.

DISCUSSION

At present, the known opioid peptides may be classified into three families; the endorphin, the enkephalin, and the dynorphin; based upon the precursors from which these neuropeptides are derived. These macromolecular precursors all contain at least one copy of the enkephalin sequence. Although there is no concrete evidence available at this point, it is possible that all three precursors may evolve from the same enkephalin gene sequence.

The peptides of the endorphin family are derived from the macromolecular precursor POMC. POMC is an interesting molecule in that its processed products include hormones such as ACTH, LPH, α-MSH, and β-MSH besides the opioid peptide β-endorphin.[3] β-Endorphin possesses very strong analgesic effect and contains the Met-enkephalin sequence at its N-terminal. The evolutionary and physiological significance of the structural association between β-endorphin and the other neuropeptides remain unclear at this point. β-Endorphin is bracketed on the N-terminal by the sequence -Lys-Arg-, thus, it is released by neuropeptide processing enzymes which typically cleave at the C-side of the paired basic amino acid sequence. However, the absence of a basic paired amino acid after the Met-enkephalin sequence makes this peptide an unlikely precursor for enkephalin. In addition, immunocytochemistry studies have shown that β-endorphin and enkephalin do not coexist in the same regions of the brain. However, as both ACTH and β-endorphin have been detected in invertebrates by immunocytochemical studies, it is likely that macromolecules similar to POMC may also exist in the invertebrates.

The presence of a number of intermediate-size peptides in bovine adrenal chromaffin cells had led to the identification of a macromolecular precursor proenkephalin.[4] Along the same line, it also raised the question whether enkephalins are the only "intended" final products of the precursor. This is especially true since some of the "intermediates" do exist at a relatively high level in the tissues. The "intermediates" have also been shown to have strong opioid activities. Nevertheless, the preliminary structure of proenkephalin showed that of six copies of the Met-enkephalin sequence, and one copy of the Leu-enkephalin sequence, only two Met-enkephalin sequences are not bracketed at the N- and C-terminal by paired basic amino acid sequence. They are Met-enkephalin-Arg^6-Phe^7 and Met-enkephalin-Arg^6-Gly^7-Leu^8. Thus, enkephalins can clearly be generated from proenkephalin neuropeptide processing enzymes.

The ratio of Met-enkephalin sequence to Leu-enkephalin sequence in proenkephalin is 6:1. On the other hand, the prodynorphin which was identified by Noda et al.[16] from cloned cDNA of porcine hypothalamus contains three copies of Leu-enkephalin sequence, but no Met-enkephalin sequences. Interestingly, the Leu-enkephalin sequences of prodynorphin are also bracketed by paired basic amino acid sequences. Similar to proenkephalin, prodynorphin in the tissues appears to be processed first to peptides of intermediate size, which have potent opioid activities and contain the enkephalin sequence. Here again, Leu-enkephalin may only be one of the "intended" products of prodynorphin. The detections of α-neo-endorphin and dynorphin in invertebrate left no doubt that proenkephalin is not the only source of Leu-enkephalin in invertebrate neural tissue. The coexistence of α-neoendorphin and dynorphin would provide very strong evidence for the presence of a precursor molecule similar to prodynorphin in invertebrates.

At this point, many interesting questions can be posed for the evolutionary relationship of these three precursors. However, very few data concerning the existence of the precursors in invertebrates are available. The presence of common enkephalin sequence in all three precursors serves to indicate that they may have a common origin. It is conceivable that the enkephalins are a very "primitive" form of neurotransmitters or neurohumors and as the organisms evolve, new neurotransmitters or hormones are generated by use of the enkephalins as a foundation. In time, these enkephalin neuropeptides seem to increase in their complexities and branch out to control numerous regulatory functions in the organisms. Thus, it is reasonable to assume that there exists in the primitive invertebrates molecules which resemble the three opioid precursors, and the isolation and identification of these precursors will provide important knowledge which will elucidate the evolution of the opioid peptides.

Recently, there have been numerous reports demonstrating the occurrence of mammalian neuropeptide-like immunoreactivity in various invertebrate nervous systems. However, much of the evidence has been based on immunocytochemistry, a technique which uses complementary amino acid structure for recognition. As a result, "complementary structures" may be demonstrated in larger molecules besides the "exact" compound. In *M. edulis*, enkephalin-like immunoreactivity is highly localized in the intraganglionic region of the pedal ganglion, which is also rich in dopamine-containing structures.[19] The present study not only supports these immunocytochemical observations, but also directly demonstrates the presence of Met-enkephalin, Leu-enkephalin, and Met-enkephalin-Arg[6]-Phe[7] as endogenous neuropeptides in *M. edulis* tissues.

The presence of Met-enkephalin, Leu-enkephalin, and Met-enkephalin-Arg[6]-Phe[7] in *M. edulis* suggests a high degree of similarity between the enkephalinergic systems of vertebrates and invertebrates. This is especially true since Met-enkephalin-Arg[6]-Phe[7] represents the C-terminal of the proenkephalin in bovine adrenal chromaffin cells. This indicates strongly the presence of a precursor in *Mytilus* that contains a C-terminal which is structurally the same as the mammalian proenkephalin. Thus, it is likely that high-molecular-weight precursors, such as proenkephalin which are found in vertebrates, are also present in invertebrate neural tissues. It should be informative to investigate the presence of such precursors in the neural tissue extract of *M. edulis*.

In conclusion, this study demonstrates that in relatively "primitive" organisms a variety of opioids can be found. The cellular communication mechanisms found in these animals are as complex as those in mammals. The importance of opioid mechanisms in evolution is thus highlighted. Indeed "simple" opioid mechanisms may have been lost in evolution and what we find today is the result of successful evolutionary competition.

ACKNOWLEDGMENTS

This work was supported by NIH-MBRS Grant #RR-018180 and ADAMHA-MARC Grant #MH-17138.

REFERENCES

1. **Hughes, J., Smith, T. W., Kosterlitz, H. W., Fothergill, L. A., Morgan, B. A., and Moores, H. R.,** Identification of two related pentapeptides from the brain with potent opiate agonist activity, *Nature (London)*, 258, 577, 1975.
2. **Hughes, J.,** Biogenesis release and inactivation of enkephalins and dynorphins, *Br. Med. Bull.*, 39, 17, 1983.
3. **Smyth, D. G.,** β-Endorphin and related peptides in pituitary, brain, pancreas and antrum, *Br. Med. Bull.*, 39, 25, 1983.
4. **Udenfriend, S. and Kilpatrick, D. L.,** Biochemistry of the enkephalin-containing peptides, *Arch. Biochem. Biophys.*, 221, 309, 1983.
5. **Stefano, G. B. and Catapane, E. J.,** Enkephalins increase dopamine levels in the CNS of a marine mollusc, *Life Sci.*, 24, 1617, 1979.
6. **Stefano, G. B., Catapane, E. J., and Kream, R. M.,** Characterization of the dopamine-stimulated adenylate cyclase in the pedal ganglia of *Mytilus edulis:* interaction with etorphine, β-endorphin, DALA and methionine enkephalin, *Cell. Mol. Neurobiol.*, 1, 57, 1981.

7. **Kream, R. M., Zukin, R. S., and Stefano, G. B.,** Demonstration of two classes of opiate binding sites in the nervous tissue of the marine mollusc *Mytilus edulis:* positive homotropic cooperativity of lower affinity binding sites, *J. Biol. Chem.*, 225, 9218, 1980.

8. **Stefano, G. B., Kream, R. M., and Zukin, R. S.,** Demonstration of opiate binding in the nervous tissue of marine mollusc *Mytilus edulis, Brain Res.*, 181, 445, 1980.

9. **Stefano, G. B.,** Aging: variation in opiate binding characteristics and dopamine responsiveness in subtidal and intertidal *Mytilus edulis* visceral ganglia, *Comp. Biochem. Physiol.*, 72C, 349, 1982.

10. **Huang, W. Y., Chang, R. C. C., Kastin, A. J., Coy, D. H., and Schally, A. V.,** Isolation and structure of pro-methionine-enkephalin: potential enkephalin precursor from porcine hypothalmus, *Proc. Natl. Acad. Sci. U. S. A.*, 76, 6177, 1979.

11. **Stefano, G. B. and Leung, M. K.,** Purification of opioid peptides from molluscan ganglia, *Cell. Mol. Neurobiol.*, 2, 347, 1982.

12. **Pert, C. D. and Snyder, S. H.,** Opiate receptor binding of agonists and antagonists affected differently by sodium, *Mol. Pharmacol.*, 10, 808, 1974.

13. **Lowry, O. H., Rosebrough, N. J., Farr, A. L., and Randall, R. J.,** Protein measurement with the Folin phenol reagent, *J. Biol. Chem.*, 193, 265, 1951.

14. **Lewis, R. V., Stein, S., Gerber, L. D., Rubinstein, M., and Udenfriend, S.,** High molecular weight opioid-containing proteins in striatum, *Proc. Natl. Acad. Sci. U.S.A.*, 75, 4021, 1978.

15. **Bloom, F., Baltenberg, E. Rossier, J., Ling, N., and Guillemin, R.,** Neurons containing β-endorphin in rat brain exist separately from those containing enkephalin: immunocytochemical studies, *Proc. Natl. Acad. Sci. U.S.A.*, 75, 1591, 1978.

16. **Noda, M., Furutani, Y., Takahashi, H., Toyosato, M., Hirose, T., Inayama, S., Nakanishi, S., and Numa, S.,** Cloning and sequence analysis of cDNA for bovine adrenal preproenkephalin, *Nature (London)*, 295, 202, 1982.

17. **Stefano, G. B. and Martin, R.,** Enkephalin-like immunoreactivity in the pedal ganglion of *Mytilus edulis* (Bivalvia) and its proximity to dopamine-containing structures, *Cell Tissue Res.*, 230, 137, 1983.

OPIOID AND RELATED NEUROPEPTIDES IN MOLLUSCAN NEURONS

Rainer Martin, Carola Haas, and Karl-H. Voigt

SUMMARY

Octopus nerves were found to contain, in agreement with biochemical studies in this volume, immunoreactive FMRFa and opioid and α-melanotropin-like substances. They occurred in the same nerve endings and neurons in the brain and periphery. In the pedal ganglion of the bivalve *Mytilus*, similar substances were detected, but each of the three peptides was expressed by a different population of neurons. In ganglia of the gastropods, *Aplysia*, and *Lymnaea*, related peptides occurred in particular neurons in all possible patterns of dissociation or association. Expression of specific sequences of a family of related peptides, and nonexpression of others, underlines the interest in precursor studies in molluscs. This peptide family in octopus nerves shows structural and functional similarities with opiomelanocortins in vertebrates.

INTRODUCTION

Evidence is accumulating that shorter neuropeptides and their receptors have been very stable in evolution. One of these old peptides, first isolated in vertebrates, is the opioid enkephalin (YGGFM/L).[1] It has now been identified in ganglia of a bivalve mollusc[2] and in analog form in cephalopod nerves.[3] Immunoreactive enkephalin-like material was found in secretory nerves of the octopus,[4] in ganglia of the pond snail *Lymnaea stagnalis*,[5] and of the clam *Mytilus edulis*.[6] Opioid receptor binding and a role of opioids comparable to that in vertebrates have been shown in cephalopod and bivalve ganglia[7] and in the terrestrial gastropod *Cepaea nemoralis*.[8]

Octopus[3] and *Mytilus*[9] nerves contain an elongated opioid peptide with the structure YGGFMRF. This heptapeptide, at least in the octopus, seems to function as an immediate precursor of the molluscan neuropeptide FMRFamide which was first found in a clam.[10] FMRFa has a marked excitatory effect on isolated heart preparations of a number of molluscan species as well as producing, in very low concentrations, contractures of certain noncardiac muscles.[11] It has potent and complex effects on *Helix* neurons[12] where it occurs in a different form.[13] Immunoreactive FMRFa-like material was recorded in *Lymnaea* ganglia[14] and in octopus nerves,[4] as well as in the rat brain.[15]

The octopus nerves which store YGGFMRFa and FMRFa were also found to contain immunoreactive α-melanotropin (MSH)- like material.[4,16] Colocalization of such peptides is interesting, since melanotropin analogs as well as an opioid are cosynthesized in vertebrates on the common proopiomelanocortin polypeptide.[17]

In the present study we report on the nature and localization of enkephalin-, FMRFa-, and α-MSH-like peptides in ganglia of *Octopus, Mytilus, Aplysia,* and *Lymnaea.* As we will show, these immunoreactive compounds in the three mollusc classes are distributed according to very different patterns of association and dissocation in neurons. Mollusc neuroendocrinology may contribute to an elucidation of the specific role of neuropeptides in identified neuronal circuits, to investigations into the evolution of peptide precursor systems, and to an understanding of the neuron-specific regulation of the expression of distinct peptide signals.

MATERIALS AND METHODS

Animals and Fixation

Slices of the brain and of the anterior vena cava from anesthetized (3% ethanol in seawater) specimens of *Octopus vulgaris,* pedal ganglia from *Mytilus edulis,* and ganglia from *Aplysia californica* were excised and immersed for 60 min into fixative. This fixative was 2% glutaraldehyde in natural seawater, and its pH was adjusted to 7.8 with NaOH. Ganglia from *Lymnaea stagnalis* were immersed for 60 min into 2% glutaraldehyde in 0.01 *M* phosphate buffer, pH 7.2. All tissues were dehydrated in ethanol and propylene oxide and embedded in Epon® 812. Treatment with heavy metals was omitted.

Semithin (0.5 μm) sections were attached by heat to glass slides and the resin was removed from the sections by Na-methoxide.[18] They were incubated with antibodies as described previously.[19] Antibody binding sites were visualized by application of the peroxidase anti-peroxidase method of Sternberger et al.[20] Also from ultrathin sections on nickel grids the resin was partially removed,[21] they were incubated first with egg albumin, then with the primary antibodies, and the protein-A gold marker was applied according to Roth.[22]

Antibodies

An antibody raised against synthetic FMRFa by E. Weber (Stanford, U.S.A.) in immunocytochemical preabsorption tests recognized C-terminally amidated FMRF and YGGFMRF, but not the desamidated forms, the enkephalins, or α-MSH (more detailed specificity tests are reported in Reference 4). It was applied on sections in dilutions of 1:20,000 to 1:30,000. In a radioimmunoassay (RIA) this antibody did not crossreact with YGGFMRF-OH, nor with the enkephalins. It crossreacted about 55% with γ-MSH, but not with α-MSH.[3]

An antibody raised against Leu-enkephalin (Immuno Nuclear Corporation, U.S.A.) in preabsorption tests was blocked by addition of Leu-enkephalin, Met-enkephalin, and longer enkephalin analogs, but not by FMRFa or by α-MSH (for details see Reference 4.) The working dilution for this antibody was 1:5000.

An antibody raised against the heptapeptide YGGFMRFa (a gift of M. Greenberg, Fla.) in preabsorption tests was blockable by the heptapeptide, the amidated as well as the desamidated form, and by FMRFa, but not by Met/Leu-enkephalin or by α-MSH (Figure 1). In a RIA it recognized the heptapeptide about 50 times better than FMRFa. In this RIA it did not crossreact with YGGFMRF-OH.[3] The working dilution of this antibody was 1:3200.

An antibody raised against synthetic α-MSH (K.H. Voigt) in preabsorption tests was blockable by authentic α-MSH and by ACTH 1-13, but not by the enkephalins or by FMRFa (for details, see Reference 16). This antibody was purified by affinity chromatography.

The specificity of the oxytocin antibody (Ferring, Malmö, Sweden) has been reported in Reference 23). The working dilution of this antibody was 1:3000.

A monoclonal antibody against β-endorphin (Ch. Gramsch, Munich) was found to recognize essentially the N-terminal tyrosine moiety of opioid peptides.[24]

Enzymatic Cleavage Tests

Pretreatment of the tissue sections with enzymes prior to incubation with different antibodies was carried out as described previously.[21]

Controls

The specificities of the antibodies were tested by addition of homologous or heterologous synthetic antigens to the antibody (40 μℓ of antiserum in working dilution + 10 μℓ of antigen in 1 mg/mℓ concentration, for 12 hr at 4°C), prior to application on tissue sections (e.g., Figure 1). Omitting one of the single steps of the peroxidase antiperoxidase procedure prevented immunoreaction. Absence of crossreactivities between the three main antibodies

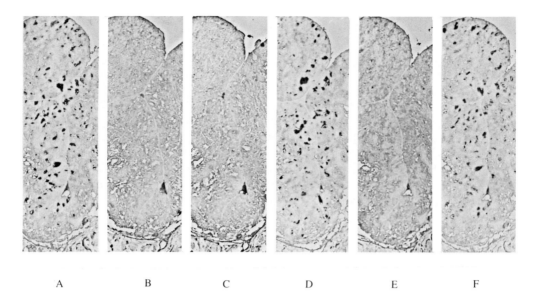

A   B   C   D   E   F

FIGURE 1. Specificity tests of the YGGFMRFa antibody on sections through the vena cava neuropil of the octopus. The immunoreactive spots in (A), (D), and (F) are peptidergic nerve terminals filled with secretory granules. In (A) the antibody has been applied without antigen (normal reaction), in (B) after addition of YGGFMRFa, (C) YGGFMRF, (D) M-enkephalin, (E) FLRFa, and (F) α-MSH. The antibody was not blocked by M-enkephalin and α-MSH (D,F); it was blocked by amidated or desamidated YGGFMRF and by FLRFa (B, C, E). (Magnification × 600.)

was indicated most clearly in the *Mytilus* ganglion in which each antibody immunostained a different population of neurons.

RESULTS

Peptidergic Neurons in the Nervous System of the Octopus

Secretory nerves that originate in the palliovisceral lobe of the posterior brain and terminate in large granule-loaded endings inside the vena cava in direct contact with the blood are at least of two different types: a population of nerve terminals with immunoreactive FMRFa-YGGFMRFa-, YGG—, and α-MSH-like peptides (Figure 2A, B, and D to F), and a population of terminals with vasopressin/oxytocin- and neurophysin-like material (Figure 2C; Reference 16). These two nerve populations appear not to account for all secretory terminals in the vena cava neuropil. Attempts to characterize other nerve endings by application of antibodies against the proopiomelanocortin fragments ACTH, 16k-fragment, β-endorphin, or against dynorphin and cholecystokinin were unsuccessful.

In the -RFa/MSH terminals, enkephalin immunoreactivity was very much enhanced by pretreatment of the sections with trypsin; this treatment at the same time abolished FMRFa reactivity. Without enzymatic cleavage enkephalin-like immunostaining was almost absent. We interpreted these observations as indication for occurrence of the heptapeptide-amide YGGFMRFa, and absence of enkephalin in pentapeptide form.[4] After treatment with trypsin, apparently all -RFa immunoreactive structures exhibited enkephalin-like immunostaining. The monoclonal YGG— antibody confirmed presence of the N-terminal opioid structure (Figure 2D). We assume that the terminals of this population all contain YGGFMRFa together with the tetrapeptide FMRFa.

Almost all brain lobes examined showed fine nerve endings with -RFa, α-MSH, and, after tryptic cleavage, enkephalin-like immunostaining in their centrally situated neuropil.

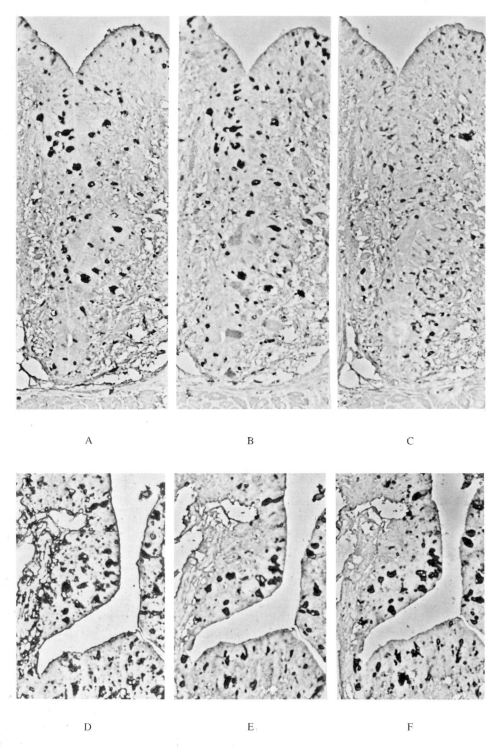

FIGURE 2.   Immunoreactivity in consecutive sections through the vena cava neuropil of the octopus, after use of the YGGFMRFa (A, E), the α-MSH (B), the oxytocin (C), the monoclonal YGG— (D), and the FMRFa (F) antibodies. All antibodies except the oxytocin antibody reacted with the same population of neurosecretory terminals. (Magnification × 500.)

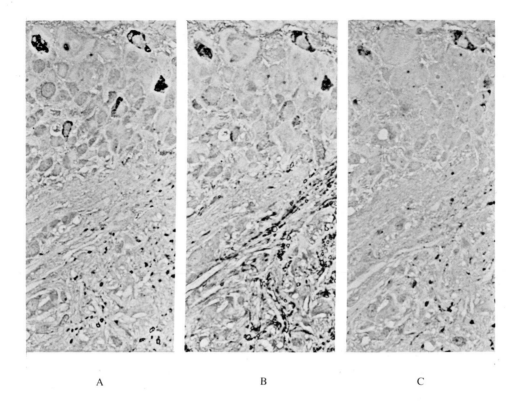

FIGURE 3. Serial sections through the superior buccal lobe of the octopus brain with (A) enkephalin-like (after tryptic cleavage), (B) FMRFa-, and (C) α-MSH-like immunoreactivity in perikarya and nerve endings. The same neurons are reactive with all three antibodies. (Magnification × 500.)

Large numbers of reactive terminals were found in the inferior and superior buccal lobes (Figure 3), the subpeduncular lobe (Figure 5) and in the optic lobe (Figures 6 and 7), they were less frequent in the pedal lobes (Figure 4), and rare or absent in the neuropils of the vertical and of the superior frontal lobe. The immunoreactive material in the terminals resided in densely packed electron opaque granules of about 90 nm diameter in the terminals. These granule populations were very uniform (Figures 7 and 8E, F).

The reactive terminals in brain lobes were too small to repeat frequently enough on 0.5-μm-thick serial sections for examination of coexistence patterns of the three peptides. However, the optic gland on the nerve stalks between the optic lobes and the central brain was shown to be innervated by one type of nerve endings only.[25] These nerve endings of the optic gland exhibited -RFa-like, α-MSH-like, and, after tryptic cleavage, enkephalin-like, but not vasopressin/oxytocin-like, immunostaining (Figure 8).

Immunoreactive perikarya in the cellular cortex of brain lobes were rather rare. Isolated cell bodies with -RFa-like immunostaining were observed in the cellular groups deep in the optic lobes, in the cortex of the superior buccal lobe (Figure 3), basal lobes, and the subpeduncular lobe (Figure 5), but not in the cortex of the vertical and superior frontal lobe. In the pedal lobes, e.g., the rare -RFa neurons were very small as compared to the motor neurons and were found in the cellular cortex in proximity to the neuropil (Figure 4). There were distinct reactive perikarya in serial sections after use of the FMRFa, the enkephalin, and the α-MSH antibody (Figure 3), but the number of immunostained cells and endings after application of the FMRFa antibody was always considerably larger. Perikarya with MSH or enkephalin immunostaining were rather rare.

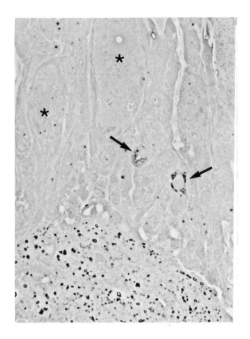

FIGURE 4.   FMRFa-like immunoreactivity in the pedal lobe of the octopus. Rare small neurons (arrows), not the large motor neurons (*), are immunoreactive. (Magnification × 400.)

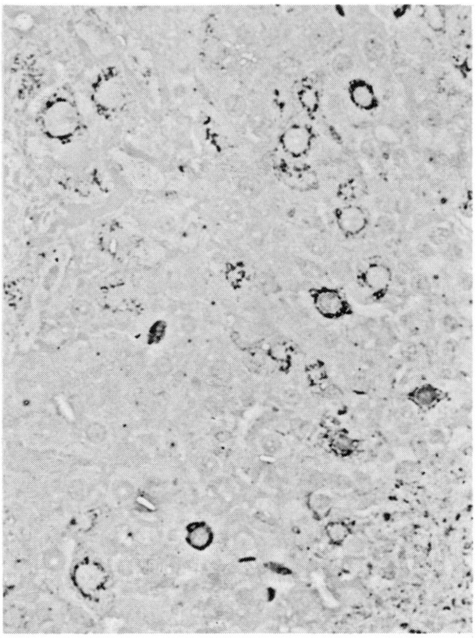

FIGURE 5.   FMRFa immunoreactive neurons in the subpeduncular lobe of the octopus. (Magnification × 400.)

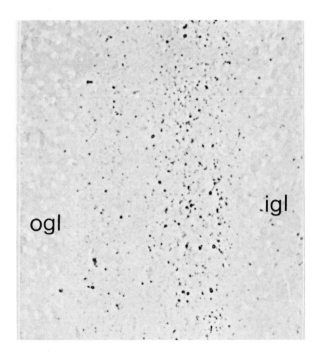

FIGURE 6.   Small FMRFa immunoreactive endings in the deep retina of the octopus optic lobe. ogl, Outer granule cell layer; igl, inner granule cell layer. (Magnification × 400.)

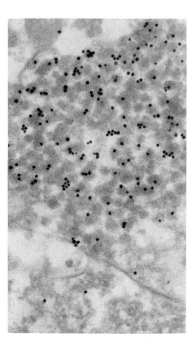

FIGURE 7. Electron micrograph of one of the FMRFa immunoreactive endings of the optic lobe (see Figure 6) with peptidergic synaptic granules. Protein-A gold labeling. (Magnification × 44,000).

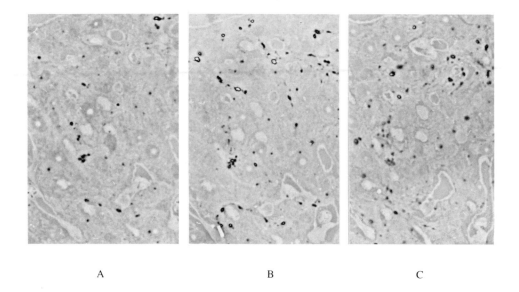

A   B   C

FIGURE 8. Light micrographs of sections through the optic gland of the octopus with (A) enkephalin immunoreactivity (after tryptic cleavage), (B) FMRFa, and (C) α-MSH immunoreactivity, but no oxytocin/vasopressin immunostaining (D). (Magnification × 600.) The optic gland has been found to be innervated by one nerve type only, i.e., the different immunoreactive substances co-occur in the same nerves. The immunoreactive material (anti-FMRFa) resides in 90-nm granules in the varicosities (E) and axons (F) of the optic gland (protein-A gold method.) (Magnification × 44,000.)

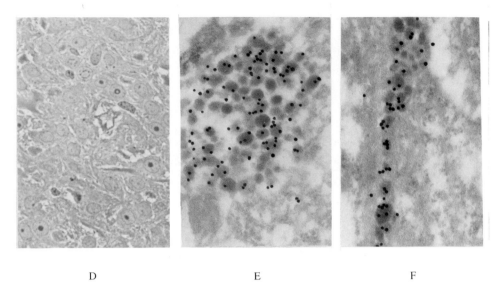

D   E   F

FIGURE 8 continued

Peptidergic Neurons in the Pedal Ganglion of *Mytilus edulis*

There were distinct differences in the distribution of the immunoreactive peptides between *Mytilus* and *Octopus* ganglia. In the posterior central region of the pedal ganglion of the mussel, a group of neuronal perikarya exhibited intense enkephalin-like immunostaining (Figure 9A). The immunoreaction was not intensified by pretreatment with enzymes. These enkephalin neurons in serial sections were different from those neurons which exhibited immunostaining with the FMRFa and with the α-MSH antibody (Figure 9A to D).

Distribution of enkephalin-, FMRFa-, and α-MSH-like material on different neuron populations was confirmed also in survey sections through the pedal ganglion. Numerous FMRFa perikarya were scattered over the whole ganglionic cortex. MSH neurons were concentrated in the posterior central region and were much less frequent in the ganglionic cortex. Enkephalin perikarya were almost exclusively found in the posterior central area. In serial sections through the ganglion the different immunoreactive neurons in no case coincided.

Very small reactive nerve endings were found in the ganglionic neuropil. FMRFa and α-MSH endings were observed all over the neuropil and in the large cerebropedal nerve, while enkephalin endings occurred in an anterior dorsal region of the neuropil only.

Peptidergic Neurons in *Aplysia Ganglia*

Distinct FMRFa-like immunoreactivity was found in all ganglia of the *Aplysia* nervous system in a limited number of perikarya in the ganglionic cortex, and in form of small nerve endings in the neuropil (Figure 10). Smaller neurons, not the typical giant neurons of *Aplysia* ganglia, were immunoreactive. α-MSH-like immunostaining in most ganglia was confined to small endings in the neuropil (Figure 11B). Only in the buccal and cerebral ganglia there were a few immunoreactive perikarya (Figure 11A). Enkephalin-like immunostaining, with the exception of a group of perikarya in the buccal ganglion, was confined to endings in the neuropil. Pretreatment of the sections with trypsin enhanced the intensity of the reaction and revealed perikarya which without enzymatic cleavage were unreactive (Figure 12A and B). There was no evidence for coexistence of the three immunoreactive substances in specific neurons, with exception of a group of perikarya in the buccal ganglion. The two bag cell clusters at the rostral ends of the abdominal ganglion which synthesize, store, and release egg-laying hormone and related peptides[26] were not antigenic to any of our antibodies.

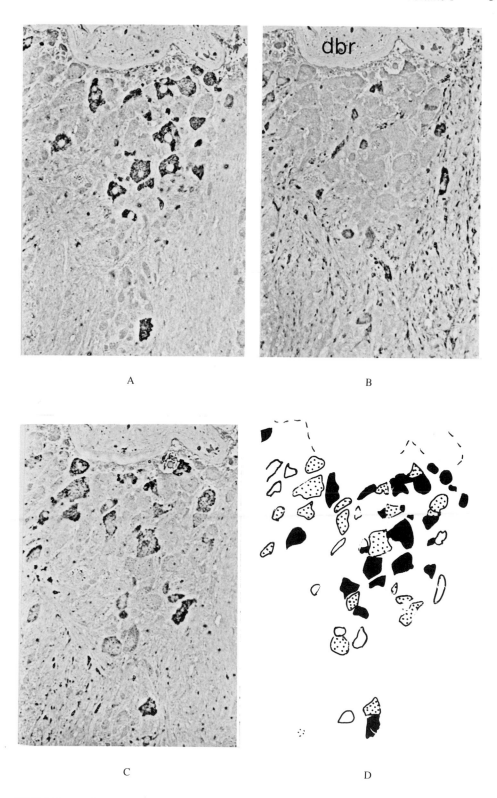

A

B

C

D

FIGURE 9.   Serial sagittal sections through the pedal ganglion of *Mytilus edulis*. The enkephalin antibody (A), the FMRFa (B), and the α-MSH antibody (C) were immunoreactive in different neuron populations (D). dbr, dorsal byssus retractor nerve; ●, enkephalin; ○, FMRFa; ◉ , α-MSH. (Magnification × 400.)

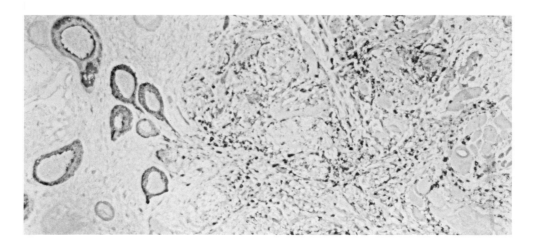

FIGURE 10.   A section through the abdominal ganglion of *Aplysia* with FMRFa immunoreactivity in perikarya and in nerve endings in the neuropil. (Magnification × 200.)

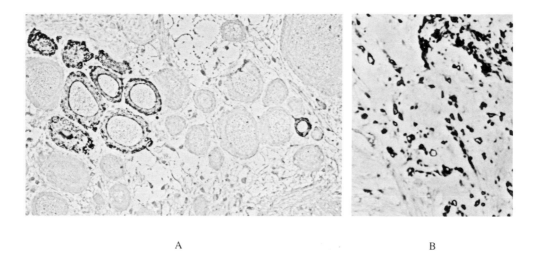

A   B

FIGURE 11.   Sections through the (A) buccal and the (B) pedal ganglion of *Aplysia* with α-MSH immunoreactivity. (Magnification (A) × 250. Magnification (B) × 500.)

Immunoreaction after use of the α-MSH, the FMRFa, or the enkephalin antibody resided always in small nerves, not in the giant nerve fibers. The peptidergic terminals were often lined up along large axons. At the fine structural level spines with peptide granules were found to protrude into large axons (Figure 13A). Also distinct synaptic contacts with membrane specializations were observed (Figure 13B). In perikarya and in nerve terminals the immunoreactive material was stored in dense granules of about 150 nm diameter (Figures 11A, 13A to C.)

Peptidergic Neurons in Ganglia of *Lymnaea stagnalis*

In large neurons of visceral and pedal ganglia of *Lymnaea stagnalis* the immunoreactive substances detected by our antibodies occurred in all possible combinations. Neurons which exhibited FMRFa and enkephalin immunoreactivity did not show α-MSH-like immunostaining (Figure 14). Enkephalin neurons with α-MSH but not FMRFa immunoreactivity were observed (Figure 15), as were neurons with α-MSH- and FMRFa- but no enkephalin-

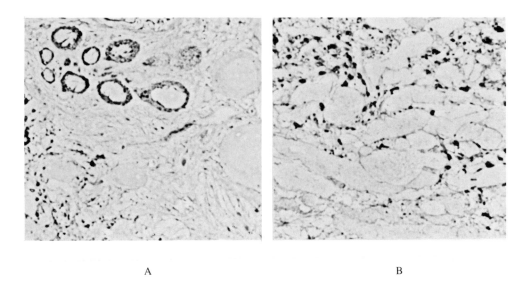

A   B

FIGURE 12.   Sections through the (A) abdominal and (B) pedal ganglion of *Aplysia* with enkephalin immuno-reactivity. The immunostaining was much enhanced by pretreatment of the sections with trypsin. (Magnification (A) × 300. Magnification (B) × 500.)

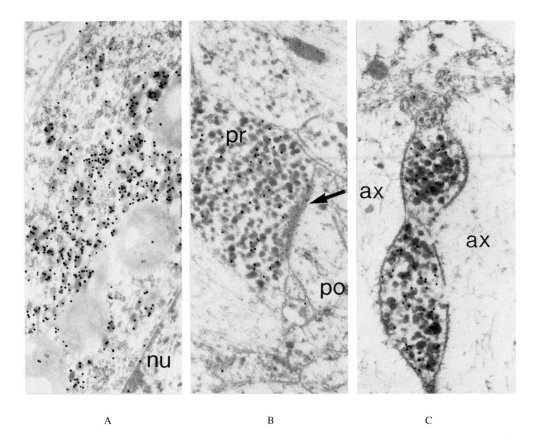

A   B   C

FIGURE 13.   In *Aplysia* perikarya (A), synaptic nerve endings (B), and varicosities (C), the immunoreaction resides in peptidergic granules of about 150 nm diameter (FMRFa antibody; protein-A gold method). nu, Nucleus; pr, presynaptic; po, postsynaptic axon; the arrow points to synaptic membranes; ax, axoplasm of a giant axon with invaginating peptidergic varicosities. (Magnification × 25,000.)

anti-enkeph.   anti-FMRFa   anti-α-MSH

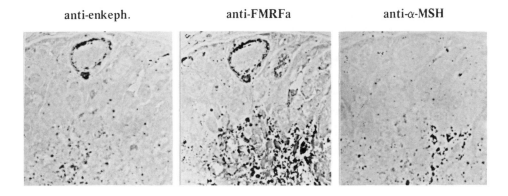

FIGURE 14

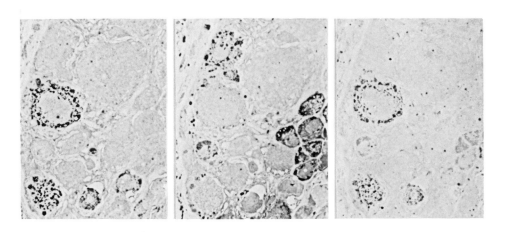

FIGURE 15

FIGURES 14 to 17.   Serial sections through ganglia of *Lymnaea stagnalis* with enkephalin (left column), FMRFa (central column) and α-MSH immunoreactivity (right column). Immunoreactive perikarya with all possible combinations of antigenic material are shown. (Magnification × 300.)

like material (Figure 16). There were also neurons with FMRFa-like immunostaining only (Figure 17).

DISCUSSION

The immunocytochemical observations of the present study agree well with biochemical characterizations of extracted peptides. In octopus nerves we found -RFa immunoreactivity associated with enkephalin-like immunoreactivity when the tissue sections were treated with endopeptidases prior to application of the enkephalin antibody. This indicates occurrence of a longer enkephalin analog which is rendered antigenic by enzymatic cleavage.[4] Enkephalin immunoreactivity without enzymatic treatment was very weak. But an antibody against the N-terminal opioid portion (YGG-) was always immunoreactive. Octopus nerves therefore appear to contain a molecule with the enkephalin sequence at the N-terminus and the -RFa structure at the C-terminus, possibly YGGFMRFa. In fact, extracts of the octopus brain when subjected to gel chromatography and HPLC fractionation showed large peaks comi-

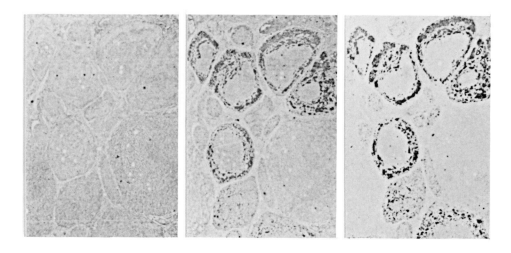

FIGURE 16

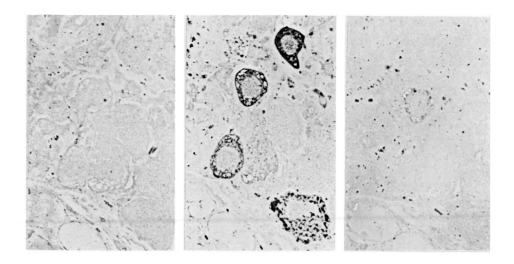

FIGURE 17

grating with FMRFa and with YGGFMRFa. The tetrapeptide was present in about three times higher concentrations on a molecular basis.[3] A peak coeluting with enkephalin as pentapeptide was detectable after tryptic digestion of the extracts.[3]

In the octopus, coexistence of FMRFa and YGGFMRFa in the same nerves was evident in section series through the vena cava neuropil, the nerves in the optic gland, and perikarya in the superior buccal lobe. These observations imply incomplete cleavage of YGGFMRFa, if the tetrapeptide is derived from the heptapeptide. However, assumption of regional diversity in the octopus brain with neurons containing the tetrapeptide only and neurons storing the heptapeptide together with the tetrapetide would not be in conflict with the biochemical and immunocytochemical data. In fact, there were many perikarya with FMRFa immuno-reactivity which even after enzymatic treatment did not exhibit enkephalin immunoreactivity.

The immunocytochemical findings also agree with biochemical data in the case of the pedal ganglion of *Mytilus*. There was an immunoreactive neuron population after use of the FMRFa antibody. In contrast to octopus nerves, a second neuron population exhibited enkephalin immunoreactivity without use of enzymes. Both enkephalin (methionine- as well

Table 1

	Octopus	Mytilus			Lymnaea		
FMRFamide	O	O			O	O	
YGGFMRFamide	Δ						
YGGFM/L			Δ			Δ	Δ
α-MSH-analog	X			X	X		X

> *Note:* Similar neuropeptides occur in representatives of three mollusc classes. In the octopus
> they coexist in the same nerves, in *Mytilus* they are distributed on three different
> neuron populations, and in *Lymnaea* varying pairs of peptides co-occur in particular
> neurons.

as leucine-enkephalin) and YGGFMRF have been found in extracts from *Mytilus* ganglia.[9] Also FMRFa most likely is present in this bivalve species. It would be important to know whether or not YGGFMRF is present in amidated form, since lack of amidation would indicate sequence differences in the respective precursor peptides. Absence of enkephalin immunoreactivity in -RFa neurons after pretreatment with trypsin may indicate that YGGFMRF is contained in the enkephalin neuron population.

To our knowledge no biochemical data on opioid and related peptides from *Lymnaea* and *Aplysia* ganglia are available at present for comparison with immunocytochemical observations. Unfortunately, biochemical data on α-MSH-like material in molluscs which we found in immunoreactive form of high intensity in all four species examined are totally lacking.

The most surprising findings in this study are wide variations in the neuronal distribution of the different peptides from mollusc class to class (Table 1). In the octopus, YGGFMRFa, FMRFa, and an α-MSH-like substance were associated in the same nerves. In *Mytilus*, enkephalin, FMRFa, and α-MSH-like material were dissociated in specific neuron populations. The most variable distribution pattern was found in *Lymnaea* ganglia in which each immunoreactive substance was associated with one of the other two in distinct neuron populations. Even more numerous combinations of peptides may occur in *Lymnaea* neurons which were found to contain a very large number of analogs to vertebrate neuropeptides.[5] The peptides in the different classes appear to be similar but not identical. The bivalve ganglion in contrast to octopus nerves stores enkephalin in pentapeptide form; a gastropod analog to FMRFa includes considerable sequence variations.[13]

We can at present only speculate about genetic mechanisms that allow expression of peptides in different combinations by specific neuron populations. The most likely explanation might be occurrence of a family of genes coding for multiple related precursors with enkephalin-, FMRFa-, and α-MSH-like sequences. Minor changes in structure between related precursors might regulate processing of particular peptide sequences in specific cells. It is then a matter of neuronal differentiation which gene is expressed in a specific neuron population. Families of related but diversified genes coding for egg-laying hormones in *Aplysia* were found in bag cell neurons.[26] Moreover, there are two or more genes for proopiomelanocortin in vertebrates.[27] Possibly, the large number of replicated DNA in polyploid gastropod neurons allows expression of more than one gene in specific neurons. There is immunocytochemical evidence for occurrence of many different FMRFa-related forms in *Lymnaea* ganglia.[28]

The octopus neuropeptides FMRFa, YGGFMRFa, and the α-MSH analog exhibit striking similarities to the proopiomelanocortin derivatives of vertebrates. Proopiomelanocortin is a promelanotropin precursor with four analogous melanotropin sequences and an opioid sequence in the C-terminal region.[29] There are also functional relationships between the pituitary opiomelanocortin system of vertebrates and the neurosecretory system of the vena cava in the octopus, since both organs appear to become activated under conditions of stress.[30] A characterization of the octopus precursor polypeptide might allow insight into the evolution of opiomelanocortins.

Moreover, molluscan ganglia appear most suitable for studies into the function of neuropeptides in defined nerve circuits. Peptidergic synapses and peptidergic spines invading giant axons as observed here in *Aplysia* and *Octopus* ganglia may represent convenient models for an examination of the role of neuropeptides in accessible nerve circuits. It would be most interesting to learn which function is exerted by an α-MSH-like peptide in molluscs.

REFERENCES

1. **Hughes, J., Kosterlitz, T. W., and Fothergill, L. A.,** Identification of two related pentapeptides from the brain with potent opiate agonist activity, *Nature (London),* 258, 577, 1975.
2. **Leung, M. K. and Stefano, G. B.,** Isolation and identification of enkephalins in pedal ganglia of *Mytilus edulis* (Mollusca), *Proc. Natl. Acad. Sci. U.S.A.,* 81, 955, 1984.
3. **Voigt, K. H. and Martin, R.,** Neuropeptides with cardioexcitatory and opioid activity in *Octopus* nerves, in *Handbook of Comparative Opioid and Related Neuropeptide Mechanisms,* Vol. 1, Stefano, G. B., Ed., CRC Press, Boca Raton, Fla., 1986.
4. **Martin, R., Frösch, D., Kiehling, C., and Voigt, K. H.,** Molluscan neuropeptide-like and enkephalin-like material co-exist in *Octopus* nerves, *Neuropeptides,* 2, 141, 1981.
5. **Schot, L. P. C., Boer, H. H., Swaab, D. F., and van Noorden, S.,** Immunocytochemical demonstration of peptidergic neurons in the central nervous system of the pond snail *Lymnaea stagnalis* with antisera raised to biologically active peptides, *Cell Tissue Res.,* 216, 273, 1981.
6. **Stefano, G. B. and Martin, R.,** Enkephalin-like immunoreactivity in the pedal ganglion of *Mytilus edulis* (Bivalvia) and its proximity to dopamine-containing structures, *Cell Tissue Res.,* 230, 147, 1983.
7. **Stefano, G. B. and Hall, B.,** Opioid inhibition of dopamine release from nervous tissue of *Mytilus edulis* and *Octopus bimaculatus, Science,* 213, 928, 1981.
8. **Kavaliers, M., Hirst, M., and Teskey, G. C.,** A functional role for an opiate system in snail thermal behavior, *Science,* 220, 99, 1983.
9. **Leung, M. and Stefano, G. B.,** Isolation of molluscan opioid peptides, *Life Sci.,* 33/I, 77, 1983.
10. **Price, D. A. and Greenberg, M. J.,** Structure of molluscan cardioexcitatory neuropeptide, *Science,* 197, 670, 1977.
11. **Greenberg, M. J. and Price, D. A.,** FMRFamide, a cardioexcitatory neuropeptide of molluscs: an agent in search of a mission, *Am. Zool.,* 19, 164, 1979.
12. **Cottrell, G. A.,** FMRFamide neuropeptides simultaneously increase and decrease K currents in an identified neurone, *Nature (London),* 296, 87, 1982.
13. **Price, D. A.,** The FMRFamide-like peptide of *Helix aspersa, Comp. Biochem. Physiol.,* 72C, 325, 1982.
14. **Schot, L. P. C. and Boer, H. H.,** Immunocytochemical demonstration of peptidergic cells in the pond snail *Lymnaea stagnalis* with an antiserum to the molluscan cardioactive tetrapeptide FMRFamide, *Cell Tissue Res.,* 225, 347, 1982.
15. **Dockray, G. J. and Williams, R. G.,** FMRFamide-like immunoreactivity in rat brain: development of a radioimmunoassay and its application in studies of distribution and chromatographic properties, *Brain Res.,* 266, 295, 1983.
16. **Martin, R., Frösch, D., and Voigt, K. H.,** Immunocytochemical evidence for melanotropin- and vasopressin-like material in a cephalopod neurohemal organ, *Gen. Comp. Endocrinol.,* 42, 235, 1980.
17. **Mains, R. E. and Eipper, B. A.,** Coordinate synthesis of corticotropins and endorphins by mouse pituitary tumor cells, *J. Biol. Chem.,* 253, 651, 1978.
18. **Mayor, H. D., Hampton, J. C., and Rosario, B.,** A simple method for removing the resin from epoxy embedded tissue, *J. Cell Biol.,* 9, 909, 1961.
19. **Weber, E., Voigt, K. H., and Martin, R.,** Pituitary somatotrophs contain Met-enkephalin-like immunoreactivity, *Proc. Natl. Acad. Sci. U.S.A.,* 75, 6134, 1978.
20. **Sternberger, L. A., Hardy, P. H., Cuculis, J. J., and Meyer, H. G.,** The unlabeled antibody enzyme method of immunocytochemistry, *J. Histochem. Cytochem.,* 18, 315, 1970.
21. **Martin, R., Schäfer, M., and Voigt, K. H.,** Enzymatic cleavage prior to antibody incubation as a method for neuropeptide immunocytochemistry, *Histochemistry,* 4, 457, 1982.

22. **Roth, J.,** Applications of immunocolloids in light microscopy. Preparation of protein A-silver and protein A-gold complexes and their application for localization of single and multiple antigens in paraffin sections, *J. Histochem. Cytochem.,* 30, 691, 1982.

23. **Martin, R., Geis, R., Holl, R., Schäfer, M., and Voigt, K. H.,** Co-existence of unrelated peptides in oxytocin and vasopressin terminals of rat neurohypophysis: immunoreactive Met-enkephalin-, Leu-enkephalin- and cholecystokinin-like substances, *Neuroscience,* 8, 213, 1983.

24. **Herz, A., Gramsch, C., Höllt, V., Meo, T., and Riethmüller, G.,** Characteristics of a monoclonal β-endorphin antibody recognizing the N-terminus of opioid peptides, *Life Sci.,* 31, 1721, 1982.

25. **Frösch, D.,** The subpedunculate lobe of the octopus brain: evidence for dual function, *Brain Res.,* 75, 277, 1974:

26. **Scheller, R. H., Jackson, J. F., McAllister, L. B., Rothman, B. S., Mayeri, E., and Axel, R.,** A single gene encodes multiple neuropeptides mediating a stereotyped behaviour, *Cell,* 32, 7, 1983.

27. **Herbert, E.,** Discovery of pro-opiomelanocortin — a cellular polyprotein, *Trends Biochem. Sci.,* 6, 184, 1981.

28. **Schot, L. P. C., Boer, H. H., and Montange-Wajer, C.,** Characterisation of multiple immunoreactive neurons in the central nervous system of the pond snail *Lymnaea stagnalis* with different fixatives and antisera adsorbed with the homologous and the heterologous antigens, Academisch Proefschrift, Vrije Universiteit, Amsterdam, 1984, 65.

29. **Nakanishi, S., Inoue, A., Kita, T., Nakamura, M., Chang, A. C. Y., Cohen, S. N., and Numa, S.,** Nucleotide sequence of cloned cDNA for bovine corticotropin-β-lipotropin precursor, *Nature (London),* 278, 423, 1979.

30. **Wells, M. H.,** Hormones and the circulation in octopus, in *Molluscan Neuro-Endocrinology,* Lever, J. and Boer, H. H., Eds., North-Holland, Amsterdam, 1983, 21.

IS THERE AUTHENTIC SUBSTANCE P IN INVERTEBRATES?

R.M. Kream, M. K., Leung, and G. B. Stefano

SUMMARY

Mytilus edulis pedal ganglia were extracted in 2 *N* acetic acid, and the supernatant fractions were concentrated by passage over C_{18} SEP-PAK® cartridges. After reconstitution, aliquots were run in a highly sensitive radioimmunoassay. Substance P-like immunoreactivity was quantitated at less than 0.2 pg/mg tissue. The immunoreactivity was not recoverable after gel filtration chromatography followed by HPLC. In addition, column fractions were inactive in both bioassay and receptor binding assays. We conclude that substance P is not present in the pedal ganglia of *M. edulis*.

INTRODUCTION

Recently, opiate systems in invertebrates have been described. Both narcotics and opioid can increase dopamine levels in various invertebrates.[1-6] In all cases this action was reversed by naloxone, suggesting the presence of opiate receptor mechanisms. Subsequently, saturable, high affinity and stereospecific opiate-binding sites were demonstrated in the neural tissues of *M. edulis*.[5-6] The properties of opiate binding to invertebrate neural tissue are similar to those reported for mammalian opiate receptors.

In addition, a naloxone reversible action of Met-enkephalin and morphine on the activity of single identified neurons in *Helix pomatia* has been reported.[7] More recently, the presence of enkephalin-like immunoreactivity in *M. edulis* pedal ganglia has been demonstrated and purification of these peptides has proceeded.[8-9]

As a result of these findings, it was of interest to determine if substance P systems are present in invertebrate neural tissues. Substance P-like immunoreactivity has been reported in *H. aspersa* neural tissues.[10] In addition, substance P was found to interact with opiates in the CNS of *H. pomatia*.[11] This study involves the detection and authentication of substance P in *M. edulis* pedal ganglia.

METHODS

Tissue Extraction

Subtidal *M. edulis* were harvested from the Long Island Sound at Northport, N.Y., during October 1983. Three pools of ganglia of 170, 143, and 113 weighing 0.349, 0.281, and 0.197 g, respectively, were extracted in 50 volumes 2 *N* acetic acid. An aliquot of the tissue extract (10%) was reserved and spiked with 100 fmol (^{3}H) substance P (New England Nuclear) to estimate recovery through the purification procedures.

The acidified extract was percolated twice over a C_{18} SEP-PAK® cartridge (Waters, Milford, Mass.) prewetted with 10 mℓ HPLC-grade methanol followed by 10 mℓ H_2O. The SEP-PAK® was washed with 10 mℓ 0.1% TFA, and 10 mℓ 10 % acetonitrile, 0.1% TFA, and then substance P was eluted in 5 mℓ 40% acetonitrile (v/v), 0.1% TFA (w/v).

For radioimmunoassay, 0.5-mℓ stock RIA sample buffer (see following) and 5-μℓ stock β-mercaptoethanol were added to the eluant fraction, and total volume was reduced to 1.0 mℓ under a stream of dry N_2. For HPLC, the eluant fraction, was reloaded over a fresh prewetted SEP-PAK® and eluted in 3 mℓ absolute ethanol percolated over the cartridge three times. Stock β-mercaptoethanol (3 μℓ) was added and the total volume was reduced to 10 μℓ under a stream of dry N_2, and then 100 μℓ 0.1% TFA was added and HPLC performed;

(^{3}H)-substance P was used to monitor recovery from the C_{18} SEP-PAK® and varied from 92.4 to 107.0%. Radioactivity was quantitated in a Beckman® LS-100 liquid scintillation spectrometer. Samples were counted in 1 mℓ 40% acetonitrile/0.1% TFA with 12 mℓ hydrofluor at a counting efficiency of 18.5% using (^{3}H)-toluene as an internal standard.

Radioimmunoassay

Generation of Antisera

Substance P (2 mg) was conjugated to succinylated thyroglobulin (4 mg) using 10 mg 1-ethyl-3-(3-dimethylaminopropyl) carbodiimide in a modification of the procedure of Mroz and Leeman.[12] Incorporation was 80% as monitored by HPLC elution of unconjugated peptide. Conjugated peptide was emulsified with an equal volume of Freund's complete adjuvant (0.7 mℓ final volume) and injected intradermally into 2 to 3 kg young female white New Zealand rabbits. From 200 to 300 μg of conjugated peptide was initially injected. Animals were twice boosted after 1-month intervals with half the initial amount of antigen. Animals were bled 2 weeks after final boost. Sera were screened for binding of radioiodinated peptide tracer, prepared as described (below).

Preparation of (^{125}I)-Labeled Bolton-Hunter Conjugated Substance P

The solvent was removed from 1 mCi of monoiodo (^{125}I)-Bolton-Hunter reagent (0.45 nmol ester) in 100-μℓ benzene in a closed conical reaction vial by drawing dry N_2 under reduced pressure throught the vial with volatile radioactivity trapped by two activated charcoal filters. Then 100 ng substance P (0.075 nmol peptide) in 1 μℓ 50% ethanol/0.5 *M* sodium borate (pH 8.0) was injected through the septum. The stoichiometry of the coupling reaction was 6:1, ester to peptide. After 30 min at 0°C, the reaction was quenched with 250 μℓ 6 *M* guanidine-HCl, 250 μℓ 0.2 *M* pyridine/1.0 *M* acetic acid and the mixture fractionated over a prewetted C_{18} SEP-PAK®. After washes with 20 mℓ 0.1% TFA, and 20 mℓ 10% acetonitrile, 0.1% TFA, (^{125}I)-Bolton-Hunter conjugated substance P was eluted in 4 mℓ absolute ethanol. Then 1% β-mercaptoethanol was added and the volume reduced for HPLC. An aliquot of the absolute ethanol eluant fraction was chromatographed isocratically on a C_{18} μ-Bondapak® analytical column with 48% methanol, 0.1% TFA as the mobile phase (Figure 1). Four peaks of radioactivity eluted at 9.8, 12.6, 18.9, and 28.7 mℓ, respectively. When assayed by quantitative immunotitration with substance P antibody (see below), only the peak at 28.7 mℓ displayed specific binding to the antibody. At a final antibody dilution of 1:240,000, approximately 5000 cpm or 50% of added (^{125}I)-labeled tracer was bound. This represented 2 fmol (^{125}I)-labeled conjugate at a specific activity of 2200 Ci/mmol and counting efficiency of 70%. The elution position of the (^{125}I)-labeled substance P derivative was equivalent to that of an independently synthesized (^{127}I)-labeled (^{3}H)-substance P derivative, was well separated from the authentic substance P (which eluted at 12 mℓ), and rechromatographed at the same position.

Coelution on HPLC of the (^{125}I)-labeled substance P derivative and the (^{3}H)-labeled monoiodinated derivative provides presumptive evidence that the (^{125}I)-labeled tracer is a monoiodinated substance P derivative. The HPLC elution position indicates the derivative to be considerably more hydrophobic than native substance P as described by Michelot et al.[13] and Liang and Cascieri.[14] The conjugation reaction was run at pH 8.0 rather than 8.5 as described previously,[13-15] to favor acylation of the amino terminal of substance P over the amino group of lysine at position three. While this may have contributed to the low yield (13%) of (^{125}I)-product, nevertheless one coupling reaction using 1 mCi (^{125}I)-Bolton-Hunter reagent produced over 100 μCi of purified (^{125}I)-labeled tracer, sufficient for 20,000 separate analyses by RIA.

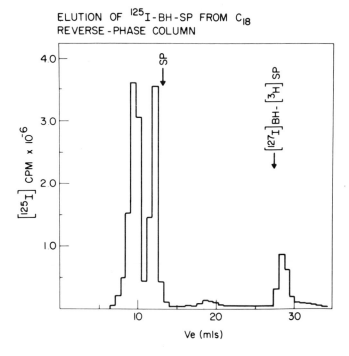

FIGURE 1. Elution profile of (^{125}I)-labeled, Bolton-Hunter (BH) substance P (SP) derivative on a C_{18} μ-Bondapak® reverse-phase column. Mobile phase was 48% methanol, 0.1% TFA pumped at 0.7 mℓ/min. Input of iodine radioactivity was 16×10^6 cpm. Radioactivity was determined in aliquots of 0.7-mℓ fractions. The (^{125}I)-labeled peak at 28 mℓ coeluted with an independently synthesized (^{127}I)-labeled, Bolton-Hunter derivatized (^{3}H)-substance P as indicated in the figure. (From Kream, R. M., Davis, B. J., Kawano, T., Margolis, F. L., and Macrides, F., *J. Comp. Neurol.*, 222, 140, 1984. With permission.)

Radioimmunoassay of Substance P

Substance P adheres to a wide variety of surfaces (see Mroz and Leeman[12] for review) and is rapidly degraded by chymotrypsin-like proteases.[16-18] Therefore, we utilized an RIA sample buffer developed in a previous study[19] that would inhibit both surface adsorption and proteolysis of radiolabeled tracer and displacer. Stock RIA sample buffer consisted of 50 mM sodium HEPES, pH 8.0/10 mM sodium EDTA/20 mg/mℓ Difco® bactopeptone/10 mg/mℓ bovine serum albumin/0.2 mg/mℓ cytochrome c/10 μg/mℓ phenyl methyl sulfonyl fluoride/10 μg/mℓ Trasylol. The stock RIA buffer was heated to 80°C for 15 min, then cooled. Any precipitated protein was removed by centrifugation. Working RIA buffer was made by dilution of stock RIA buffer 1:1 with H_2O. Unfractionated lyophilized antiserum was reconstituted with H_2O, then diluted tenfold with stock RIA buffer, and stored at −20°C. Antiserum was used at a final dilution of 300,000 in the RIA, as determined by quantitative immunotitration (data not shown).

The RIA standard curve consisted of serial dilutions of a 10 ng/mℓ solution of synthetic substance P in working RIA buffer. Ten standard concentrations were run in duplicate (200-μℓ aliquots ranging from 2 to 1000 pg per tube). Extracted tissue samples or HPLC eluant fractions were reconstituted in working RIA buffer and duplicate 200-μℓ aliquots were assayed. For each tissue extract, three dilutions were assayed in duplicate. Sequential 20-μℓ aliquots of antibody and (^{125}I)-Bolton-Hunter conjugated substance P (each diluted in working buffer) were added yielding a total assay volume of 240 μℓ with 10,000 cpm of radioiodinated tracer per tube. Incubation was for 30 hr at 4°C followed by precipitation of

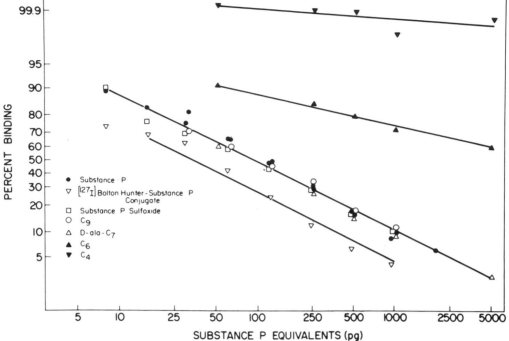

FIGURE 2.   A logit-log plot of a substance P RIA calibration. Points represent the mean of duplicate analyses. For substance P two independent determinations are presented. Total binding of (^{125}I)-labeled tracer was 50% in the absence of added displacer. Substance P sulfoxide was prepared by incubation of substance P with 0.72% hydrogen peroxide for 30 min, at room temperature. The sulfoxide was equipotent with native substance P (IC$_{50}$ or 50% displacement of tracer was 95 pg) as were the C$_9$ and D-Ala-C7 peptides. The (^{125}I)-labeled substance P derivative was three times as potent as substance P in the assay (IC$_{50}$ = 36 pg substance P equivalents). The C$_6$ and C$_4$ peptides were at least two orders of magnitude less potent than substance P. (From Kream, R. M., Davis, B. J., Kawano, T., Margolis, F. L., and Macrides, F., *J. Comp. Neurol.*, 222, 140, 1984. With permission.)

bound from free tracer by addition of 1000 µℓ absolute ethanol. After centrifugation, supernatants were aspirated and the pellets counted in a Packard® Gamma Counter at greater than 70% efficiency. Nonspecific binding was measured as precipitated tracer in the absence of antiserum and total binding was measured as precipitated tracer in the absence of added displacer. For unknown samples, substance P values were interpolated from a logit-log linear plot of the standard dose-response displacement curve (Figure 2).

High-Performance Liquid Chromatography

HPLC was performed using two Waters® B510 pumps, Model 680 Gradient Programmer, ISCO® UA-5 Absorbance Detector with Type 9 Optical Unit, ISCO® HPLC Autoinjector System, and ISCO® Foxy programmable fraction collector. Data were reduced using a Hewlett-Packard ® Model 3390A reporting integrator. Samples were injected through a 100-nℓ sample loop on a Rheodyne valve. Isocratic elution was performed with either 48% methanol, 0.1% TFA (w/v), or 54% methanol, 0.1% TFA at a flow rate of approximately 0.5 mℓ/min. Substance P and its analogs were monitored at 214 nm.

Gradient peptide elution was performed on a 8 × 100 mm µ-Bondapak® Radial Pak column using a convex or linear gradient of 27 to 32% acetonitrile in 0.1% trifluoroacetic acid (TFA). Ion-exchange HPLC was performed on a 8 × 100 mm Radial Pak Partisil SCX-10 cation exchange column using a linear gradient of 0.02 *M* KH$_2$PO$_4$, 0.05 *M* KCL in 25% acetonitrile, 0.1% TFA, to 0.2 M KH$_2$PO$_4$, 0.5 M KCl in 25% acetonitrile, 0.1% TFA. Fractions were monitored at 214 nm, and collected using preset time windows programmed into the Foxy.

Gel Filtration Chromatography

Extracts were fractionated on a column of Ultrogel® A$_c$A202 (LKB, total volume of 280 mℓ). Elution was with 2 N acetic acid. The column was calibrated with BSA (67 kdaltons), cytochrome (13 kdaltons), aprotonin (7 kdaltons), β-endorphin (3 kdaltons), and substance P (1.5 kdaltons).

Biochemical Reagents and Solutions

Tritiated substance P (2-L-prolyl-3,4-^{3}H(N)) (27.6 Ci/mmol) in ethanol containing 1% β-mercaptoethanol and (^{125}I)-Bolton-Hunter reagent (*N*-succinimidyl 3,4-hydroxy,5-(^{125}I)-iodophenylpropionate) (2200 Ci/mmol) in benzene were obtained from New England Nuclear (Boston, Mass.). Nonradioactive Bolton-Hunter reagent (*N*-succinimidyl 3,4-hydroxyphenyl propionate) and anhydrous TFA were purchased from Pierce Chemical Co. (Rockford, Ill.). Synthetic substance P (1-11) and five substance P carboxyl-terminal peptides, nonapeptide (C$_{9;3-11}$), (D-Ala^4)-heptapeptide (C$_{7;5-11}$), hexapeptide (C$_{6;6-11}$), tetrapeptide (C$_{4;8-11}$), and D-Ala^4)-heptapeptide (C$_{7;5-11}$), hexapeptide (C$_{6;6-11}$), tetrapeptide (C$_{4;8-11}$), and D-Ala-Met-enkephalinamide (DALA), were obtained from Peninsula Laboratories (San Carlos, Calif.). Purity of the peptides was monitored by HPLC (see below). Peptides were stored as milligrams per milliliter solutions in absolute ethanol containing 1% β-mercaptoethanol. Lactoperoxidase (purified grade) and phenylmethylsulfonyl fluoride (PMSF) were purchased from Calbiochem (La Jolla, Calif.). The TPCK-treated trypsin, α-chymotrypsin, bradykinin acetate, thyroxine, cytochrome c, bovine serum albumin EDTA, and bactopeptone were from Sigma Chemical Co. (St. Louis, Mo.). Neurotensin was from Boehringer-Mannheim (West Germany). The β-mercaptoethanol, 30% hydrogen peroxide, and HPLC-grade solvents were obtained from Fisher (Fairlawn, N.J.).

RESULTS

Radioimmunoassay

A highly sensitive radioimmunoassay was used to determine if authentic substance P was present in *M. edulis* pedal ganglia. A typical standard curve yielded parameters similar to those reported in a previous study.[19] At a final dilution of 1:300,000 absence of added displacer, approximately 3000 cpm or 30% of added tracer was specifically bound. The sensitivity of the assay defined as 10% displacement of added tracer was approximately 4 pg substance P, and the IC$_{50}$ or 50% displacement occurred at 30 pg substance P. Interassay sensitivity, based on over 20 independently performed RIAs, varied from 4 to 8 pg substance P, whereas interassay IC$_{50}$ ranged from 30 to 50 pg substance P. Interassay values for nonspecific binding were less than 5%, yielding specific to nonspecific binding ratios of greater than 10:1. The C-terminal fragments of substance P C-terminal nonapeptide (substance P 3-11) and C-terminal (D-Ala-^{4})-heptapeptide (substance P 5-11) were equipotent with native substance P. The C-terminal hexapeptide (substance P 6-11) was three orders of magnitude weaker (IC$_{50}$ + 20,000 pg) with a nonparallel displacement curve. The C-terminal tetrapeptide (substance P 8-11) displayed only 15 to 20% displacement of the tracer at 1 μg. L-thyroxine and the peptides neurotensin, bradykinin, and D-Ala^2-Met^5-enkephalinamide were all found to be ineffective at 1 μg. Thus, no C-terminal substance P fragment smaller than the D-Ala-heptapeptide was effectively recognized in our RIA. Finally, the related tachykinins, physalemin and eledoisin, displayed only 0.01% crossreactivity in the assay, eliminating significant contribution by these peptides.

Quantitation of Substance P

Total substance P in the tissue pools of 170, 143, 113 ganglia were 29.5, 22.3, and 14.6 pg, respectively. Specific activities were 0.1, 0.1, and 0.1 pg/1 mg tissue, respectively. Specific activities were corrected for 10% tissue weight committed to recovery measurements. Recoveries of (^{3}H) substance P were greater than 90% through C_{18} SEP-PAK® purification and radioimmunoassay. We thus eliminate the possibility that authentic substance P is unrecoverable by our procedures.

Chromatography of Substance P-like Immunoreactivity

Extracts were fractionated by gel chromatography in 2 N acetic acid. Fractions of 5 mℓ were collected, lyophilized, and reconstituted in 0.5 mℓ H_2O. Either 125 μℓ or 25% of each total fraction was run in duplicate in the radioimmunoassay. No substance P-like immunoreactivity was recovered in the elution position of synthetic substance P. Similarly, no substance P-like immunoreactivity was recovered in any of the column fractions. In addition, none of the column fractions exhibited bioactivity in the guinea pig ileum myenteric plexus-longitudinal muscle preparation or in inhibition of binding of (^{125}I)-Bolton-Hunter conjugated substance P to brain membranes (adaptation of the method of Kream et al.[5]). We therefore conclude that substance P is not present in pedal ganglia in significant concentrations.

Additional Studies

Invertebrate ganglia were homogenized in extraction solution consisting of 0.1 M acetic acid, 20 mM HCl, 1 μg each of PMSF and pepstatin A per mℓ, and 1% 2-mercaptoethanol. The high-molecular-weight materials in the homogenates were precipitated with TCA. The supernatants were extracted with ether to remove lipids and TCA. The aqueous layers were lyophilized and redissolved in HPLC buffer consisting of 10 mM ammonium acetate at pH 4.0. HPLC analyses of these samples were carried out with a reverse-phase column. The column was eluted with a solvent system consisting of the HPLC buffer and linear gradient of 5 to 25% 2-propanol in 30 min. The eluant was monitored at 260 nm. Preliminary studies on ganglia from various invertebrates such as mussel (*M. edulis*), sea hare (*Aplysia*), and cockroach (*Leucophaea maderae*) showed substance P was not present in any of the ganglia at a level which is detectable by this procedure.

DISCUSSION

Salanki et al.[11] have demonstrated that substance P, depending on the physiological state of *H. pomatia,* can alter the normal rhythmic bursting pattern of the Br-type neuron by prolonging the bursting period. Substance P appears to increase the bursting period, without altering the firing frequency, by eliminating subsequent interburst intervals. This action is achieved 10 to 15 sec following its application. The responsiveness of the Br-type neuron to substance P was found to depend on the physiological state of the cell and, in turn, on the state of the organism itself.[11] Organisms whose Br neurons were examined upon arousal for substance P effects, were unresponsive. This also was true in examining substance P effects on cells which had been treated with opioids. In fresh preparations, where the Br neuron was found to be continuously firing, substance P proved to be ineffective. Possible explanations for the above are too numerous to mention. For example, the Br neuron may be under the influence of endogenous opioids or substance P, thereby making it unresponsive to exogenous agent administration. Other transmitters, other than peptides, may influence its activity and have a higher intrinsic "priority" over the cell's activity pattern. Despite the restricted bath application of the agents, the activities obtained are quite specific because the cells in the immediate vicinity of the Br-type neuron were unresponsive to the treatments. This study strongly suggests that substance P or a substance P-like substance has its own receptor system in this organism. Substance P and Met-enkephalin antagonize each other's activity by separate mechanisms. The study suggests that many of the peptide activities found in mammals occur in invertebrates as well.

In contrast, we have found that the authentic undecapeptide is not present in *M. edulis* pedal ganglia. This raises the possibility that an invertebrate tachykinin is present in lieu of substance P. If so, concentrations are so low that we were unable to detect this putative peptide by bioassay or receptor binding techniques. Future studies must utilize thousands of ganglia in order to yield reliable activity data. A final possibility is that substance P or a related tachykinin is seasonally expressed. Such phenomena have been previously described for *M. edulis* opioid systems.[20] It would be of interest to reevaluate levels of substance P-like immunoreactivity at other time points during the year.

ACKNOWLEDGMENTS

Supported by New England Medical Center GRS Grant 19-80741 (R.M.K.) and NIH-MBRS Grant RR 08180 and ADAMHA Grant 17138 (G.B.S. and M.K.L.). We acknowledge the assistance given by ADAMHA Fellow G. Gonzalez.

REFERENCES

1. **Stefano, G. B. and Catapane, E. J.,** Enkephalins increases dopamine levels, *Soc. Neurosci. Abstr.*, 4, 283, 1978.
2. **Stefano, G. B. and Catapane, E. J.,** Enkephalins increase dopamine levels in the CNS of a marine mollusc, *Life Sci.*, 24, 1617, 1979.
3. **Osborne, N. W. and Neuhoff, J.,** Are there opiate receptors in the invertebrates?, *J. Pharm. Pharmacol.*, 31, 481, 1979.
4. **Stefano, G. B. and Hiripi, L.,** Methionine enkephalin and morphine alter monoamine and cyclic nucleotide levels in the cerebral ganglia of the fresh water bivalve, *Anodonta cygnea, Life Sci.*, 25, 291, 1979.
5. **Kream, R. M., Zukin, R. S., and Stefano, G. B.,** Demonstration of two classes of opiate binding sites in the nervous tissue of the marine mollusc *Mytilus edulis:* positive homotropic cooperativity of lower affinity binding sites, *J. Biol. Chem.*, 225, 218, 1980.
6. **Stefano, G. B., Kream, R. M., and Zukin, R. S.,** Demonstration of stereospecific opiate binding in the nervous tissue of the marine mollusc *Mytilus edulis, Brain Res.*, 181, 445, 1980.
7. **Stefano, G. B., Vadasz, I., and Hiripi, L.,** Methionine enkephalin inhibits the bursting activity of the Br-type neuron in *Helix pomatia, Experientia*, 36, 666, 1980.
8. **Stefano, G. B. and Martin, R.,** Enkephalin-like immunoreactivity in the pedal ganglia of *Mytilus edulis* (Bivalvia) and its proximity to dopamine-containing structures, *Cell Tissue Res.*, 230, 147, 1983.
9. **Stefano, G. B. and Leung, M.,** Purification of opioid peptides from molluscan ganglia, *Cell Mol. Neurobiol.*, 2, 347, 1982.
10. **Osborne, W. N., Cuello, A. C., and Dockray, G. J.,** Substance P and cholecystokinin-like peptides in *Helix* neurons and cholecystokinin and serotonin in giant neuron, *Science*, 216, 409, 1982.
11. **Salanki, J., Vehovseky, A., and Stefano, G. B.,** Interaction of substance P and opiates in the CNS of *Helix pomatia* L., *Comp. Biochem. Physiol.*, 75c, 387, 1983.
12. **Mroz, E. A. and Leeman, S. E.,** Substance P, in *Methods of Hormone Radioimmunoassay*, Jaffe, B. M. and Behrmann, H. R., Eds., Academic Press, New York, 1979, 121.
13. **Michelot, R. H., Gozlan, J. C., Beaujouvan, M. J., Torrens, B. Y., and Glowinski, J.,** Synthesis and biological activities of substance P iodinated derivatives, *Biochem. Biophys. Res. Commun.*, 95, 491, 1980.
14. **Liang, T. and Cascieri, M. A.,** Specific binding of an immunoreactive and biologically active (^{125}I)-labeled N(I) acylated substance P derivative to parotid cells, *Biochem. Biophys. Res. Commun.*, 96, 1793, 1980.
15. **Bolton, A. G. and Hunter, W. M.,** The labelling of proteins to high specific radioactivities by conjugation to a (^{125}I)-containing acylating agent, *Biochem. J.*, 133, 529, 1973.
16. **Lee, C. M., Sanberg, B. E. B., Hanley, M. R., and Iversen, L. L.,** Purification and characterization of a membrane-bound substance P degrading enzyme from human brain, *Eur. J. Biochem.*, 114, 315, 1981.

17. **Harmar, A. and Keen, P.,** Chemical characterization of substance P-like immunoreactivity in primary affect neurons, *Brain Res.,* 220, 203, 1981.
18. **Harmar, A., Schofield, J. G., and Keen, P.,** Substance P biosynthesis in dorsal root ganglia. An immunochemical study of (^{35}S) methionine and (^{3}H)-proline incorporation *in vitro, Neuroscience,* 6, 1917, 1981.
19. **Kream, R. M., Davis, B. J., Kawano, T., Margolis, F. L., and Macrides, F.,** Substance P and catecholaminergic expression in neurons of the hamster olfactory bulb, *J. Comp. Neurol.,* 222, 140, 1984.
20. **Stefano, G. B., Kream, R. M., and Zukin, R. S.,** Seasonal variation of stereospecific enkephalin binding and dopamine responsiveness in *Mytilus edulis* pedal ganglia, in *Neurotransmitters in Invertebrates,* Rozsa, K. S., Ed., Pergamon Press, London, 1980, 22 and 453.

IMMUNOCHEMICAL EVIDENCE ON THE OCCURRENCE OF OPIOID- AND GASTRIN-LIKE PEPTIDES IN TISSUES OF THE EARTHWORM *LUMBRICUS TERRESTRIS*

Kenneth V. Kaloustian and Peter J. Rzasa

SUMMARY

The presence of Met- and Leu-enkephalin-like, β-endorphin-like, and gastrin-like immunoreactive peptides in tissues of the earthworm, *Lumbricus terrestris,* was demonstrated by means of radioimmunoassay technique. All of these peptides are common to both nerve as well as gut tissues. Quantification of regional gut levels of these peptides revealed peak concentrations in regions coinciding with the area of the gut containing high levels of digestive enzyme activities, and it is assumed that they may play a role in digestive functions. Treatment of the gut with either Met-enkephalin, Leu-enkephalin, or β-endorphin decreases contraction and presumably increases the transient time spent by food in gut for successful digestion. The dual localization of all of these peptides in earthworm nerve and gut tissues follows a similar pattern observed with a variety of peptidal hormones in vertebrates and raises interesting phylogenetic implications. The presence of the enkephalins and β-endorphin-like immunoreactivity in seminal vesicles and the enkephalins and gastrin-like immunoreactivity in body wall may indicate a more ubiquitous distribution and function of these peptides in the earthworm.

INTRODUCTION

The variety and wide distribution of peptides originally found in vertebrate tissues that are also present in invertebrate tissues support the assumption that invertebrates possess substances that are structurally related to vertebrate peptides and may play an important role in regulatory processes.[1-6] From such phylogenetic studies, a better understanding can be gained as to the origin, heterogeneity, and function of peptidergic hormonal systems and might prove particularly effective in elucidating the relationship between peptidal hormones which are dually localized in both nerve and gut tissues.

While the distribution of opioid-like peptides[7-16] and gastrin/CCK-like peptides[7,9,12,17-23] in invertebrates has been strongly documented, the earthworm *Lumbricus terrestris* has been used frequently as a model for such studies. Most of the research with the earthworm has emphasized the nervous system since these tissues consist of neurons loaded with large electron-dense granules similar to those of the peptide-hormone-producing neurosecretory cells of mammals.[24] Hence, it is not surprising that in the earthworm ample immunochemical evidence exists for β-endorphin- and enkephalin-like peptides in supra- and subpharyngeal ganglia and ventral nerve fiber,[25] ACTH-like peptide in supra- and subpharyngeal ganglia,[26] substance P-like peptide in subpharyngeal ganglia,[26] pancreatic polypeptide-like peptide in supra- and subpharyngeal ganglia, circumpharyngeal connectives, and ventral nerve fiber,[27] and vasoactive intestinal peptide-like substance in subpharyngeal and ventral nerve fiber.[27] Rzasa et al.[28-29] and Lyons et al.[30] have further expanded the study on earthworm tissue distribution of Met- and Leu-enkephalin-like, gastrin-like, and β-endorphin-like immunoreactivity and have established the presence of such peptides not only in nerve but also other tissues as well and raised interesting questions on the presence and probable functions of these peptides in different tissues. The purpose of this paper is to report evidence from our previous studies on immunochemical evidence for earthworm tissue distribution of opioid and gastrin-like peptides[28-30] and elaborate on some functions for these peptides from our current work.

MATERIALS AND METHODS

Earthworms were purchased from a biological supply company and maintained at 14°C in moistened soil supplemented with leaves. For fasting studies, animals were starved for 7 days while maintained individually in 125-mℓ Erlenmeyer flasks with 2 mℓ of distilled water at 14°C.

Tissue Extraction

Animals were anesthetized in 10% ethanol solution. The gut from the gizzard to the anus was excised, cut longitudinally, and rinsed with a stream of distilled water in order to remove intestinal content. The gut was divided into 6 sections; gizzard, section 1; end of gizzard to beginning of clitellum, section 2; clitellum, section 3; and equal divisions of the postclitellum gut section designated as sections 4 to 6.

Ventral nerve fiber was removed by careful dissection of the fiber from the base of the subpharyngeal ganglion to gut section 6. The supra- and subpharyngeal ganglia were removed and collected with the intact connectives. Seminal vesicles were collected by detaching the gland from the underlying tissues, free of nerve fibers. Body wall tissues were collected by scraping underlying intestinal tissues followed by rinsing with distilled water.

At least eight extracts of each individual tissue were prepared. In the case of ventral nerve fiber and supra- and subpharyngeal ganglia, samples were pooled from 20 individual animals.

Tissue Preparation

All tissue extracts for gastrin assay were prepared by the method of Straus et al.[31] while that of Met- and Leu-enkephalin and β-endorphin assay were prepared by the method of Duka et al.[32] With the gastrin assay intestinal tissues were treated with pronase and chymotryspin-free trypsin to elucidate peptidal nature of the earthworm active gastrin components. The protease inhibitor Trasylol (500 units/mℓ) was added to the tissue homogenates for both Met- and Leu-enkephalin assays and the pH adjusted to 7.0. In all cases, the final supernatant was frozen at -20°C until used. No significant change in immunoreactivity was detected when samples were frozen. Rabbit hypothalamus was prepared as a control for the enkephalins.

Column Chromatography
Gastrin

Intestinal tissue extracts were gel filtered at 24°C on a column (0.9 $\times$ 100 cm of Sephadex G-50-Superfine) by using 0.02 M Tris buffer at pH 8.0. The flow rate of the column was adjusted to 6 mℓ/hr and fractions of 0.8 mℓ were collected and subjected to radioimmunoassay. Calibration of the column was accomplished using pure human gastrin G-17.

β-Endorphin

In order to separate β-endorphin from LPH, tissue extracts were gel filtered at 5°C on a column (1.5 $\times$ 30 cm of Sephadex G-75-Superfine) by using 0.1 M phosphate buffer, containing 0.05 M NaCl, at pH 6.0. The flow rate was adjusted to 1 mℓ/min and 0.5-mℓ fractions were collected. The column was calibrated with 1% blue dextran and ^{125}I β-endorphin. β-LPH location was approximated by the blue dextran region.

Radioimmunoassay

Radioimmunoassay were performed using commercial kits from Becton-Dickinson (Orangeburg, N.Y.) for gastrin; Immuno Nuclear Corporation (Stillwater, Minn.) for Met- and Leu-enkephalin; and New England Nuclear (Boston, Mass.) for β-endorphin. All RIA analyses were performed in duplicate. Radioactive counting was performed in a Searle 1185 gamma scintillation counter and the results expressed as percent bound vs. concentration. All values reported were obtained by evaluation of experimental data against standard curves.

Verification of cross reactivities of Met-enkephalin antiserum to Leu-enkephalin and vice versa was performed by running simultaneous and separate assays with the homologous and crossreacting peptides. Results were expressed as the ratio of the enkephalin standard concentration to the crossreacting peptide concentration at 50% of maximum binding. Since porcine gastrin (G-17) has been shown to display some homology with Met-enkephalin,[33] the crossreactivity of enkephalin antisera with gastrin (G-17) was also determined.

Recovery for both Met- and Leu-enkephalin, gastrin, and β-endorphin from gut tissues was determined by spiking the tissues with respective standards preceding tissue extraction and RIA analysis. In each case, baseline tissue concentrations of respective peptides were subtracted before recoveries were determined.

RESULTS

In the gut of the earthworm, highest activities for all peptides studied were measured in sections 2 to 5 with Met-enkephalin values peaking in section 3 and that of gastrin in section 4 for both feeding and fasting conditions. Fasting gut gastrin levels were significantly higher than those of feeding (at 99% level of confidence) with the exception of section 1 where both values were relatively equal (Table 1). Gut β-endorphin activities were detectable when sections 3 to 5 were pooled, however, the values were significantly lower than that of the nerve tissues and seminal vesicles. Leu-enkephalin values in the gut sections were negligible compared to those of Met-enkephalin and furthermore, they may actually represent Met-enkephalin activities to a degree since crossreacting studies (Table 2) indicate that the crossreactivity of Leu-enkephalin antiserum with Met-enkephalin to be 26%, while that of Met-enkephalin antiserum with Leu-enkephalin was determined to be only 3.2%.

Gel filtration of the intestinal tissue extracts revealed immunoreactive gastrin-like activity as three separate peaks with one of the peaks corresponding to that of human gastrin (G17I) taken from a standard run (fractions 100 to 120) (Figure 1). Incubation of the intestinal extracts for gastrin assay with pronase destroyed all activity, suggesting peptide nature of the active components. Treatment with chymotrypsin-free trypsin yielded no change in immunoreactivity suggesting that, similar to mammaliam gastrin,[34] earthworm gastrin-like peptides do not contain a lysine or arginine residue, at least in regions of the molecule that elicit antigenicity.

In nerve tissues peak activities for both Met- and Leu-enkephalin were detected in the supra- and subpharyngeal ganglia with moderately lower concentrations in the ventral nerve fibers. For β-endorphin, however, the supra- and subpharyngeal ganglia and ventral nerve fiber showed equal values (Table 1). Although we were not able to detect gastrin-like immunoreactivity in nerve tissues with the RIA techniques, we have, however, been able to show immunoreactivity using peroxidase-antiperoxidase (PAP) techniques (unpublished data) thus agreeing with the work of Larson and Vigna[35] who have shown similar values for earthworm gut gastrin/CCK to those of Rzasa et al.[29] as well as immunoreactivity in nerve tissues.

Our results of rabbit hypothalmic Met-enkephalin-like immunoreactivity of 279 pmol/g and Leu-enkephalin-like immunoreactivity of 47.0 pmol/g were approximately half those of Hughes et al.[36] perhaps owing to different methodologies and antiserum employed. However, in both studies the ratio of Met- to Leu-enkephalin-like immunoreactivity was essentially the same (5.9).

Enkephalin-like and β-endorphin-like immunoreactivity but not gastrin-like immunoreactivity were observed in seminal vesicles. In the case of β-endorphin, seminal vesicles contained the highest immunoreactivity. With the exception of β-endorphin, immunoreac-

Table 1
TISSUE MET- AND LEU-ENKEPHALIN-LIKE, GASTRIN-LIKE, AND β-ENDORPHIN-LIKE IMMUNOREACTIVITY IN *LUMBRICUS TERRESTRIS*

Tissue	Met-enkephalin (ng/g ± S.E.M.)[a]	Leu-enkephalin (ng/g ± S.E.M.)	Gastrin, feeding (pg/g ± S.E.M.)	Gastrin, fasting (pg/g ± S.E.M.)	β-endorphin (pg/mℓ ± S.E.M.)[b]
Supra- and subpharyngeal ganglia	22.60 ± 1.69	16.66 ± 3.37	Not detected	Not detected	15.5 ± 5.5
Ventral nerve fiber	13.84 ± 0.72	3.78 ± 0.40	Not detected[c]	Not detected[c]	15.1 ± 3.2
Seminal vesicles	13.34 ± 0.24	4.82 ± 1.20	Not detected	Not detected	19.1 ± 5.8
Body wall	6.11 ± 1.04	2.59 ± 0.83	94 ± 2	94 ± 2	Not detected
Gut sections					
1	4.24 ± 0.37	1.04 ± 0.55	180 ± 26	164 ± 41	9.9 ± 1.4[d]
2	30.12 ± 1.97	2.58 ± 0.52	403 ± 49	458 ± 77	
3	33.22 ± 4.56	2.30 ± 0.61	380 ± 41	454 ± 62	
4	30.14 ± 3.22	0.59 ± 0.07	832 ± 83	1072 ± 179	
5	23.28 ± 1.98	0.69 ± 0.32	658 ± 163	906 ± 112	
6	14.21 ± 2.15	1.28 ± 0.45	Not detected	Not detected	

[a] Values in gram wet tissue.
[b] Values in mℓ tissue fraction.
[c] Immunoreactivity detected using peroxidase-antiperoxidase techniques.
[d] Values pooled from gut sections 2, 3, and 4.

Table 2
CROSSREACTIONS OF ENKEPHALIN AND
GASTRIN ANTISERA WITH DIFFERENT PEPTIDES
FROM EXPERIMENTAL DETERMINATIONS

	Crossreactivity (%)[a]		
Peptides	Met-enkephalin antiserum	Leu-enkephalin antiserum	Gastrin antiserum
Met-enkephalin	100.0	26.0	1.0
Leu-enkephalin	3.2	100.0	1.0
Gastrin (G-17)	1.0	1.0	100.0

[a] The crossreactivities have been calculated from the concentration of the unlabeled peptides which decrease the binding of ^{125}I-Met-enkephalin, ^{125}I-Leu-enkephalin, and ^{125}I-gastrin (G-17) to the respective antisera by 50%.

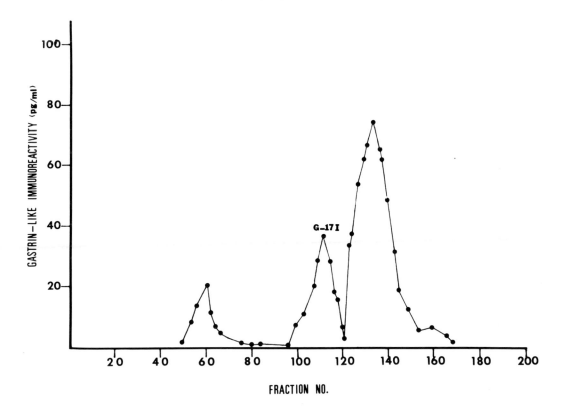

FIGURE 1.   Separation of immunoreactive gastrin-like components of *L. terrestris* intestinal tissues by gel filtration. Samples were applied to Sephadex G-50 Superfine Column (0.9 × 100 cm) and eluted with 0.02 *M* Tris buffer, pH 8.0, with a flow rate of 6 mℓ/hr. Fractions of 0.8 mℓ were collected and assayed. The column was calibrated using pure human gastrin G-17.

tivity for the other peptides was also detected in the body wall (Table 1). Recently, using PAP techniques, we have shown gastrin-like immunoreactivity in nerve fibers of the body wall tissues (unpublished data). It is probable that immunoreactivity measured for both Met- and Leu-enkephalin in body walls may be due to the nerve fibers that innervate these tissues.

Recovery studies for over 60% for both Met- and Leu-enkephalin, 81% for gastrin, and over 90% for β-endorphin indicate satisfactory recovery values which exclude the possibility that the peptides we are measuring are being deactivated by specific enzymes leading to losses in activities and to high nonspecific binding.

Meaningful comparisons of gastrin-, enkephalin-, or β-endorphin-like immunoreactivity in various tissue specimens require that competition for antibody binding sites increase systematically with tissue concentration in a consistent fashion in all tissues examined. From Figures 2A, C, and E, it can be seen that competition for Met-enkephalin, Leu-enkephalin, and β-endorphin antibody binding increased in a linear and parallel manner with large amounts of nerve tissue extract when plotted on a logit-log basis. Parallelism was also observed with intestinal tissues and body wall for Met-enkephalin (Figure D) and intestinal tissues for gastrin (Figure G). This parallelism between standard and tissue extracts may suggest a homologous structure. For non-nerve tissues, however, parallelism was not observed with both Leu-enkephalin (Figure B) and β-endorphin (Figure F) as well as that of seminal vesicles for Met-enkephalin (Figure D).

DISCUSSION

The results of the present study have demonstrated the presence of Met- and Leu-enkephalin-, β-endorphin, and gastrin-like immunoreactivity in tissues of the earthworm *L. terrestris*. A high degree of structural homology was observed between Met- and Leu-enkephalin and β-endorphin standards vs. the tissue extracts from supra- and subpharyngeal ganglia and ventral nerve fiber (Figures 2A, C, and E) suggesting similar structures to those of the standards. In all cases studied, seminal vesicles lacked parallelism with the standards (Figures 2B and D), especially with Leu-enkephalin (Figure 2B). This is unusual since immunoreactivity for all peptides measured in seminal vesicles, with the exception of gastrin where no activity was measured, was moderately high, equal in values to the ventral nerve fiber for both Met- and Leu-enkephalin and exceeding all other tissue activities for β-endorphin. These results, however, agree with reports of enkephalin-like immunoreactivity in ganglia innervating sex organs of the leech[37] and smooth muscle fibers of the human prostate and seminal vesicles.[38] In addition, opiate receptors have also been detected in mouse vas deferens.[39]

The body wall of the earthworm is highly innervated as evidenced by its many physiological functions, such as photosensitivity, respiration, and tactile responses. Relatively low levels of immunoreactivity measured for Met- and Leu-enkephalin and gastrin in body wall tissues are probably due to nerve tissue terminals embedded in body walls, since using PAP techniques for gastrin, we were able to detect immunoreactivity only in the nerve tissues and not in other tissues. It is tempting to indicate that these compounds may function as neurotransmitters in body wall tissues for specific physiological functions.

Regional differences in earthworm gut enkephalin immunoreactivity are similar to differences observed in the vertebrate intestine. In guinea pigs, highest activity has been observed in duodenum with decreasing amounts in succeeding intestinal sections.[40] In humans,[41] highest activities of Met- and Leu-enkephalin are detected in antral mucosa with subsequent linear decrease in activity towards colonal tissues. In the earthworm, highest levels of immunoreactivity for all the peptides studied peaked at regions of the gut containing high digestive enzyme activities[42] with immunoreactivity decreasing significantly or being absent, as is the case of β-endorphin, in gut section 6. We have measured the actions of Met- and Leu-enkephalin and β-endorphin or earthworm gut contraction. Gut sections treated with these peptides show significant decrease in contractile properties. It appears that they induce their effect by their action on dopamine (unpublished data), a finding similar to that of Stefano and Hall.[43] We presume that this allows an increased transient time for food in areas of the gut with high digestive enzyme activities, hence enhancing digestive processes.

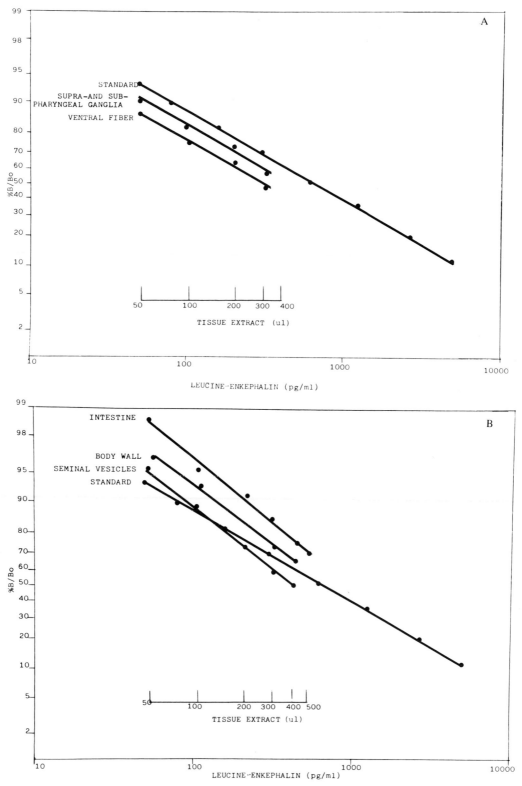

FIGURE 2.   Dose-response curves for (A and B) Leu-enkephalin, (C and D) Met-enkephalin, (E and F) β-endorphin, and (G) gastrin standards against *L. terrestris* tissue extracts. Results are expressed as the logit transformation of the response vs. log dose of Leu- and Met-enkephalin, β-endorphin, and gastrin standards or volume of tissue extracts assayed per tube.

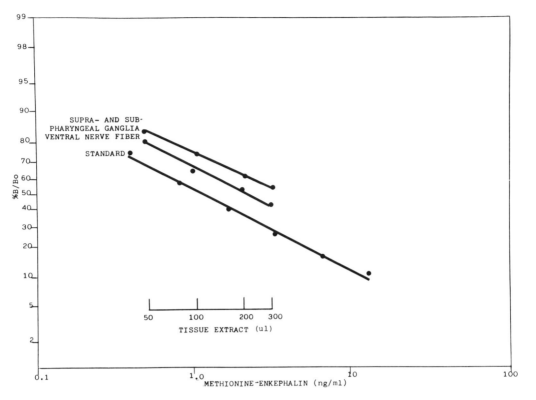

FIGURE 2C

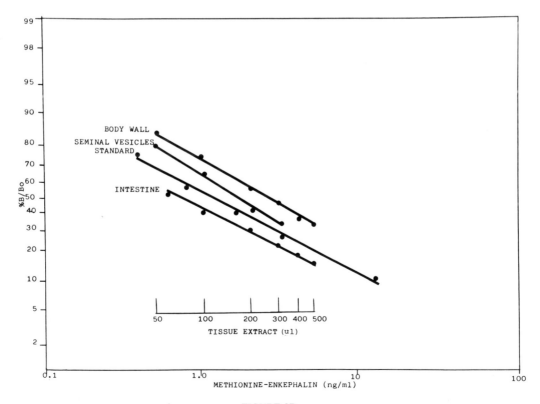

FIGURE 2D

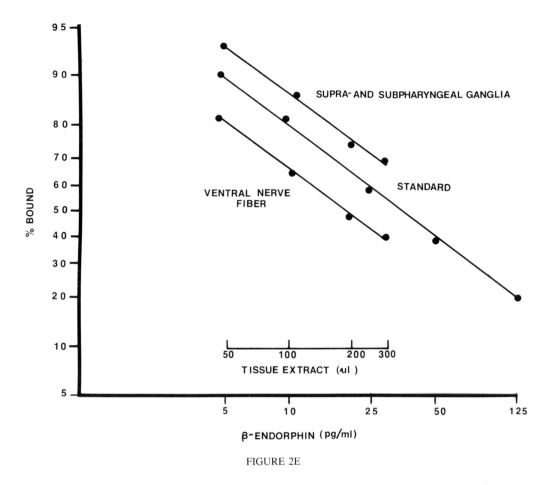

FIGURE 2E

The high gastrin-like immunoreactivity measured in intestinal tissues of fasting earthworms is unclear. Since earthworm gut tissues regress in size and volume upon starvation, increased gastrin activity may play a role in providing acidic environment for tissue degradation. Earthworm tissues contain a high ratio of proteins with the body wall composed mainly of collagen. The catabolism of these proteins during starvation becomes extremely important for survival. Kaloustian[44] has reported collagenase-type enzyme activation in starving and estivating earthworm that has a requirement for acidic pH which gastrin may provide. Using PAP techniques, we have observed intense staining for gastrin in tissues of fasting earthworms. We are currently attempting to elucidate whether the staining is specific for gastrin-like immunoreactivity or nonspecific staining due to actions of other proteins or products of proteolytic digestion in fasting tissues.

The dual localization of these peptides in both nerve and gut tissues in the earthworm follows a similar pattern observed for many peptidal hormones in vertebrates.[45-47] This dual localization in earthworms may indicate that gut and nerve regulating peptides arose more or less in parallel in animals much earlier than originally anticipated and provides an opportunity to further substantiate their regulatory properties.

In conclusion, studies leading to the elucidation of stereospecific binding for these peptides should not only strengthen our knowledge for the presence of such peptides in invertebrates, but possibly enhance our understanding of their regulatory properties. It should not be surprising if such receptors are identified in earthworm tissues considering Stefano et al.[15] have reported the presence of stereospecific opiate binding in the nervous tissue of *Mytilus edulis* with properties similar to those of opiate receptors in rat brain. Additional studies by

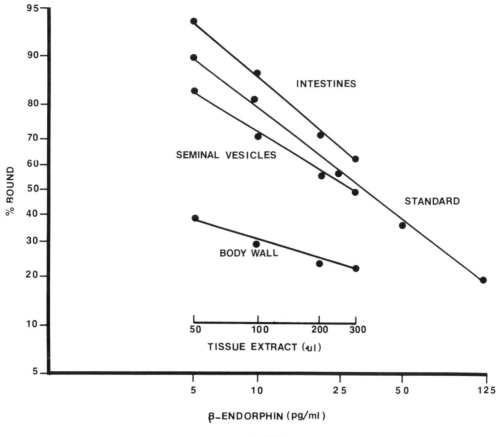

FIGURE 2F

Kream et al.[48] have shown the presence of at least two types of opioid receptors in *Mytilus edulis* which are analogous to those of vertebrate brain opioid receptors mu and delta. Future work in our laboratory will involve studies to identify receptors for these peptides.

ACKNOWLEDGMENTS

The authors express their gratitude to Dr. Irwin Beitch, Quinnipiac College, for performing the PAP technique. The authors also thank Mrs. Shahnaz Siddiqui and Ms. Judy Berman for their technical assistance with earthworm intestinal contractions. This work was supported by grants through Quinnipiac Faculty Research and by a grant from Sigma Xi.

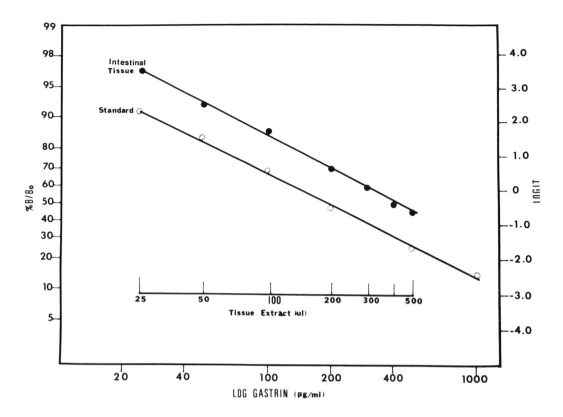

FIGURE 2G

REFERENCES

1. **Davidson, J. K., Falkmer, S., Mehrota, B. K., and Wilson, S.,** Insulin assays and light microscopical studies of digestive organs in protostomian and deuterostomian species and in coelenterates, *Gen. Comp. Endocrinol.,* 17, 388, 1971.
2. **LeRoith, D., Lesniak, M. A., and Roth, J.,** Insulin in insects and annelids, *Diabetes,* 30, 70, 1981.
3. **Duve, H., Thorpe, A., and Lazurus, N. R.,** Isolation of material displaying insulin-like immunological biological activity from the brain of the blowfly *Calliphora vomitoria, Biochem. J.,* 184, 221, 1979.
4. **Pliselskaya, E., Kazakov, V. K., Solitilakays, L., and Leinson, L. G.,** Insulin producing cells in the gut of freshwater bivalve molluscs *Anodonta cygnea* and *Unio pictorum* and the role of insulin in the regulation of their carbohydrate metabolism, *Gen. Comp. Endocrinol.,* 35, 133, 1978.
5. **Maier, V., Kroder, A., and Groner, E.,** Glucagon-like activity (GLI) in the intestine of *Porcus domesticus* and *Astacus fluviatilis, Acta Endocrinol. (Copenhagen) Suppl.* 193, (Abstr.) 41, 1975.
6. **Tager, H. S., Markese, J., Kramer, K. J., Spiers, R. S., and Childs, C. N.,** Glucagon-like and insulin-like hormones of the insect neurosecretory system, *Biochem. J.,* 156, 515, 1976.
7. **Van Noorden, S., Fritsch, H. A. R., Grillo, T. A. I., Polak, J. M., and Pearse, A. G. E.,** Immunocytochemical staining for vertebrate peptides in the nervous system of a gastropod mollusc, *Gen. Comp. Endocrinol.,* 40, 375, 1980.
8. **Zipser, B.,** Identification of specific leech neurones immunoreactive to enkephalin, *Nature (London),* 283, 857, 1980.
9. **El-Salhy, M., Abou-El-Ela, R., Falkmer, S., Grimelius, L., and Wilander, E.,** Immunohistochemical evidence of gastro-enteropancreatic neurohormonal peptides of vertebrate type in the nervous system of the larva of a dipteran insect, the hoverfly, *Eristalis aeneus, Regulatory Peptides,* 1, 187, 1980.
10. **Gros, C., Lafon-Cazal, M., and Dray, F.,** Presence de substances immunoreactivement apparentées aux enkephalines chez un insectem *Locusta migratoris, C. R. Acad. Sci. Paris,* 287D, 647, 1978.

11. **Remy, C. and Dubois, M. P.,** Immunohistological evidence of methionine enkephalin-like material in the brain of the migratory locust, *Cell and Tissue Res.,* 213, 271, 1981.
12. **Schot, L. P. C., Boer, H. H., Swaab, D. F., and Van Noorden, S.,** Immunocytochemical demonstration of peptidergic neurons in the central nervous system of the pond snail *L. stagnalis* with antisera raised to biologically active peptides of vertebrates, *Cell Tissue Res.,* 216, 273, 1981.
13. **Stefano, G. B., Rozsa, K. S., and Hiripi, L.,** Actions of met-enkephalin and morphine on single neuronal activity in *Helix pomatia* L., *Comp. Biochem. Physiol.,* 66, 193, 1979.
14. **Remy, C. H. and Dubois, M. P.,** Localisation par immunofluorescence de peptides analogues a lα-endorphine dans les ganglion infra-oesophagiens du *Lombricide dendrobaena* subrubicunda Eisen., *Experientia,* 35, 137, 1979.
15. **Stefano, G. B., Kream, R. M., and Zukin, R. S.,** Demonstration of stereospecific opiate binding in the nervous tissue of the marine mollusc *Mytilus edulis, Brain Res.,* 181, 440, 1980.
16. **Leung, M. K. and Stefano, G. B.,** Isolation and identification of enkephalins in pedal ganglia of *Mytilus edulis* (Mollusca), *Proc. Natl. Acad. Sci. U.S.A.,* 81, 955, 1984.
17. **Kramer, K. J., Speirs, R. D., and Childs, C. N.,** Immunochemical evidence for a gastrin-like peptide in insect neuroendocrine system, *Gen. Comp. Endocrinol.,* 32, 423, 1977.
18. **Yui, R., Fujita, T., and Ito, S.,** Insulin-gastrin-pancreatic polypeptide-like immunoreactive neurons in the brain of the silkworm, *Bombyx mori, Biomed. Res.,* 1, 42, 1980.
19. **Grimmelikhuijzen, C., Sundler, F. S., and Rehfeld, J. F.,** Cholecystokinin-like immunoreactivity in nerve cells of coelenterates, *Histochemistry,* 69, 61, 1980.
20. **Duve, H. and Thorpe, A.,** Gastrin-cholecystokinin (CCK)-like immunoreactive neurons in the brain of the blowfly, *Calliphora erythrocephala* (Diptera), *Gen. Comp. Endocrinol.,* 43, 381, 1981.
21. **Fritsch, H. A. R., Van Noorden, S., and Pearse, A. G. E.,** Localization of somatostatin and gastrin-like immunoreactivity in the gastrointestinal tract of *C. intestinalis, Cell Tissue Res.,* 186, 181, 1978.
22. **Van Noorden, S. and Pearse, A. G. E.,** Localization of immunoreactivity to insulin, glucagon, and gastrin in the gut of *Amphioxus (Branchiostoma) lanceolatum,* in *Evolution of Pancreatic Islets,* Grillo, T. A. I. and Epple, A., Eds., Oxford, 1977, 46.
23. **Straus, E. and Yalow, R. S.,** Gastrointestinal peptides in the brain, *Fed. Proc. Fed. Am. Soc. Exp. Biol.,* 38, 2320, 1979.
24. **Scharrer, B.,** Peptidergic neurons: facts and trends, *Gen. Comp. Endocrinol.,* 34, 50, 1978.
25. **Alumets, J. Håkanson, R., Sundler, F., and Thorell, J.,** Neuronal localization of immunoreactive enkephalin and β-endorphin in the earthworm, *Nature (London),* 279, 805, 1979.
26. **Aros, B., Wenger, V. B., and Vigh-Teickmann, I.,** Immunohistochemical localization of substance P and ACTH-like activity in the central nervous system of the earthworm *Lumbricus terrestris* L., *Acta Histochem.,* 66, 262, 1980.
27. **Sundler, F., Håkanson, R., Alumets, J., and Walles, B.,** Neuronal localization of pancreatic polypeptide (PP) and vasoactive intestinal peptide (VIP) immunoreactivity in the earthworm *Lumbricus terrestris, Brain Res. Bull.,* 2, 811, 1977.
28. **Rzasa, P. J., Kaloustian, K. V., and Prokop, E. K.,** Immunochemical evidence for met-enkephalin-like and leu-enkephalin-like peptides in tissues of the earthworm *Lumbricus terrestris, Comp. Biochem. Physiol.,* 77C, 345, 1984.
29. **Rzasa, P. J., Kaloustian, K. V., and Prokop, E. K.,** Immunochemical evidence for a gastrin-like peptide in the intestinal tissues of the earthworm *Lumbricus terrestris, Comp. Biochem. Physiol.,* 71A, 631, 1982.
30. **Lyons, N., Kaloustian, K. V., and Rzasa, P. J.,** Immunochemical evidence for β-endorphin-like compound in tissues of the earthworm *Lumbricus terrestris,* 10th New England Endocrine Conference, Boston, 1984.
31. **Straus, E., Yalow, R. S., and Gainer, H.,** Molluscan gastrin: concentration and molecular forms, *Science,* 190, 687, 1975.
32. **Duka, T., Holt, V., Przewlocki, R., and Wesche, D.,** Distribution of methionine- and leucine-enkephalin within the rat pituitary gland measured by highly specific radioimmunoassays, *Biochem. Biophys. Res. Commun.,* 85, 1119, 1978.
33. **Rehfeld, J. F.,** Four basic characteristics of the gastrin-cholecystokinin systems, *Am. J. Physiol.,* 240, G225, 1981.
34. **Gregory, H., Hardy, P. M., and Jones, D. S.,** The antral hormone gastrin; structure of gastrin, *Nature (London),* 204, 931, 1964.
35. **Larson, B. A. and Vigna, S. R.,** Species and tissue distribution of cholecystokinin/gastrin-like substances in some invertebrates, *Gen. Comp. Endocrinol.,* 50, 469, 1983.
36. **Hughes, J., Kosterlitz, H. W., and Smith, T. W.,** The distribution of methionine-enkephalin and leucine-enkephalin in the brain and peripheral tissues, *Br. J. Pharmacol.,* 61, 639, 1977.
37. **Zipser, B.,** Identification of specific leech neurones immunoreactive to enkephalin, *Nature,* 283, 857, 1980.

38. **Vaalasti, A., Linnoila, I., and Hervonon, A.,** Immunohistochemical demonstration of VIP, met⁵- and leu⁵-enkephalin immunoreactive nerve fibers in the human prostate and seminal vesicles, *Histochemistry,* 66, 89, 1980.

39. **Henderson, G., Hughes, J., and Kosterlitz, W.,** A new example of a morphine-sensitive neuro-effector junction: adrenergic transmission in the mouse vas deferens, *Br. J. Pharmacol.,* 46, 764, 1972.

40. **Linnoila, R. I., DiAugustine, R. P., Miller, R. J., Chung, K. J., and Cuatrecasas, P.,** An immunohistochemical and radioimmunological study of the distribution of (Met⁵) and (Leu⁵)-enkephalin in the gastrointestinal tract, *Neuroscience,* 3, 1187, 1978.

41. **Polak, J. M., Bloom, S. R., Sullivan, S. N., Facer, P., and Pearse, H. G. E.,** Enkephalin-like immunoreactivity in the human gastrointestinal tract, *Lancet,* 1, 972, 1977.

42. **Laverack, M. S.,** *The Physiology of Earthworms,* Pergamon Press, Oxford, 1963.

43. **Stefano, G. B. and Hall, B.,** Opioid inhibition of dopamine release from nervous tissue of *Mytilus edulis* and *Octopus bimaculatus, Science,* 213, 928, 1981.

44. **Kaloustian, K. V.,** A collagenolytic type enzyme from the posterior body wall tissues of the estivating earthworm, *Allolobophora caliginosa* (Savigny), *Comp. Biochem. Physiol.,* 68A, 669, 1981.

45. **Loonen, J. M. and Soudijn, W.,** Peptides with a dual function: central neuroregulators and gut hormones, *J. Physiol.,* 75, 831, 1979.

46. **Dockray, G. J. and Gregory, R. A.,** Relations between neuropeptides and gut hormones, *Proc. R. Soc. London Ser. B,* 210, 151, 1980.

47. **Rehfeld, J. F.,** Immunochemical studies on cholecystokinin. II. Distribution and molecular heterogeneity in the central nervous system and small intestine of man and hog, *J. Biol. Chem.,* 253, 4022, 1978.

48. **Kream, R. M., Zukin, R. S., and Stefano, G. B.,** Demonstration of two classes of opiate binding sites in the nervous tissues of the marine mollusc *Mytilus edulis, J. Biol. Chem.,* 255, 9218, 1980.

Evolutionary History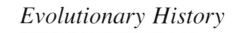

STRUCTURE AND FUNCTION OF THE HEAD ACTIVATOR IN *HYDRA* AND IN MAMMALS

H. Chica Schaller and
Heinz Bodenmüller

SUMMARY

From the fresh-water coelenterate *Hydra,* we recently isolated and sequenced a peptide which acts as a head inducing morphogen. This peptide, called head activator, has the sequence pGlu-Pro-Pro-Gly-Gly-Ser-Lys-Val-Ile-Leu-Phe and is responsible in *Hydra* for head-specific growth and differentiation processes. It affects growth by triggering cells to divide and cellular differentiation by influencing the determination of cells such as the differentiation of interstitial stem cells to nerve cells. A peptide with the same amino acid sequence as the head activator was also isolated from mammalian brain and intestine. Immunological methods indicate that the head activator also occurs in other animals and in other locations. It is especially abundant in frog skin and in neuronal or neuroendocrine cells or tumors. The head activator may be the first example of a neuropeptide that conserved its structure and maybe its trophic function all through evolution from coelenterates to humans.

INTRODUCTION

Within the animal kingdom, cells specialized for nervous function at first appear among coelenterates. We have chosen to work with *Hydra,* because its nervous system and its general organization seemed to be simple. The animal consists of only two cell layers, ectoderm and endoderm, made up of epitheliomuscular cells and of stem cells which give rise to two differentiation products, nerve cells and nematocytes. The nervous system is organized as a continuous net with a higher density in the head and in the foot region.

The nervous system in *Hydra* serves two functions: it is necessary for the coordination of fast processes like body movements, and it controls slow processes like cellular growth and differentiation. So far, not much is known how and with which transmitter substances the fast interneuronal communication system in *Hydra* works. Accumulating evidence suggests that conventional transmitters such as acetylcholine or catecholamines are not present in nerve cells of *Hydra.* It seems more likely that peptides may play that role since more and more of them are found coexisting or separately in nerve cells of hydra[1,2] (see contribution by Grimmelikhuijzen). We know more about the second function of the nervous system and about the substances which control cellular growth and differentiation and thus pattern formation in *Hydra.* To understand pattern formation, we want to know why a certain structure is induced at a definite position and time during development, and why it subsequently inhibits the formation of an identical such structure nearby.

We could show that four substances are responsible for induction and inhibition in *Hydra:* an activator and an inhibitor of head formation and an activator and an inhibitor for foot formation.[3] All four substances in normal *Hydra* are products of nerve cells. The gradients in biological properties are caused by a graded distribution of these four substances along the body axis of the animal. The two activators are peptides, the inhibitors resemble in their chemical properties conventional transmitter-like substances. The inhibitors are hydrophilic molecules, the activators are more hydrophobic. Thus, the head activator at concentrations higher than $10^{-3}M$ is insoluble in water, soluble in organic solvents, and has a very high surface adhesivity. The head inhibitor is easily soluble in water and does not seem to stick to hydrophobic surfaces. This difference in hydrophobicity or surface adhesivity resulting in completely different diffusion rates provides a plausible explanation for the biological finding that the action of the inhibitors is long range, whereas that of the activators is short range.

From the degree of purity and the molecular weight, we can calculate that all four substances act at very low concentrations. According to the present state of purity, the two foot factors and the head inhibitor act at concentrations below 10^{-8} *M* the head activator at 10^{-13} *M*. This also means that a *Hydra* needs very little of these substances and explains our difficulties in obtaining in pure form workable quantities from *Hydra* for a chemical analysis. For example, we were able to isolate from 3×10^6 *Hydra* (3 kg), accumulated over several years, 0.5 μg of the pure head activator. Fortunately, we discovered that other coelenterates, in particular sea anemones, which are 10^4-fold larger and abundant in the ocean, contain the same set of substances as *Hydra* in similar concentrations. Meanwhile we have processed 200 kg of the sea anemone, *Anthopleura elegantissima*, and used it so far for the chemical analysis of the head activator.[4]

ISOLATION AND SEQUENCE ANALYSIS OF THE HEAD ACTIVATOR

To obtain workable quantities of the peptide we had to start with relatively large amounts of tissue. From the 200 kg of *Anthopleura* we isolated 20 nmol of head activator and from 3 kg of *Hydra* 500 pmol. We also isolated the head activator from other animal sources, especially mammals, where we extracted 3000 pieces of rat intestine and obtained 20 nmol and 2 human hypothalami, which yielded 2 nmol of head activator. In general, we could enrich the head activator 10^8- to 10^9-fold with a 10 to 20% yield. Purification steps included organic solvent extraction, molecular sieve and ion-exchange chromatography, and HPLC methods.[4,5] The amounts of head activator available for a sequence analysis were relatively minute and required use and improvement of micromethods. To determine the amino acid composition of the head activator the purified peptide was hydrolyzed, dansylated, and the dansylated amino acids were separated for a qualitative analysis on micro-thin-layer plates and for quantitation on HPLC. We found that the head activator contained the following amino acids: Gly (2), Glu (1), Ile (1), Leu (1), Lys (1), Phe (1), Pro (2), Ser (1), Val (1). The sequence analysis was complicated by the fact that the head activator, like many other biologically active peptides, had no free amino terminus, that its amino end was blocked by pyroglutamic acid. Therefore, most of the sequence was obtained by enzymatic degradation methods and in part by a chemical degradation of the biologically inactive Glu-peptide which contained glutamic acid instead of pyroglutamic acid as amino terminal amino acid. The head activator was found to have the sequence pGlu-Pro-Pro-Gly-Gly-Ser-Lys-Val-Ile-Leu-Phe.[4] No difference in sequence was found for the head activator from the sea anemone, *Anthopleura elegantissima*, from *Hydra attenuata*, from rat intestine, and from bovine and human hypothalamus.[4,5] This sequence of the head activator was confirmed by synthesis.[6] We could show that there is no difference between the purified native and the synthetic peptides in chemical, enzymatic, and biological properties. We and others also produced a series of peptides with minor modifications. So far, none of those was as biologically active as the native molecule. Together with the evolutionary conservation of the head activator from *Hydra* to mammals, we interpret this to mean that the native molecule, at least in regards to its action on *Hydra* receptors, has very strict conformational requirements.

The chemistry of the head activator predicts or explains the following outstanding properties: (1) it is very lipophilic leading to a relatively low solubility in aqueous solvents; and (2) in common with many other biologically active peptides, it has a very high surface adhesivity resulting in absorption in vitro to glass, plastic, etc. and in vivo probably to membranes and to hydrophobic parts of proteins rendering it relatively immobile or pre-

venting diffusion to distant areas. The head activator occurs in two distinguishable conformations, one with an apparent molecular weight of about 1500 daltons and one of about
700 daltons. Preliminary evidence supports the idea that the low-molecular-weight conformation is a monomeric peptide, whereas the larger one is dimeric. The freshly isolated,
relatively impure, and dilute native peptide has the monomeric conformation. The synthetic
molecule and the purified head activator in physiological media convert to the more stable
dimeric conformation. The two conformations are interconvertible. Since only the monomer
is biologically active, we think that this conversion may be of physiological relevance.

EFFECTS OF THE HEAD ACTIVATOR ON MORPHOGENESIS, CELLULAR GROWTH, AND DIFFERENTIATION

The head activator and its counterpart, the head inhibitor, are present in *Hydra* in an
inactive structure-bound form. The head activator is stored in neurosecretory granules.[7] This
has the advantage that the release can be regulated. The release of the head factors is controlled
by the head inhibitor which acts as a release-inhibiting factor for its own release and for
that of the head activator as well.[8] The head activator has no effect on its own release nor
on that of the head inhibitor. Thus release of both factors is triggered by the absence of the
head inhibitor. The head inhibitor, as a small hydrophilic molecule, is able to diffuse once
released into the intercellular space over long distances. The head activator, due to its
hydrophobic character, will adhere to membranes of neighboring cells or to the extracellular
matrix and will therefore by restricted locally. The distribution of sources for both substances
within *Hydra* is relatively similar, both having their maximal concentration in the head area
(hypostome). Thus although both substances may be released from the same location, at the
release site there will be a local predominance of head activator over head inhibitor and at
a greater distance predominance of head inhibitor over head activator.

At the cellular level, head activator and head inhibitor act antagonistically: the head
activator stimulates cells to divide and is necessary for head-specific differentiation; the head
inhibitor blocks such processes. Thus an interstitial stem cell will become a head-specific
nerve cell if the concentration of the head activator is higher than that of the head inhibitor.
The stem cell will remain stem cell or enter another differentiation pathway (e.g., to nematocytes) if head inhibitor prevails.[8-11] This means that close to a common release site
stem cells differentiate to nerve cells, whereas further away they enter another differentiation
pathway. The new nerve cells will then produce head activator and/or head inhibitor thus
reinforcing or amplifying the head-specific influence. Such a feedback mechanism may
ensure that wherever head activator predominates over head inhibitor, head-like structures
will be induced and maintained. Release control and events at the cellular level have recently
been incorporated into a computer simulation of head regeneration in *Hydra*.[12]

ACTION OF THE HEAD ACTIVATOR IN MAMMALS

The head activator was also isolated and sequenced from human and bovine hypothalami
and from rat intestines. Immunological methods (HPLC-RIA or ELISA)[13,14] indicate that
a head-activator-like peptide also occurs in other animals (drosophila, artemia, frog, rabbit,
mouse, guinea pig) and in other locations.

The sequence of the head activator differs from that of other biologically active peptides.
The only structural homology we observed is with bradykinins, in the sequence X-Pro-Pro-
Gly-X-Ser-. Bradykinins are known as smooth muscle stimulants. The head activator at a
concentration of 10^{-5} *M* did stimulate uterus and ileum contraction in rats.[16] This may imply
a very weak structural and functional homology. From the localization of the head activator
in the brain and intestinal tract, from its appearance early in mammalian development, and

in analogy to its action in Hydra, the following speculation may be allowed. In the brain it may act as a transmitter-like messenger. Some preliminary evidence suggests that like other hypothalamic peptides it may have a regulatory effect on the pituitary.[17] The high concentration in the upper part of the intestine may hint at a regulatory function in the digestive system. Thus we found that the levels of head activator in human blood increase immediately after ingestion of a meal. One consequence of this may be that exocrine pancreas secretion measured as amylase release is stimulated.[15] In embryonal life and maybe also in the adult the head activator may act on some neuronal stem cell either as growth factor or as differentiation signal. Its presence in tumors or tumor cell lines derived from early stages of neuronal differentiation hints at such a possible trophic or mitotic role in nerve cell development. This may be the first example that a neuropeptide conserved its structure, and perhaps also its growth or differentiation controlling functions during evolution from coelenterates to humans.

ACKNOWLEDGMENTS

H. C. Schaller was a recipient of a Heisenberg fellowship. We are supported by the Deutsche Forschungsgemeinschaft (Scha 253/8).

REFERENCES

1. **Grimmelikhuijzen, C. J. P., Dierickx, K., and Boer, G. J.,** Oxytocin/vasopressin-like immunoreactivity is present in the nervous system of hydra, *Neuroscience,* 7, 3191, 1982.
2. **Grimmelikhuijzen, C. J. P.,** Coexistence of neuropeptides in hydra, *Neuroscience,* 9, 837, 1983.
3. **Schaller, H. C., Schmidt, T., and Grimmelikhuijzen, C. J. P.,** Separation and specificity of action of four morphogens from hydra, *Wilhelm Roux's Archives,* 186, 139, 1979.
4. **Schaller, H. C. and Bodenmüller, H.,** Isolation and amino acid sequence of a morphogenetic peptide from hydra, *PNAS,* 78, 7000, 1981.
5. **Bodenmüller, H. and Schaller, H. C.,** Conserved amino acid sequence of a new neuropeptide, the head activator, from coelenterates to humans, *Nature,* 293, 579, 1981.
6. **Birr, C., Zachmann, B., Bodenmüller, H., and Schaller, H. C.,** Synthesis of a new neuropeptide, the head activator from hydra, *FEBS Lett.,* 131, 317, 1981.
7. **Schaller, H. C. and Gierer, A.,** Distribution of the head-activating substance in hydra and its localisation in membraneous particles in nerve cells, *J. Embryol. Exp. Morph.,* 29, 39, 1973.
8. **Kemmner, W. and Schaller, H. C.,** Actions of head activator ahd head inhibitor during regeneration in hydra, *Differentiation,* 26, 91, 1984.
9. **Schaller, H. C., Bodenmüller, H., and Kemmner, W.,** Struktur und Funktion morphogener Substanzen aus Hydra, *Verh. Dtsch. Zool. Ges.,* 81, 1982.
10. **Schaller, H. C.,** Action of the head activator as a growth hormone in hydra, *Cell Differ.,* 5, 1, 1976.
11. **Schaller, H. C.,** Action of the head activator on the determination of interstitial cells in hydra, *Cell Differ.,* 5, 13, 1976.
12. **Kemmner, W.,** A model of head regeneration in hydra, *Differentiation,* 26, 83, 1984.
13. **Bodenmüller, H. and Zachmann, B.,** A radioimmunoassay for the head activator from hydra, *FEBS Lett.,* 159, 237, 1983.
14. **Schaller, H. C., Bodenmüller, H., Zachmann, B., and Schilling, E.,** Enzyme-linked immunosorbent assay for the neuropeptide head activator, *Eur. J. Biochem.,* 138, 365, 1984.
15. **Feuerle, G. E., Bodenmüller, H., and Báca, J.,** The neuropeptide head activator stimulates amylase secretion from rat pancrease in vitro, *Neuroscience Letters,* 38, 287, 1983.
16. **Ganten, D.,** unpublished.
17. **Kordon, P.,** unpublished.

THE ENKEPHALINS AND FMRFamide-LIKE PEPTIDES: THE CASE FOR COEVOLUTION

Michael J. Greenberg, Scott M. Lambert, Herman K. Lehman, and David A. Price

INTRODUCTION: BROWNSTEIN'S CONJECTURE

The opioid pentapeptide, Met-enkephalin, was chemically identified in mammalian brain in 1975,[1] and the "molluscan cardioactive peptide" FMRFamide was identified in clam ganglion extracts 2 years later.[2] Given the obvious functional and structural differences between these peptides, there was no reason *a priori* for associating them, particularly since their reactive sites are, respectively, at the amino and carboxy terminals. Consequently, the occurrence of the dipeptide -Phe-Met- in both Met-enkephalin and FMRFamide went unnoticed.

In 1979, the opioid heptapeptide Met-enkephalin-Arg[6]-Phe[7] (YGGFMRF) was discovered in bovine adrenal chromaffin cells.[3] Shortly afterward, Brownstein[4] noted that the sequence of the new opioid contained that of FMRFamide, and he included the molluscan peptide in a list of possible opioid peptide precursors (Table 1 in Reference 4).

SOME PHYLOGENETIC CONSIDERATIONS

The coincidence of a tetrapeptide sequence in a mollusc and a mammal can be viewed either as a chance occurrence or as an example of homology. Regarding chance, the probability that a given tetrapeptide from one species will appear as a C-terminal sequence in another is rather high — about 60%; in fact, close analogs of -Phe-Met-Arg-Phe- actually occur in a variety of known vertebrate hormones and precursors.[5] The high probability of coincidence suggests that the shared tetrapeptide is an example of evolutionary convergence, and the general dissimilarity between molluscs and vertebrates supports that notion.

Although the argument for convergence is compelling, Greenberg et al.[6] took the more heuristic view that " . . . the tetrapeptide sequence Phe-Met-Arg-Phe, common to both FMRFamide and the opioid YGGFMRF, implies some degree of homology between the two classes of neuropeptides and their complementary receptors. One speculation might be that they coevolved from an ancestral peptide YGGFMRFamide and its receptor." Can this idea be described more concretely?

The hypothetical scheme in Figure 1 illustrates one way in which the FMRFamide-like and enkephalin-containing (EC-) peptides may have coevolved. The starting postulate is that, at some point, YGGFMRFamide appeared as a secreted molecule, binding to a complementary receptor site to produce some regulatory effect.* The original location of the "message" (i.e., the effective binding site on the molecule) cannot, of course, be specified; in Figure 1, the primitive peptide and receptor are portrayed as being, respectively, amphiactive and bivalent. In any event, we can speculate that two new types of bivalent receptors soon evolved: each type bound to both the N- and C-terminals of the heptapeptide; but whereas the effective (message) site for one type was on the C-terminal (FMRFamide-like receptors), the other kind bound effectively to the N-terminal (enkephalin-like receptors). Subsequent evolution led, presumably, to the present condition: an array of FMRFamide-like and EC-peptides (together with their receptors) with conserved message sequences at the C- and N-terminals, respectively, and with various address sequences at the opposite ends of the molecule. Consistent with this scheme, we could think of YGGFMRF and its receptor as representing a stage in the loss of the C-terminal address (Figure 1).

---

* The possibilities, that YGGFMRFamide might have been secreted as a functionless by-product of processing before a receptor evolved, or that the receptor evolved first, are not rationally discussable.

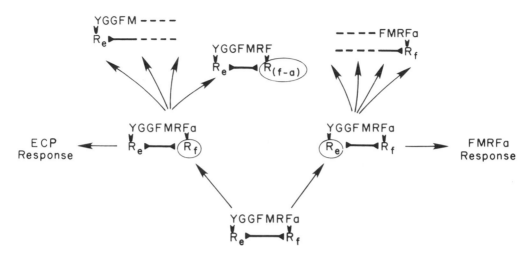

FIGURE 1.   A hypothetical scheme describing the coevolution of the enkephalin-containing- (ECP) and FMRFamide-like peptides, and their respective receptors, from a common, ancestral peptide and receptor. The primitive
receptor (at the bottom) is shown as bivalent with an EC-binding site at the N-terminal (R_c) and a FMRFamide-
binding site at the C-terminal (R_f). The primitive amphimorphic ligand is taken to be Tyr-Gly-Gly-Phe-Met-Arg-
Phe-NH$_2$ (YGGFMRFa). In evolution, binding sites may retain their messages, or serve only as addresses (circled).
In the mammalian heptapeptide Met-enkephalin-Arg^6-Phe^7 (YGGFMRF), the address-$R_{(f-a)}$ has been shortened. The
diverse set of recent EC- and FMRFamide-like peptides (and their receptors) have, according to this scheme, mostly
lost all traces of their respective ancestral address sites. Further description in the text.

Both the FMRFamide-like and EC-peptides have been detected, primarily by immunological techniques, at all levels of the animal kingdom, from the most primitive eumetazoans,
to the insects, cephalopods, and mammals (Figure 2). If we accept the validity of these
identifications, then the phylogenetic distribution in Figure 2 provides a time frame for the
evolutionary sequence proposed above. That is, if the opioid and FMRFamide-like peptides
had a common origin, it must have occurred very early, in some primitive flagellate.
Similarly, if YGGFMRF (from mammalian adrenal gland) and its receptor are relics of this
common origin, then similar relics should occur frequently among the 25 to 30 invertebrate
phyla. In the remainder of this essay, we critically examine a few likely relics that have
turned up among molluscs.

EVIDENCE FROM MOLLUSCS OF A COMMON PRECURSOR

The best evidence that the FMRFamide-like and EC-peptides had a common origin would
be a molluscan gene, encoding a precursor protein which would include the sequences of
both peptide families, and which would be similar in its overall sequence to one of the
known vertebrate opioid precursors. A good example of the latter would be preproenkephalin
A, which appears to be conserved in vertebrates, from amphibians to mammals.[7] A FMRFamide gene has been isolated, but the precursor it encodes includes no EC sequences.[7a] Under
these circumstances, three alternate sorts of data would be helpful: (1) immunocytochemical
evidence that a single cell is producing both enkephalin- and FMRFamide-like peptides; (2)
complete chemical identification of amphimorphic neurosecretory peptides (e.g.,
YGGFMRFamide) containing the sequences of both peptide families; and (3) the demonstration, by radioreceptor or pharmacological assay, of bivalent receptors activated by either
or both types of peptides, but responding optimally to YGGFMRFamide.

By themselves, none of these evidences would be overwhelmingly strong; they would
become much more compelling in concert.

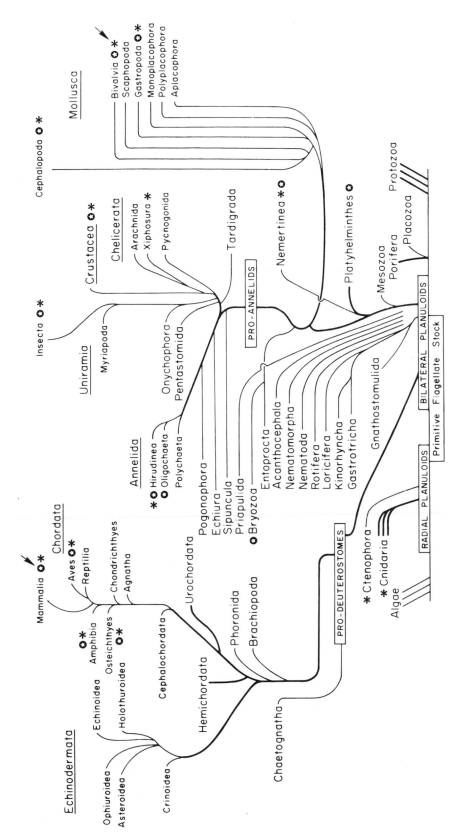

FIGURE 2. Animal phylogenesis showing every group in which enkephalin-containing (EC) peptides (white star) and FMRFamide-like peptides (asterisk) have been detected, usually by immunological or pharmacological techniques. Peptides have been chemically identified only in their home phylum; taxon of origin indicated by an arrow.

One Cell Produces Members from Both Peptide Families

Where immunocytochemical staining for both FMRFamide and enkephalins has been done, the two classes of peptides have been observed in different cells. For one example, the distributions of immunocytochemical (ic-) FMRFamide and ic-Met-enkephalin among and within the ganglia of *Lymnaea stagnalis* are completely distinct (compare Figure 5 in Reference 8 with Figure 1 in Reference 9). Most of the ic-Met-enkephalin is in the pedal ganglia, each of which contains about 0.7 pmol enkephalin, as measured by radioreceptor assay.[10] In contrast, the number of ic-FMRFamide-staining cells is especially high in the visceral and parietal ganglia which contain little ic-Met-enkephalin.

Another cogent example of separate distributions is in the pedal ganglion of *Mytilus edulis,* where both ic-FMRFamide and ic-enkephalin have been demonstrated, but in different, nonoverlapping cell populations.[11]

Colocalization of ic-FMRFamide and ic-Met-enkephalin has been reported only in cell bodies and terminals of the neurosecretory system of the vena cava in *Octopus bimaculatus.*[12] However, the demonstration required that serial sections be stained with antibodies to FMRFamide and Met-enkephalin, both in untreated tissues, and in tissues previously treated with trypsin or carboxypeptidase B. The enzyme digestion eliminated ic-FMRFamide, but enhanced ic-Met-enkephalin.[12]

We have examined a tissue-free analog of this observation: the binding of synthetic peptides in a radioimmunoassay (RIA) for Leu-enkephalin (Figure 3). The procedure, similar to that of Lewis et al.,[13] was to treat a test peptide (e.g., YGGFLRFamide) with trypsin to remove the C-terminal Phe-NH$_2$, and then with carboxypeptidase B to cleave off the arginyl residue. The expected product of digestion is Leu-enkephalin (YGGFL). Therefore, if the binding of the peptide to Leu-enkephalin antibody is measured before and after digestion, (details of the RIA in the legend to Figure 3), the unsurprising result is that proteolysis of YGGFLRF-amide (having produced Leu-enkephalin) enhances binding (Figure 3). But how specific is this procedure?

We repeated the experiment, but used synthetic Glp-Asp-Pro-Phe-Leu-Arg-Phe-NH$_2$ (pQDPFLRFamide) as the substrate. This FMRFamide-like peptide was recently identified in the ganglia of *Helix aspersa.*[14] At the concentrations tested, pQDPFLRFamide does not crossreact with the Leu-enkephalin antibody (Figure 3). After digestion, however, a small amount of binding occurred; more would not be expected since the product, pQDPFL, and the antigen, YGGFL, have in common only the terminal dipeptide. Nevertheless, if a peptide such as pQDPFLRFamide were present in high concentrations, as it is in some neuronal cell bodies or terminals or *Helix,*[15] then digestion of sections of these tissues with trypsin and carboxypeptidase B would reveal ic-Leu-enkephalin.

In summary, the immunohistochemical evidence that Met-enkephalin and FMRFamide occur in sequence on the same precursors in cells of *Octopus* still needs considerable support. Such support may be forthcoming, as discussed briefly, below.[16]

The Occurrence of Amphimorphic Peptides

Of the FMRFamide-like peptides detected by immunological or pharmacological techniques in various animals, only three have been completely sequenced: FMRFamide itself was detected in *Aplysia* head ganglia;[17] pQDPFLRFamide occurs in *Helix* neurons, both central and peripheral;[14,15] and the pentapeptide, Leu-Pro-Leu-Arg-Phe-NH$_2$, was found in chicken brain.[18] Thus, there is still no definitive evidence of a peptide containing both enkephalin and FMRFamide-like sequences. However, the data of Voigt et al.[16] strongly suggest that YGGFMRFamide is the FMRFamide-like peptide in the neurohemal organ of the vena cava of *Octopus.*

Most of the enkephalins and endorphins widely detected by immunohistochemistry in the invertebrates (Figure 2) have not been characterized. Recently, however, the pedal ganglia

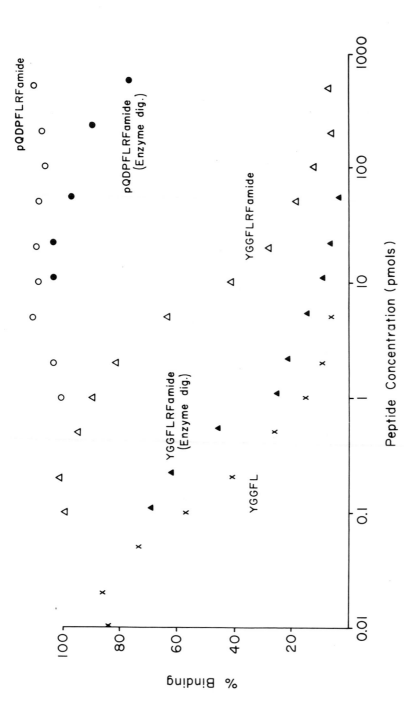

FIGURE 3.   Leu-enkephalin radioimmunoassay (RIA): the displacement of ^{125}I-Leu-enkephalin (trace) from a Leu-enkephalin antiserum (Peninsula Laboratories) by Leu-enkephalin (YGGFL: standard) and two analogs: YGGFLRFamide and pQDPFLRFamide. Both analogs were assayed in the natural state, and after sequential digestion with trypsin and carboxypeptidase B. The final concentration of digest: 5 nmol/45 μ*ℓ*. RIA: each tube contained 100 μ*ℓ* of diluted antiserum, 50 μ*ℓ* of diluted trace (7000 cpm), and 50 μ*ℓ* of sample or buffer, final peptide concentration (abscissa), pmol/ 200 μ*ℓ*; bound and free trace separated with charcoal. The peptides are named according to their approved one-letter abbreviations: Y-Tyr; G-Gly; F-Phe; L-Leu; R-Arg; D-Asp; P-Pro; pQ-pyroGlu.

of *Mytilus edulis* were reported to contain Met- and Leu-enkephalin, and Met-enkephalin-Arg[6]-Phe[7] (YGGFMRF). Even though the amide is lacking, and YGGFMRF is inactive in FMRFamide bioassays,[6,19] the finding of the opioid heptapeptide in molluscs still supports the notion that this amphimorphic sequence evolved very early and has been strongly conserved.

Bivalent Receptors

Where the effects of both FMRFamide-like and EC-peptides have been tested, molluscan somatic and cardiac muscles usually have not responded to opioids, but only to FMRFamide and its amidated analogs, and these responses were not blocked by naloxone (e.g., bivalve heart, radula protractor, anterior byssus retractor muscle (ABRM);[6] *Octopus* heart[20]). [Two exceptions are noteworthy. First, the heartbeat of the venerid clam, *Protothaca staminea*, was inhibited by Met-enkephalin, and the response blocked by naloxone; but this enkephalin effect was not consistently repeatable.[21] Second, morphine (enkephalins not tested) enhances the contracture of a *Mytilus* ABRM treated with ACh.[22]]

Although it is extended by a Met-enkephalin sequence, YGGFMRFamide is not different in its pharmacology from other N-terminal-extended analogs. The effects of all such analogs upon molluscan somatic and heart muscles are usually qualitatively similar to those of FMRFamide, although the relative potencies may vary. A complex example is the biphasic action of FMRFamide and YGGFMRFamide on the tentacle retractor muscle of *Helix aspersa:* whereas both peptides are about equipotent inducers of rhythmicity, YGGFMRFamide is more potent as an inhibitor of tone and rhythmical activity.[23] The enkephalins have no effect on the tentacle retractor muscle.[15,23]

The clam rectum is an unusual preparation, in that it responds consistently to both FMRFamide and the EC peptides. The effects of FMRFamide are an increase in tone and the induction of rhythmical activity; the EC-peptides reduce tone and inhibit rhythmicity. YGGFMRFamide is amphiactive on the clam rectum, its effect (excitation, inhibition, or both) varying with preparation and dose. As a contracture agent, YGGFMRFamide is about equipotent to FMRFamide; but as an EC-peptide, it is the most potent analog tested, about 100 times more active than any of the enkephalins. None of these effects is blocked by naloxone ($<10^{-4}M$) which acts as a weak agonist.[24,25]

The critical finding with the clam rectum, that the most active enkephalin was YGGFMRFamide (even compared with YGGFMRF), suggested the occurrence of an opioid receptor with a binding site for a ligand having -Met-Arg-Phe-NH_2 at its C-terminal. Therefore, since high concentrations (About 10 pmol per ganglion) of authentic mammalian opioid peptides (i.e., Met- and Leu-enkephalins and Met-enkephalin-Arg[6]-Phe[7]) have been characterized in *Mytilus* ganglion,[26,27] we supposed that these same peptides, or even YGGFMRFamide, might turn up in ganglion extracts of *Mercenaria* assayed by techniques used to detect these same enkephalins in mammalian tissues.

No Enkephalins Detectable in Clam Ganglia

The starting material was the acetone-extracted ganglia from 1000 *Mercenaria*. These were further extracted in acid[3] and chromatographed on Sephadex G-15. The fractions were tested in an opiate radioreceptor assay employing cell membranes isolated from rat brain, and in two RIAs featuring antibodies to, respectively, Leu- and Met-enkephalin (methods in Figure 4 legend).

Two prominent peaks of ligand binding were detected with the radioreceptor assay. Leu-enkephalin immunoreactivity (ir) appeared only in the second peak (about 30 pmol/mℓ), as did relatively low levels of ir-Met-enkephalin (about 2.7 pmol/mℓ).

No step in the extraction procedure excluded sodium, so it remained in the supernatant fractions with the extracted peptides. However, sodium decreases the affinity of mammalian opiate receptors for their ligands, largely by increasing the dissociation rate of the peptide-

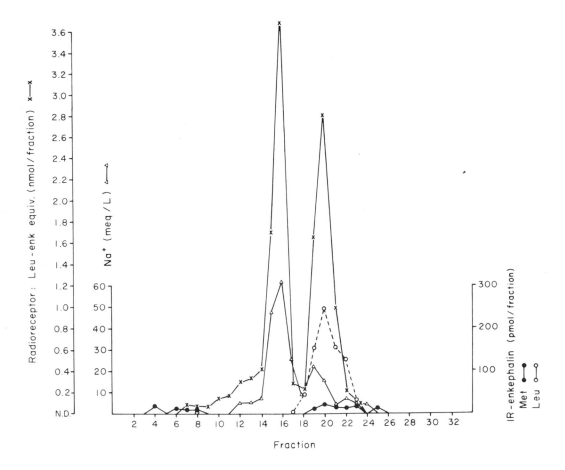

FIGURE 4. Chromatographic separation of a clam ganglion extract, and assay of the fractions for enkephalin-containing (EC) peptides and sodium content. Ganglia from 1000 clams were extracted (see text), the sample applied to a column of Sephadex G-15, eluted with 0.1 N acetic acid, and 8-mℓ fractions collected. Aliquots were tested for enkephalin activity by radioimmunoassay using antisera to Met-enkephalin (Penninsula Laboratories) and Leu-enkephalin (Immuno-Nuclear Corp.); method in Figure 3 legend. Opiate activity was measured by radioreceptor assay using ³H-D-Ala-Met-enkephalinamide binding to rat brain membranes by the method of Pert and Snyder,[28] except that incubations were carried out at 0°C for 90 min in the presence of 20 µg/mℓ bacitracin to minimize sample proteolysis. Radioreceptor assays are expressed in terms of Leu-enkephalin molar equivalents. The sodium content of the fractions was determined by flame photometry.

receptor complex.[28] Thus, if sodium occurred in a fraction, the equilibrium binding of trace (³H-D-Ala-Met-enkephalinamide) would be decreased, an effect indistinguishable, in the radioreceptor assay, from competitive binding by an opiate peptide or alkaloid. Sodium concentration was measured with a flame photometer in those fractions comprising the two radioreceptor peaks; most of it eluted in the first peak (Figure 4).

A fraction containing enkephalins can be desalted (and the sodium removed) by loading it onto a Sep-Pak® and washing first with water, and then with methanol; all detectable enkephalin activity elutes in the methanol. This procedure was carried out on the active fractions (illustrated in Figure 4). No ir-Leu-enkephalin was detected in the methanol washes from any fraction, and the radioreceptor assay detected only small amounts of binding, none of it in the most propitious fraction (#20, Figure 4).

Two further attempts were made to demonstrate the presence of opioid peptides. First, aliquots of the fractions were treated with leucine aminopeptidase (20µg/mℓ) which should cleave the Tyr¹-Gly² bond of enkephalins. However, this treatment had no effect on the small amounts of EC-peptides detected by radioreceptor assay or RIA. Second, we assumed that Met-enkephalin-Arg^6-Phe^7 might be present in high concentrations and might not have been detected by the Met-enkephalin RIA. Therefore, aliquots of the active fractions (14 to 22 in Figure 4) were treated successively with trypsin (100 μg/mℓ, 22°C, 2 hr) and carboxypeptidase B (15 μg/mℓ, 22°C, 2 hr). However, rather than enhancing Met-enkephalin levels, the digestion abolished them completely.

In summary, we had expected to recover and measure at least 1 nmol of EC-peptide in those pooled bivalve ganglia, but instead we found very low levels of substances that did not appear to behave like authentic enkephalins. This result could reflect seasonal or age-related variations in enkephalin levels, a phenomenon reported in *Mytilus*, [20,30] but about which nothing is known in *Mercenaria*.

CONCLUSION

Among invertebrates, opioid peptides have been chemically identified only in molluscs, and these molecules have not been variants, but authentic members of the vertebrate peptide family.[26,27] Only two FMRFamide-like peptides are known in molluscs, and none of these has been shown unequivocally to occur in vertebrates or any other phylum. The evidence supporting a common origin for the enkephalins and FMRFamide-like peptides is scattered and weak. As we wrote in 1981,[6] "Such speculations can be better evaluated when the phylogenetic distribution of opioid and FMRFamide-like peptides, and the specificities of their target tissue receptors, are better known." Four years later, there is no need to change that assessment.

REFERENCES

1. **Hughes, J., Smith, T. W., Kosterlitz, H. W., Fothergill, L. A., Morgan, B. A., and Morris, H. R.,** Identification of two related pentapeptides from the brain with potent opiate agonist activity, *Nature (London),* 258, 577, 1975.
2. **Price, D. A. and Greenberg, M. J.,** The structure of a molluscan cardioexcitatory neuropeptide, *Science,* 197, 670, 1977.
3. **Stern, A. S., Lewis, R. V., Kimura, S., Rossier, J., Gerber, L. D., Brink, L., Stein, S., and Udenfriend, S.,** Isolation of the opioid heptapeptide Met-enkephalin[Arg^6, Phe^7] from bovine adrenal medullary granules and striatum, *Proc. Natl. Acad. Sci. U.S.A.,* 76, 6680, 1979.
4. **Brownstein, M. J.,** Opioid peptides: search for the precursors, *Nature (London),* 287, 678, 1980.
5. **Price, D. A.,** FMRFamide: assays and artifacts, in *Molluscan Neuroendocrinology,* Lever, J. and Boer, H. H., Eds., North-Holland, Amsterdam, 1983, 184.
6. **Greenberg, M. J., Painter, S. D., and Price, D. A.,** The amide of the naturally occurring opioid [Met]enkephalin-Arg^6-Phe^7 is a potent analog of the molluscan neuropeptide FMRFamide, *Neuropeptides,* 1, 309, 1981.
7. **Kilpatrick, D. L., Howells, R. D., Lahm, H.-W., and Udenfriend, S.,** Evidence for a proenkephalin-like precursor in amphibian brain, *Proc. Natl. Acad. Sci. U.S.A.,* 80, 5772, 1983.
7a. **Schaefer, M., Picciotto, M. R., Kreiner, T., Kaldany, R.-R., Taussig, R., and Scheller, R. H.,** *Aplysia* neurons express a gene encoding multiple FMRFamide neuropeptides, *Cell,* 41, 457, 1985.
8. **Schot, L. P. C., Boer, H. H., Swaab, D. F., and Van Noorden, S.,** Immunocytochemical demonstration of peptidergic neurons in the central nervous system of the pond snail *Lymnaea stagnalis* with antisera raised to biologically active peptides of vertebrates, *Cell Tissue Res.,* 216, 273, 1981.
9. **Schot, L. P. C. and Boer, H. H.,** Immunocytochemical demonstration of peptidergic cells in the pond snail *Lymnaea stagnalis* with an antiserum to the molluscan cardioactive tetrapeptide FMRFamide, *Cell Tissue Res.,* 225, 347, 1982.
10. **Lambert, S. M.,** unpublished data, 1984.
11. **Stefano, G. B. and Martin, R.,** Enkephalin-like immunoreactivity in the pedal ganglion of *Mytilus edulis* (Bivalvia) and its proximity to dopamine-containing structures, *Cell Tissue Res.,* 230, 147, 1983.
12. **Martin, R., Frösch, D., Kiehling, C., and Voigt, K. H.,** Molluscan neuropeptide-like and enkephalin-like material coexists in octopus nerves, *Neuropeptides,* 2, 141, 1981.

13. **Lewis, R. V., Stern, A. S., Kimura, S., Rossier, J., Stein, S., and Udenfriend, S.,** An about 50,000-dalton protein in adrenal medulla: a common precursor of [Met]- and [Leu]enkephalin, *Science,* 208, 1459, 1980.

14. **Price, D. A., Cottrell, G. A., Doble, K. E., Greenberg, M. J., Jorenby, W., Lehman, H. K., and Riehm, J. P.,** A novel FMRFamide-related peptide in *Helix:* pQDPFLRFamide, *Biol. Bull.,* 169, 256, 1985.

15. **Lehman, H. K.,** The distribution and actions of a FMRFamide-like peptide, *Soc. Neurosci. Abstr.,* 9, 78, 1983.

16. **Voigt, K. H. and Martin, R.,** Neuropeptides with cardioexcitatory and opioid activity in octopus nerves, in *Handbook of Comparative Opioid and Related Neuropeptide Mechanisms,* Vol. 1, Stefano, G. B., Ed., CRC Press, Boca Raton, Fla., 1986.

17. **Lehman, H. K., Greenberg, M. J., and Price, D. A.,** The FMRFamide-like peptide of *Aplysia* is FMRFamide, *Biol. Bull.,* 167, 460, 1984.

18. **Dockray, G. J., Reeve, J. R., Jr., Shively, J., Gayton, R. J., and Barnard, C. S.,** A novel active pentapeptide from chicken brain identified by antibodies to FMRFamide, *Nature (London),* 305, 328, 1983.

19. **Painter, S. D., Morley, J. S., and Price, D. A.,** Structure-activity relations of the molluscan neuropeptide FMRFamide on some molluscan muscles, *Life Sci.,* 31, 2471, 1982.

20. **Voigt, K. H., Kiehling, C., Frösch, D., Schiebe, M., and Martin, R.,** Enkephalin-related peptides: direct action on the octopus heart, *Neurosci. Lett.,* 27, 25, 1981.

21. **Koch, R. A.,** personal communication, 1984.

22. **Bianchi, C. P. and Wang, Z.,** Morphine enhancement of the cholinergic response of anterior byssus retractor muscle of *Mytilus edulis,* in *Handbook of Comparative Opioid and Related Neuropeptide Mechanisms,* Vol. 2, Stefano, G. B., Ed., CRC Press, Boca Raton, Fla., 1986.

23. **Cottrell, G. A., Greenberg, M. J., and Price, D. A.,** Differential effects of the molluscan neuropeptide FMRFamide and the related Met-enkephalin derivative YGGFMRFamide on the *Helix* tentacle retractor muscle, *Comp. Biochem. Physiol.,* 75C, 373, 1983.

24. **Doble, K. E. and Greenberg, M. J.,** The clam rectum is sensitive to FMRFamide, the enkephalins and their common analogs, *Neuropeptides,* 2, 157, 1982.

25. **Greenberg, M. J., Painter, S. D., Doble, K. E., Nagle, G. T., Price, D. A., and Lehman, H. K.,** The molluscan neurosecretory peptide FMRFamide: comparative pharmacology and relationship to the enkephalins, *Fed. Proc. Fed. Am. Soc. Exp. Biol.,* 42, 82, 1983.

26. **Leung, M. and Stefano, G. B.,** Isolation of opioid peptides, *Life Sci.,* 33 (Suppl. 1), 77, 1983.

27. **Leung, M. K. and Stefano, G. B.,** Isolation and identification of enkephalins in pedal ganglia of *Mytilus edulis* (Mollusca), *Proc. Natl. Acad. Sci. U.S.A.,* 81, 955, 1984. *958*

28. **Pert, C. B. and Snyder, S. H.,** Opiate receptor binding of agonists and antagonists affected differentially by sodium, *Mol. Pharmacol.,* 10, 868, 1974.

29. **Stefano, G. B., Kream, R. M., Zukin, R. S., and Catapane, E. J.,** Seasonal variation of stereospecific enkephalin binding and pharmacological activity in marine molluscs nervous tissue, in *Advances in Physiological Science, Neurotransmitters in Invertebrates,* S-Rózsa, K., Ed., Pergamon Press, London, 1981, 22 and 453.

30. **Stefano, G. B.,** Decrease in the number of high-affinity opiate binding sites during the aging process in *Mytilus edulis* (Bivalvia), *Cell Mol. Neurobiol.,* 1, 343, 1981.

FMRFamide-LIKE PEPTIDES IN THE PRIMITIVE NERVOUS SYSTEMS OF COELENTERATES AND COMPLEX NERVOUS SYSTEMS OF HIGHER ANIMALS

C. J. P. Grimmelikhuijzen*

SUMMARY

Antisera to the molluscan cardioexcitatory peptide FMRFamide intensely stained neuronal networks or nerve rings in all coelenterate species investigated. These antisera did not crossreact with FMRFamide-related peptides such as Met-enkephalin-Arg[6]-Phe[7] (YGGFMRF), and the carboxyterminus of gastrin or cholecystokinin (WMDFamide). Most antisera, however, seriously crossreacted with peptides belonging to the family of pancreatic polypeptide, which have an arginine and an amidated aromatic amino acid (RYamide) in common positions with FMRFamide. Antisera, especially selected for their inability to react with avian and bovine pancreatic polypeptide, persisted in staining all neuronal structures in coelenterates, showing that the peptide in this phylum is genuinely FMRFamide-like. These "specific" antisera to FMRFamide also stained neurons in the central nervous system of vertebrates such as fishes and mammals. In guinea pig brain, only one group of perikarya was stained, which was mainly located in the nucleus dorsomedialis and extended to the nucleus paraventricularis and nucleus periventricularis hypothalami. These perikarya projected to several regions of the hypothalamus, mesencephalon, medulla, and spinal cord.

INTRODUCTION

The most primitive nervous systems in the animal kingdom are found within species belonging to the phylum of coelenterates. Coelenterates were probably the first animal group, during evolution, in which a nervous system emerged. The best-known coelenterate is certainly *Hydra*, a small polyp living in fresh water (Figure 1). Like all other coelenterates *Hydra* consists of only two cell layers, ectoderm and endoderm, between which lies an acellular layer called mesoglea. The nervous system in *Hydra* is organized as a simple net, which is mainly located in the ectoderm of the animal, and which is denser in the mouth (hypostome) and foot regions (basal disk and peduncle). From sections of *Hydra*, early microscopists[1] subdivided the neurons in sensory cells (long perikarya contacting the outer surface and orientated perpendicular to the mesoglea, with processes at their basal parts) and ganglion cells (round perikarya lying in the basal part of the ectoderm). Later, electronmicroscopists discovered that neurons in *Hydra* formed normal chemical synapses with dense core synaptic vesicles in the range of 100 nm.[2-4] Furthermore, Westfall and co-workers found that all neurons in *Hydra*, sensory as well as ganglion cells, were multifunctional, each having motorneuronal, interneuronal, sensory, as well as neurosecretory properties.[5,6] Recently it was observed that neurons in the head and foot regions of *Hydra* formed gap junctions, which were lying mainly adjacent to the chemical synapses.[7]

The phylum of coelenterates is subdivided in three classes: the Hydrozoa (such as *Hydra*), the Scyphozoa (such as true jellyfishes), and the Anthozoa (such as sea anemones and corals). Aside from these groups (also called Cnidaria), a fourth group exists, the Acnidaria (Ctenophora or combjellies), which is often taken as a separate phylum. Most species belonging to the classes of Hydrozoa and Scyphozoa have a life cycle including a polyp

---

* Used one-letter abbreviations of amino acids: D, aspartic acid; F, phenylalanine; G, glycine; H, histidine; K, lysine; L, leucine; M, methionine; R, arginine; T, threonine; V, valine; W, tryptophan; Y, tyrosine.

FIGURE 1.   *Hydra attenuata*, a fresh-water polyp of about 0.5 cm in length. (From Grimmelikhuijzen, C. J. P., *Evolution and Tumorpathology of the Neuroendocrine System*, Falkmer, S., Håkanson, R., and Sundler, F., Eds., Elsevier Biomedical Press, Amsterdam, 1984, 39. With permission.)

form (sessile) and medusa form (swimming). The higher mobility of the medusa forms is reflected by a nervous system, which is much more developed than that of the polyp. In hydrozoan medusae, neurons are grouped together to form nerve rings at the outer and inner side of the velum (a circular sheet of swimming muscles at the lower end of the bell). Neurons belonging to each of these rings are electrically coupled and fire synchronously.[8-10] In scyphozoan medusae, no such nerve rings are present, but here the neurons form discrete ganglia at the bases of the tentacles.[10,11] Species belonging to the Anthozoa have only polyp forms, and their nervous system is not essentially different from that in the hydrozoan polyps. In Anthozoa most muscles occur in the endoderm, with the consequence that the nervous net is predominantly present in this inner cell layer.[12] From Ctenophora it is known that they have a diffuse ectodermal nerve net, which appears to control the swimming movements of the combs.[13]

Ultrastructural and physiological data leave no doubt that most synapses in the coelenterate nervous systems are chemical.[2-10,13-17] Until recently, however, no clear information was available on the neurotransmitter substances in these animals.[18] For *Hydra*, we have not been able to find acetylcholine, using a very sensitive bioassay on leech muscle (K. Wächtler and C. J. P. Grimmelikhuijzen, unpublished). After applying various variants of the Falck-Hillarp technique, I could not demonstrate any catecholamines or serotonin in the nervous system of *Hydra* (unpublished). After using a series of antisera to serotonin (obtained from H. Steinbusch, Amsterdam), it became clear that no serotonin was present in this animal (unpublished). These data together strongly suggest that none of the "classical" neurotransmitters is present in *Hydra*.

MATERIALS AND METHODS

Animals, antisera, and immunocytochemical procedures have been extensively described in previous reports.[19-23]

RESULTS

The Coelenterate Nervous System is Peptidergic

The morphology and size of many of the synaptic vesicles in the coelenterate nervous system (dense cored, 100-nm range), is very similar to that found in peptidergic neurons of phylogenetically higher animals. In order to investigate whether *Hydra* neurons contain peptides related to those of more evolved animals, I have carried out immunocytochemistry on sections of fixed *Hydra* using antisera to many of the known neuropeptides from molluscs, crustaceans, insects, amphibians, and mammals. In collaboration with other laboratories we have tested hydra extracts in peptide radioimmunoassays.

Negative results (using several established antisera) were found for ACTH, angiotensin I and II, adipokinetic hormone, avian and bovine pancreatic polypeptide, bradykinin, calcitonin, corticotropin-releasing factor, β-endorphin, follicle-stimulating hormone, crustacean hyperglycemic hormone, insulin, Leu-enkephalin, β-lipotropin, Met-enkephalin, LHRH, litorin, motilin, α-MSH, nerve-growth factor, neurophysin I and II, physalemin, proctolin, renatensin, secretin, somatostatin, thyrotropin-releasing hormone, vasoactive intestinal polypeptide, and xenopsin.[24] Positive results were found for peptides related to gastrin/cholecystokinin, substance P, neurotensin, bombesin/gastrin-releasing peptide, FMRFamide, and oxytocin/vasopressin.[19,20,24-30] From the immunocytochemical and radioimmunological data, however, it could not be excluded that the "different" substance P and bombesin immunoreactivities were, in fact, caused by one single ancestral *Hydra* peptide, related to both substance P and bombesin (both peptides have already the sequence LMamide in common).[24] By using a special double-labeling technique, it was found that the *Hydra* bombesin-like material coexists in some (but not all) oxytocin-positive neurons, whereas oxytocin- and FMRFamide-like material were never found together.[20] In the following, I want to confine myself to FMRFamide-like peptides as FMRFamide is related to the opioids (cf. Met-enkephalin-Arg⁶-Phe⁷, with the sequence YGGFMRF[31,32]), the main topic of this *Handbook*.

FMRFamide-Like Peptides in the Nervous System of *Hydra*

With a radioimmunoassay for FMRFamide we found 8 pmol/g wet weight of immunoreactive material in *Hydra* extracts (in collaboration with G. Dockray, Liverpool). After gel chromatography, this material eluted shortly after the position of synthetic FMRFamide, suggesting that the FMRFamide-like peptide in *Hydra* has a low molecular weight.[29]

With immunocytochemistry FMRFamide-like material was found in a cluster of sensory cells around the mouth opening, in sensory and ganglion cells of the tentacles, and in a group of densely packed ganglion cells in the peduncle (Figure 2). Numerous FMRFamide antisera all gave identical staining, and therefore only two antisera with good staining properties (codes: 114I and 117I) were characterized in more detail (see Table 1). Solid-phase as well as liquid-phase absorptions showed that the antisera had a high affinity to FMRFamide. Solid-phase absorptions excluded any affinity to the FMRFamide-related peptide Met-enkephalin-Arg⁶-Phe⁷ (YGGFMRF), which shares a four amino-acid sequence with the molluscan peptide. The antisera did not recognize the carboxyterminus of gastrin or cholecystokinin (WMDFamide), which has two amino acids, including an amidation in common positions with FMRFamide, nor γ_2-MSH (YVMGHFRWDRFG), which shares the internal sequence RF. No affinity was also found to a variety of unrelated peptides such as angiotensin, bombesin, gastrin-releasing peptide, glicentin, glucagon, neurophysin, neuro-

FIGURE 2.   FMRFamide-like material in the nervous system of *Hydra attenuata*. Ec, ectoderm; En, endoderm. (a) Cryostat section through the hypostome showing a cluster of strongly immunoreactive sensory cells around the mouth opening. (Magnification × 800.) (b) Partly oblique section through two tentacles, showing a network of immunoreactive perikarya and processes in the basal part of the ectoderm. (Magnification × 450.) (From Grimmelikhuijzen, C. J. P., *Histochemistry*, 72, 199, 1983. With permission.)

tensin, oxytocin, somatostatin, substance P, vasoactive intestinal polypeptide, and vasopressin (see Table 1 and Reference 21). The antisera, however, showed some affinity to the sequence RFamide, which is contained in both FMRFamide and γ_1-MSH (YVMGH-FRWDRFamide).[33,34]Furthermore, some affinity was found to the carboxyterminus of bovine pancreatic polypeptide (LTRPRYamide), which has an arginine and an amidated aromatic amino acid in common positions with FMRFamide. This raises the question whether the coelenterate FMRFamide-like peptide could, in fact, be more pancreatic polypeptide-like (or γ_1-MSH-like). An answer to this comes from the liquid-phase absorptions (Table 1), which shows that antisera, which are raised against and have a high affinity to FMRFamide, have a much higher affinity to the *Hydra* antigen than to LTRPRYamide indicating that the *Hydra* peptide is more FMRFamide-like (the same holds for γ_1-MSH, RFamide, FLRFamide, and FMKFamide). More straightforward data are presented in Table 2, which shows that it is possible to obtain FMRFamide antisera, which do not have any crossreactivity to bovine (and avian) pancreatic polypeptide, but still have a high affinity to the hydra antigen.

FMRFamide-Like Peptides Generally Occur in the Nervous System of Coelenterates
    With a variety of antisera to FMRFamide (among them 114I and 117I), immunoreactive networks or nerve rings were found in all hydrozoan, scyphozoan, anthozoan, and ctenophoran species investigated.[19,24] In gastrozooids of the colonial hydrozoan *Hydractinia*, for example, strong immunoreactivity is occurring in the whole nervous net of body wall and

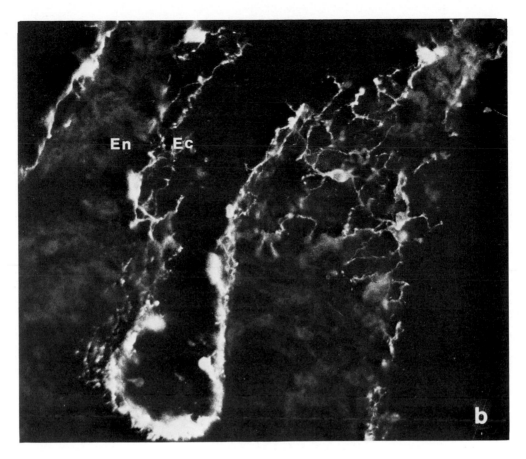

FIGURE 2b

tentacles and especially in sensory cells around the mouth opening (Figure 3a). In the hydrozoan medusa *Leuckarthiara*, FMRFamide-like peptides occur in the tentacle-nerve net and outer nerve ring (Figure 3b and c). In the sea anemone *Metridium*, FMRFamide immunoreactivity occurs in the nerve nets of the tentacles and oral disk, in fibers associated with the endodermal muscle bands, and in sensory cells, which are grouped together at the bases of the mesenteries (Figure 4).

FMRFamide-Like Peptides in Higher Animals

From the preceding review, it is obvious that FMRFamide-like peptides must play an important role in the primitive nervous systems of coelenterates. When FMRFamide-like peptides are to be demonstrated by immunocytochemical means in higher animals, one must always be aware of the limitations of this method, as FMRFamide antisera will generally crossreact with peptides of the pancreatic polypeptide family (see Table 2). This fact was and is still not realized in most of the reports on FMRFamide immunoreactivity in molluscs, crustaceans, insects, fishes, birds, and mammals[35-42] (see, however, the recent report of Veenstra and Schooneveld on insects[43]). After solid-phase absorption of several FMRFamide antisera (including 114I and 117I) with the carboxyterminus of bovine pancreatic polypeptide and double labeling with FMRFamide and bovine pancreatic polypeptide antisera, we concluded that most neuronal structures in the rodent central nervous system, which were previously described to contain FMRFamide immunoreactivity,[39,44] in fact contained pancreatic polypeptide-like material[21] (most probably neuropeptide Y[45]). These data are in good

Table 1
STAINING OF *HYDRA* NEURONAL STRUCTURES AFTER LIQUID- OR SOLID-PHASE ABSORPTION OF ANTISERA 114I OR 117I (DILUTED 1:1000)

Liquid-phase absorption with	Staining with	
	114I	117I
1 μg/mℓ FMRFamide	−	−
1 μg/mℓ FLRFamide	++	+++
10 μg/mℓ FLRFamide	+	++
100 μg/mℓ FLRFamide	−	+
1 μg/mℓ FMKFamide	+++	+++
10 μg/mℓ FMKFamide	++	+++
100 μg/mℓ FMKFamide	−	+++
10 μg/mℓ RFamide	+++	+++
100 μg/mℓ RFamide	++	++
100 μg/mℓ Famide	+++	+++
100 μg/mℓ YGGFMRF	+++	+++
100 μg/mℓ YVMGHFRWDRFG	+++	+++
100 μg/mℓ YVMGHFRWDRFamide	+++	+++
100 μg/mℓ LYRPRYamide	+++	+++

Solid-phase absorption with	Staining with	
	114I	117I
100 μg/mℓ FMRFamide	−	−
100 μg/mℓ FLRFamide	−	−
100 μg/mℓ FMKFamide	−	−
100 μg/mℓ RFamide	+	−
100 μg/mℓ Famide	+++	+++
100 μg/mℓ WMDFamide	+++	+++
100 μg/mℓ YGGFMRF	+++	+++
100 μg/mℓ YVMGHFRWDRFG	+++	+++
100 μg/mℓ YVMGHFRWDRFamide	+	+
100 μg/mℓ LTRPRYamide	−	+
100 μg/mℓ Gastrin-releasing peptide	+++	+++
100 μg/mℓ Neurotensin	+++	+++
100 μg/mℓ Oxytocin	+++	+++
100 μg/mℓ Substance P	+++	+++
100 μg/mℓ Vasopressin	+++	+++

Note: Maximal staining is expressed as + + +; moderate as + +; little as +; and absolutely no staining as −.

Table 2
STAINING OF *HYDRA* NEURONAL STRUCTURES AND RAT OR CHICKEN PANCREATIC ENDOCRINE CELLS WITH ANTISERA TO FMRFamide

Antiserum	Hydra	Rat pancreas	Chicken pancreas	Antiserum	Hydra	Rat pancreas	Chicken pancreas
Rabbit 114I-VI	+++	+++	+++	Guinea pig 547	+++	++	+++
115I-III	+++	+++	+++	548	+++	+++	+++
116I-VI	+++	+++	+++	549	++	-	+
117I-VI	+++	+++	+++	550	+++	-	-
118I-VII	+++	+++	+++	551	+++	-	+
				552	+++	+++	+++
Guinea pig 536	+++	+	++	624	+++	+++	+++
537	+++	-	+	625	-	-	-
538	+++	-	+	626	+++	+++	+++
539	+++	+	+	627	+	-	-
540	+++	-	-	628	+	-	-
541	+++	-	+	629	+++	-	+
542	+++	++	++	630	+	-	+
543	+++	++	++	631	++	-	-
544	+++	-	-	632	-	-	-
545	+	-	+	633	++	-	++
546	++	-		634	+++	+++	+++
				635	+	-	-

Note: From double labeling with FMRFamide and pancreatic polypeptide antisera, the pancreatic endocrine cells were identified as D-cells. Note that some FMRFamide antisera recognize avian pancreatic polypeptide (carboxyterminus VTRHRYamide) better than bovine pancreatic polypeptide (carboxyterminus LTRPRYamide). This is due to the histidine (H) in position 34 of avian pancreatic polypeptide, which, like the methionine (M) of FMRFamide but unlike the proline (P) in position 34 of bovine pancreatic polypeptide, is an α-amino acid. Staining intensities are as in Table 1.

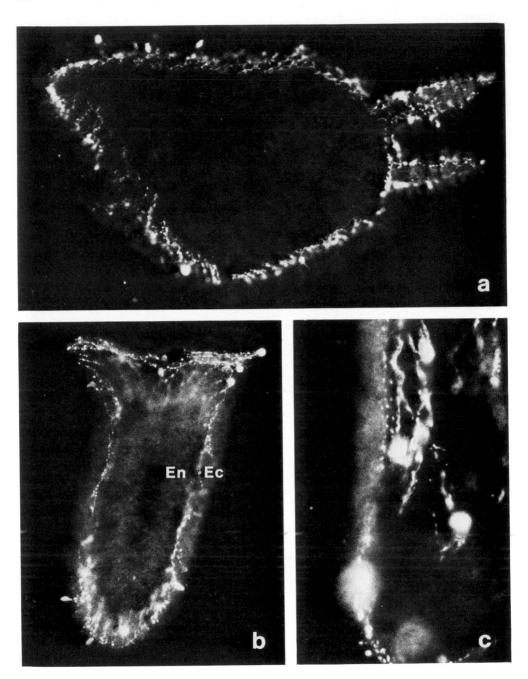

FIGURE 3.   FMRFamide-like material in the nervous system of several hydrozoans. Ec, ectoderm; En, endoderm. (a) Longitudinal cryostat section through a gastrozooid of *Hydractinia echinata*, showing an immunoreactive nervous net in the ectoderm of tentacles (right) and body wall. (Magnification × 160.) (b) Section through a tentacle and outer nerve ring (top) of the medusa *Leuckarthiara nobilis*, showing the immunoreactive nervous net in the ectoderm of the tentacle and perikarya and fibers in the outer nerve ring. (Magnification × 300.) (c) The outer nerve ring of *Leuckarthiara nobilis*. (Magnification × 800.)

agreement with a very recent report of Moore and co-workers[46] who observed a crossreaction of FMRFamide, avian pancreatic polypeptide, and neuropeptide Y antisera in a secondary visual pathway in rat brain. In the guinea pig brain, only one group of perikarya (mainly located in the nucleus dorsomedialis hypothalami) kept being stained after solid-phase ab-

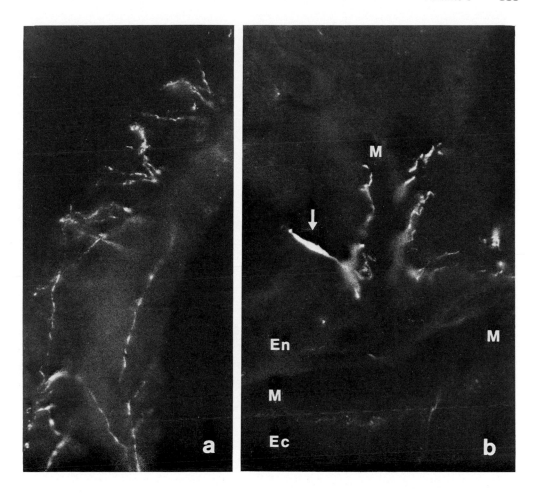

FIGURE 4.   FMRFamide-like material in the nervous system of the sea anemone *Metridium senile*. Ec, ectoderm; En, endoderm; M, mesoglea. (a) Cross section through a mesentery, showing many immunoreactive fibers at both sides of the mesentery (Magnification × 650.) (b) Cross section through the basis of a mesentery, showing a strongly immunoreactive sensory cell (arrow) connected to processes running along the mesoglea. (Magnification × 650.)

sorption of the FMRFamide antisera with pancreatic polypeptide and, in double-labeling experiments, was stained by FMRFamide and not by pancreatic polypeptide antisera.[21]

From the preceding summary it will be clear that antisera which do not show a cross-reactivity to peptides of the pancreatic polypeptide family (Table 2) are very useful. Such antisera stained perikarya in the ventromedial region of the hypothalamus of cyclostomes (Figure 5), showing that "genuine" FMRFamide-like peptides are still present in primitive vertebrates (which is in accordance with the data of Jirikowski and co-workers[23]). The same antisera, when applied on sections of guinea pig central nervous system, intensely stained only one group of perikarya which was mainly located in the nucleus dorsomedialis and extended to the nucleus paraventricularis and nucleus periventricularis hypothalami (see Figure 6). These perikarya projected to several regions of the hypothalamus, mesencephalon, medulla, and spinal cord.[22]

DISCUSSION

The primitive nervous system of coelenterates produces peptides, and among these ancient neuropeptides, FMRFamide-like material is most abundant. From the effects of FMRFamide

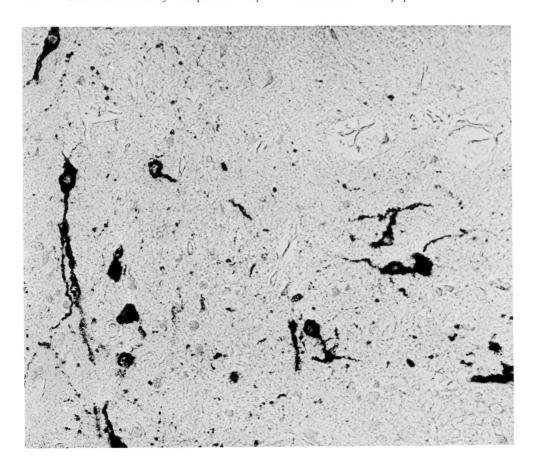

FIGURE 5. FMRFamide-like material in a group of perikarya in the posterior part of the ventromedial hypothalamus of the hagfish *Eptatretus burgeri* (Cyclostomata). These cells were stained by the "specific" antiserum 544 (see Table 2 and Reference 22 for the staining technique). The staining did not disappear after solid-phase absorption of the antiserum with α_1-MSH. The data with antiserum 544 agree well with those of Jirikowski et al.[23] (Magnification × 650.)

on molluscan heart,[47] on molluscan noncardiac smooth muscle,[48,49] and on an identified neuron of a molluscan cerebral ganglion,[50] it is clear that the peptide can have neurotransmitter-like actions. This makes it likely that the FMRFamide-like peptide in coelenterates plays a similar role.

Specific antisera show that FMRFamide-like peptides are also present in the more complex nervous systems of lower and higher vertebrates. In guinea pigs, only one group of perikarya in the dorsomedial region of the hypothalamus produces FMRFamide-like material. FMRFamide was shown to have an excitatory effect on medullary reticular neurons.[51] This is in good agreement with our findings that the FMRFamide-immunoreactive perikarya in the hypothalamus project to this region.[22]

The presence of an ancient neuropeptide in a phylogenetically old part of the brain is intriguing, and I would like to speculate that the group of "antique" neurons in the nucleus dorsomedialis hypothalami is involved in some basic and essential brain process.

ACKNOWLEDGMENTS

I would like to thank Drs. G. J. Dockray, J. Jirikowski, H. C. Schaller, L. P. C. Schot,

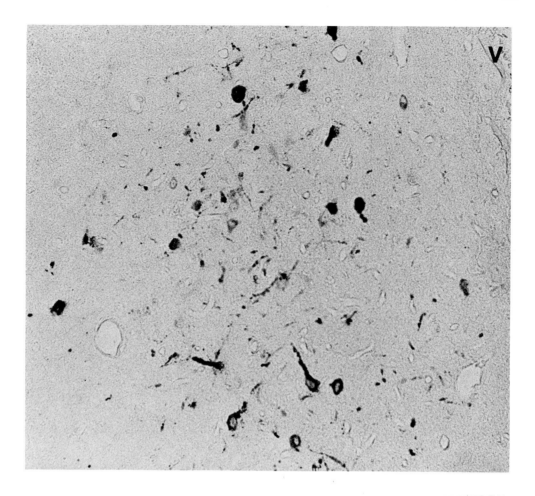

FIGURE 6. FMRFamide-like material in perikarya of the dorsomedial region of the hypothalamus of colchicine-treated guinea pigs. V, third ventricle. Staining was carried out by the "specific" antiserum 544 (see Table 2). See also Reference 22 for a complete mapping. (Magnification × 400.)

F. Sundler, J. Triepel, and K. Wächtler for collaboration in some of the projects described in this review; Mrs. K. Derendorf for typing the manuscript, and the Deutsche Forschungsgemeinschaft for financial support (Gr. 762/4). I am the recipient of a Heisenberg fellowship (Gr 762/1).

REFERENCES

1. **Hadži, J.**, Über das Nervensystem von Hydra, *Arb Zool. Inst. Univ. Wien*, 17, 225, 1909.
2. **Lentz, T. L. and Barrnett, R. J.**, Fine structure of the nervous system of hydra, *Am. Zool.*, 5, 341, 1965.
3. **Davis, L. E., Burnett, A. L., and Haynes, J. F.**, Histological and ultrastructural study of the muscular and nervous systems in hydra, *J. Exp. Zool.*, 167, 295, 1968.
4. **Westfall, J. A., Yamataka, S., and Enos, P. D.**, Ultrastructural evidence of polarized synapses in the nerve net of hydra, *J. Cell. Biol.*, 51, 318, 1971.
5. **Westfall, J. A.**, Ultrastructural evidence for a granule-containing sensory-motor-interneuron in *Hydra littoralis*, *J. Ultrastruct. Res.*, 42, 268, 1973.

6. **Westfall, J. A. and Kinnamon, J. C.,** A second sensory-motor-interneuron with neurosecretory granules in hydra, *J. Neurocytol.,* 7, 365, 1978.

7. **Westfall, J. A., Kinnamon, J. C. and Sims, D. E.,** Neuroepitheliomuscular cell and neuro-neuronal gap junctions in hydra, *J. Neurocytol.,* 9, 725, 1980.

8. **Spencer, A. N.,** The parameters and properties of a group of electrically coupled neurons in the central nervous system of a hydrozoan jellyfish, *J. Exp. Biol.,* 93, 33, 1981.

9. **Spencer, A. N. and Arkett, S. A.,** Radial symmetry and the organization of central neurones in a hydrozoan jelly fish, *J. Exp. Biol.,* 110, 69, 1984.

10. **Anderson, P. A. V. and Schwab, W. E.,** Recent advances and model systems in coelenterate neurobiology, *Prog. Neurobiol. (Oxford),* 19, 213, 1982.

11. **Romanes, G. J.,** Preliminary observations on the locomotor system of medusae, *Phil. Trans. R. Soc. London,* 166, 269, 1877.

12. **Batham, E. J., Pantin, C. F. A., and Robson, E. A.,** The nerve net of the sea anemone *Metridium senile:* the mesenteries and the column, *Q. J. Microsc. Sci.,* 101, 487, 1960.

13. **Tamm, S. L.,** Ctenophora, in *Electrical Conduction and Behaviour in Simple Invertebrates,* Shelton, G. A. B., Ed., Clarendon Press, Oxford, 1982, 266.

14. **Horridge, G. A.,** Non-motile sensory cilia and neuromuscular junctions in a ctenophore independent effector organ, *Proc. R. Soc. London Ser. B,* 162, 333, 1965.

15. **Hernandez-Nicaise, M. L.,** Specialized connections between nerve cells and mesenchymal cells in ctenophores, *Nature (London),* 217, 1075, 1968.

16. **Jha, R. K. and Mackie, G. O.,** The recognition, distribution and ultrastructure of hydrozoan nerve elements, *J. Morphol.,* 123, 43, 1967.

17. **Westfall, J. A.,** Ultrastructural evidence for neuromuscular systems in coelenterates, *Am. Zool.,* 13, 237, 1973.

18. **Martin, S. M. and Spencer, A. N.,** Neurotransmitters in coelenterates, *Comp. Biochem. Physiol.,* 74c, 1, 1983.

19. **Grimmelikhuijzen, C. J. P.,** FMRFamide immunoreactivity is generally occurring in the nervous systems of coelenterates, *Histochemistry,* 78, 361, 1983.

20. **Grimmelikhuijzen, C. J. P.,** Coexistence of neuropeptides in hydra, *Neuroscience,* 9, 837, 1983.

21. **Triepel, J. and Grimmelikhuijzen, C. J. P.,** A critical examination of the occurrence of FMRFamide immunoreactivity in the brain of guinea pig and rat, *Histochemistry,* 80, 63, 1984.

22. **Triepel, J. and Grimmelikhuijzen, C. J. P.,** Mapping of neurons in the central nervous system of the guinea pig by the use of antisera specific to FMRFamide, *Cell Tissue Res.,* 237, 575, 1984.

23. **Jirikowski, G., Erhardt, G., Grimmelikhuijzen, C. J. P., Triepel, J., and Patzner, A.,** FMRFamide-like immunoreactivity in the brain and pituitary of the hagfish *Eptatretus burgeri* (Cyclostomata), *Cell Tissue Res.,* 237, 363, 1984.

24. **Grimmelikhuijzen, C. J. P.,** Peptides in the nervous systems of coelenterates, in *Evolution and Tumorpathology of the Neuroendocrine System,* Falkmer, S., Håkanson, R., and Sundler, F., Eds., Elsevier Biomedical Press, Amsterdam, 1984, 39.

25. **Grimmelikhuijzen, C. J. P., Sundler, F., and Rehfeld, J. F.,** Gastrin/CCK-like immunoreactivity in the nervous system of coelenterates, *Histochemistry,* 69, 61, 1980.

26. **Grimmelikhuijzen, C. J. P., Balfe, A., Emson, P. C., Powell, D., and Sundler, F.,** Substance P-like immunoreactivity in the nervous system of hydra, *Histochemistry,* 71, 325, 1981.

27. **Grimmellkhuijzen, C. J. P., Carraway, R. E., Rökaeus, Å., and Sundler, F.,** Neurotensin-like immunoreactivity in the nervous system of hydra, *Histochemistry,* 72, 199, 1981.

28. **Grimmelikhuijzen, C. J. P., Dockray, G. J., and Yanaihara, N.,** Bombesin-like immunoreactivity in the nervous system of hydra, *Histochemistry,* 73, 171, 1981.

29. **Grimmelikhuijzen, C. J. P., Dockray, G. J., and Schot, L. P. C.,** FMRFamide-like immunoreactivity in the nervous system of hydra, *Histochemistry,* 73, 499, 1982.

30. **Grimmelikhuijzen, C. J. P., Dierickx, K., and Boer, G. J.,** Oxytocin/vasopressin-like immunoreactivity is present in the nervous system of hydra, *Neuroscience,* 7, 3191, 1982.

31. **Stern, A. S., Lewis, R. V., Kimura, S., Rossier, J., Gerber, L. D., Brink, L., Stein, S., and Udenfriend, S.,** Isolation of the opioid heptapeptide Met-enkephalin (Arg[6], Phe[7]) from bovine adrenal medullary granules and striatum, *Proc. Natl. Acad. Sci. U.S.A.,* 76, 6680, 1979.

32. **Rossier, J., Audigier, Y., Ling, N., Cros, J., and Udenfriend, S.,** Met-enkephalin-Arg[6]-Phe[7], present in high amounts in brain of rat, is an opioid agonist, *Nature (London),* 288, 88, 1980.

33. **Nakanishi, S., Inoue, A., Kita, T., Nakamura, M., Chang, A. C. Y., Cohen, S. A., and Numa, S.,** Nucleotide sequence of cloned cDNA for bovine corticotropin-β-lipotropin precursor, *Nature (London),* 278, 423, 1979.

34. **Böhlen, P., Esch, F., Shibasaki, T., Baird, A., Ling, N., and Guillemin, R.,** Isolation and characterization of a γ_1-melanotropin-like peptide from bovine neurointermediate pituitary, *FEBS Lett.,* 128, 67, 1981.

35. **Boer, H. H., Schot, L. P. C., Veenstra, J. A., and Reichelt, D.,** Immunocytochemical identification of neural elements in the central nervous system of a snail, some insects, a fish and a mammal with an antiserum to the molluscan cardio-excitatory tetrapeptide FMRFamide, *Cell Tissue Res.*, 213, 21, 1980.

36. **Dockray, G. J., Vaillant, C., and Williams, R. G.,** New vertebrate brain-gut peptide related to a molluscan neuropeptide and an opioid peptide, *Nature (London)*, 293, 656, 1981.

37. **Dockray, G. J., Vaillant, C., Williams, R. G., Gayton, R. J., and Osborne, N. N.,** Vertebrate brain-gut peptides related to FMRFamide and Met-enkephalin-Arg^{6}Phe^7, *Peptides*, 2, 25, 1981.

38. **Martin, R., Frösch, D., Kiehling, C., and Voight, K. H.,** Molluscan neuropeptide-like and enkephalin-like material coexist in octopus nerves, *Neuropeptides*, 2, 141, 1981.

39. **Weber, E., Evans, C. J., Samuelson, S. J., and Barchas, J. D.,** Novel peptide neuronal system in rat brain and pituitary, *Science*, 214, 1248, 1981.

40. **Schot, L. P. C. and Boer, H. H.,** Immunocytochemical demonstration of peptidergic cells in the pond snail *Lymnea stagnalis* with an antiserum to the molluscan cardioactive tetrapeptide FMRFamide, *Cell Tissue Res.*, 225, 347, 1982.

41. **Cardot, J. and Fellman, D.,** Immunofluorescent evidence of an FMRFamide-like peptide in the peripheral nervous system of the gastropod mollusc *Helix aspersa*, *Neurosci. Lett.*, 43, 167, 1983.

42. **Jacobs, A. and Van Herp, F.,** Immunocytochemical localization of a substance in the eyestalk of the prawn, *Palaemon serratus*, reactive with an anti-FMRF-amide rabbit serum, *Cell Tissue Res.*, 235, 601, 1984.

43. **Veenstra, J. A. and Schooneveld, H.,** Immunocytochemical localisation of neurons in the nervous system of the Colorado potato beetle with antisera against FMRFamide and bovine pancreatic polypeptide, *Cell Tissue Res.*, 235, 303, 1984.

44. **Williams, R. G. and Dockray, G. J.,** Immunocytochemical studies of FMRFamide-like immunoreactivity in rat brain, *Brain Res.*, 276, 213, 1983.

45. **Tatemoto, K., Carlquist, M., and Mutt, V.,** Neuropeptide Y — a novel brain peptide with structural similarities to peptide YY and pancreatic polypeptide, *Nature (London)*, 296, 659, 1982.

46. **Moore, R. Y., Gustafson, E. L., and Card, J. P.,** Identical immunoreactivity of afferents to the rat suprachiasmatic nucleus with antisera against avian pancreatic polypeptide, molluscan cardioexcitatory peptide and neuropeptide Y, *Cell Tissue Res.*, 236, 41, 1984.

47. **Greenberg, M. J. and Price, D. A.,** Cardioregulatory peptides in molluscs, in *Peptides: Integrators of Cell and Tissue Function*, Bloom, F. E., Ed., Raven Press, New York, 1980, 107.

48. **Painter, S. D.,** FMRFamide catch contractures of a molluscan smooth muscle: pharmacology, ionic dependence and cyclic nucleotides, *J. Comp. Physiol.*, 148, 491, 1982.

49. **Cottrell, G. A., Schot, L. P. C., and Dockray, G. J.,** Identification and probable role of a single neurone containing the neuropeptide *Helix* FMRFamide, *Nature (London)*, 304, 638, 1983.

50. **Cottrell, G. A.,** FMRFamide neuropeptides simultaneously increase and decrease K$^+$ currents in an identified neurone, *Nature (London)*, 269, 87, 1982.

51. **Gayton, R. J.,** Mammalian neuronal actions of FMRFamide and the structurally related opioid Met-enkephalin-Arg^6-Phe^7, *Nature (London)*, 298, 275, 1982.

FMRFamide: ACTIONS ON IONIC CURRENTS IN NEURONS AND MECHANICAL EFFECTS ON MUSCLES

G. A. Cottrell and N. W. Davies

SUMMARY

FMRFamide and related peptides have at least three separate actions on *Helix* neurons: (1) a selective increase in potassium conductance (inhibitory); (2) an increase in sodium conductance (excitatory); and (3) a reduction in a voltage-dependent potassium conductance (excitatory, modulatory). Using a computer simulation, it is shown that each of the major effects of FMRFamide peptides on different molluscan muscles (excitatory, inhibitory, and biphasic) may be explained in terms of the above actions on neuronal membranes.

INTRODUCTION

The molluscan neuropeptide Phe-Met-Arg-Phe-NH$_2$ or FMRFamide[9] and related synthetic and natural peptides have potent and diverse actions on molluscan cardiac and smooth muscles[6,8,9] and neurons.[2,3] "FMRFamide-like" pharmacological activity appears to reside in the sequence Phe-Met-(or-Leu-) Arg-Phe-NH$_2$, but some FMRFamide activity is also present where one or both Phe is substituted with Try. Peptides with N-terminal extensions have similar, but as will be seen in some cases described below, also more complex actions than the tetrapeptides.

In this presentation, we attempt to account for the different types of pharmacological effects observed on various muscle preparations in terms of three well-defined ionic permeability changes induced by FMRFamide-peptides in identified gastropod neurons.[2,3] These are (1) an increase in permeability to potassium ions (inhibitory); (2) an increase in permeability to sodium ions (excitatory); and (3) a decrease in permeability to potassium ions (modulatory or excitatory if membrane potential is depolarized). This response is mediated via different potassium channels to 1. We have recently obtained evidence suggesting that the response (1) may be subdivided into two types, with perhaps different receptors associated with each type of channel, but here for simplicity we shall consider that there is only one type of response involving an increase in potassium ion permeability.

The paper is divided into three sections: (1) a short description of some of the effects induced by FMRFamide and related peptides on different molluscan muscles; (2) a description of the three main ionic responses induced by FMRFamide in gastropod neurons; and (3) an attempt to account for the different types of mechanical responses observed in the muscles in terms of the presence of one or more of the receptor and ion channel complexes using a computer simulation.

MECHANICAL RESPONSES INDUCED IN MOLLUSCAN MUSCLES BY FMRFamide AND RELATED PEPTIDES

Various smooth muscles are induced to contract on exposure to FMRFamide, sometimes in very low concentrations, e.g., the radular protractor muscle of *Busycon contrarium*.[7] Some other muscles are relaxed by FMRFamide. In some cases, the muscle responds with a smooth contraction (see Figure 1); in others, rhythmical phasic contractions result (see Figure 2). Often there is a considerable delay, lasting several seconds or even a minute, between exposure to the peptide and the response. In some muscles, e.g., the tentacle retractor muscle of *Helix,* both excitation and inhibition may be observed depending on

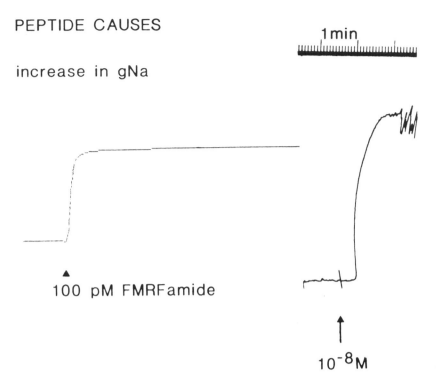

FIGURE 1.   The response of an isolated pharangeal retractor muscle of *Busycon contrarium* to exposure to 10^{-8} *M* FMRFamide is shown on the right and the simulated response shown on the left.

concentration of the peptide (see, e.g., Figure 3). This dual effect is seen most clearly with the "Met-enkephalin analog", YGGFMRFamide. (No activity is, however, seen on the muscles of either YGGFM or YGGFMRF.)

FMRFamide is a potent cardioexcitatory compound in several species (see Figure 4). It is also, however, cardioinhibitory in some other species (Figure 4).

IONIC RESPONSES INDUCED BY FMRFamide

Voltage clamp experiments on identified neurons in *Helix*[3] and other species, see below, have demonstrated potent, selective actions of FMRFamide on the permeability of neuronal membranes.

Selective Increase in Potassium Ion Permeability (Inhibition)

This response results in hyperpolarization of the membrane[1] and thus inhibition. It has also been observed in neurons in *Planorbis* and *Lymnaea* (Figure 5). As shown in Figure 5, alteration in the extracellular potassium concentration produces a change in reversal potential of the response as predicted by the Nernst equation suggesting, as in *Helix,* that the response results from an increase in permeability to potassium ions alone. This response is not rapidly desensitized by repeated application of the peptide. It is reduced by exposure to tetraethylammonium bromide, which is known to block potassium currents.

Increase in Permeability to Sodium Ions (Excitation)

On some neurons, iontophoresis of FMRFamide produces a rapid and quickly recovering excitation with spike firing. This response is desensitized with repeated application of the

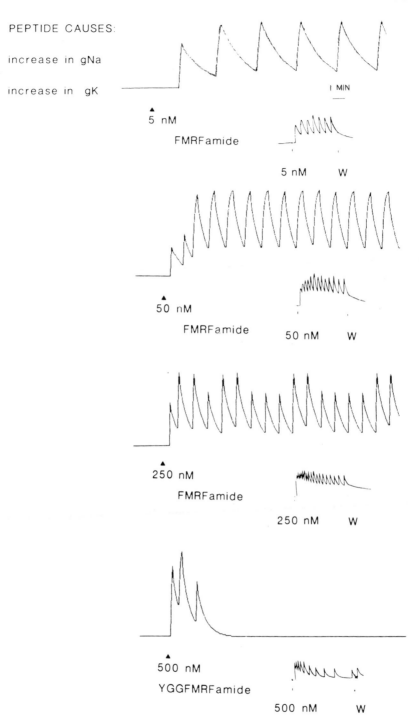

FIGURE 2. Responses and predicted responses of the tentacle retractor muscle of *Helix aspersa* to different concentrations of FMRFamide. Each experimental result, taken from Cottrell et al.[4] is the smaller of each pair of records. The predicted responses clearly parallel the experimental data. It was assumed the responses resulted from an increase in gNa and an increase in gK.

PEPTIDE CAUSES

increase in gK

decrease in gK

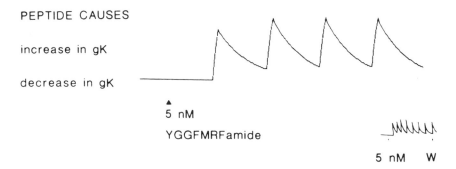

▲
5 nM

YGGFMRFamide

5 nM   W

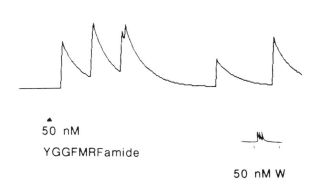

▲
50 nM

YGGFMRFamide

50 nM W

▲
250 nM

YGGFMRFamide

250 nM

FIGURE 3.   Actual responses (smaller record in each pair) and predicted responses of the *Helix* tentacle retractor muscle to YGGFMRFamide. Because of the potent action of YGGFMRFamide on the "reduction in gK" response, but lack of effect on the gNa response of the neurons (see text), we have attempted to explain the effects of this peptide in terms of changes in conductance to K alone, i.e., a combination of suppressing the voltage-dependent gK and an activation of the gK resulting in hyperpolarization. Each of the different responses to the peptide may be accounted for with these changes in membrane permeability, i.e., (1) the induction of the rhythmical phasic contractions, (2) the increase in frequency of "beating" with increased dose, and (3) the inhibition of "beating" and relaxation with higher doses.

PEPTIDE CAUSES

increase in gNa

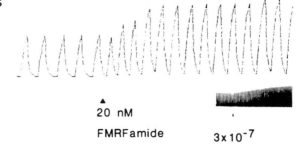

▲
20 nM
FMRFamide 3×10^{-7}

increase in gNa

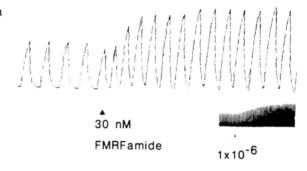

▲
30 nM
FMRFamide 1×10^{-6}

increase in gK

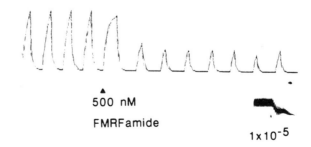

▲
500 nM
FMRFamide 1×10^{-5}

FIGURE 4.   Examples (smaller trace in each pair of recordings) of the responses of hearts of two species of mollusc to FMRFamide, *Dinocardium robustum* (above), which is excited by the peptide, and *Pseudochama exogyra* which is inhibited.[8] The simulations shown were based on the assumption that the responses of *Dinocardium* result from an action of FMRFamide on increasing gNa and gK, whereas that of *Pseudochama* results from an action increasing gK alone, presumably because the other receptor-ion channel complex is absent. The mechanical response of *Dinocardium* and other species which are excited by FMRFamide could also be explained in terms of the action resulting in a decrease in the voltage-dependent gK (see, e.g., Figure 3). Presumably the presence of all three receptor channel complexes could result in very complex types of responses.

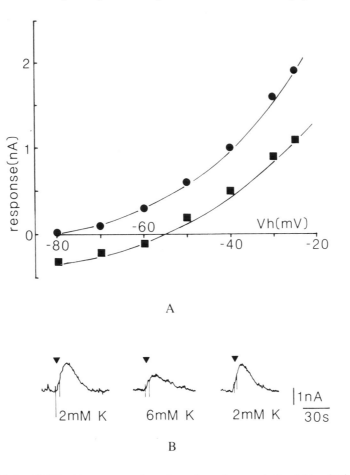

FIGURE 5.   FMRFamide responses of a voltage-clamped neuron in the right parietal ganglion of *Lymnaea*. (A) Relationship between response and holding potential in 2 mM external K (■) and 6 mM K (●). The reversal potential of the response has been shifted by 26 mV in the positive direction during exposure to 6 mM K. (B) Current responses to FMRFamide observed in the same neuron at a holding potential of −30 mV. From left to right: control (2 mM K), high K, wash.

peptide. Under voltage clamp, the size of the induced inward current can be shown to be dependent on the extracellular concentration of sodium ions. Although it is clear that sodium ions are the major charge carrier, it is possible, but not established, that one or more other ion(s) also makes a small contribution to the overall response. (Cf. ACh-activated channels at the neuromuscular junction.)

Suppression of Voltage-Dependent Potassium Channels (Modulation-Excitation)

In some *Helix* neurons, a FMRFamide-induced inward current response has been observed which is markedly voltage dependent.[2] This response may also be observed with current-clamp recording. At the resting potential level no response is seen. When the membrane is artificially depolarized, however, by current injection, application of FMRFamide produced further depolarization.

The response is suppressed by extracellular cobalt ions, suggesting some dependence on calcium ions, but it is also influenced in a Nernstian manner by altering the extracellular potassium ion concentration. These observations suggest that the peptide suppresses a voltage-dependent potassium current, perhaps the potassium current induced by intracellular calcium ions. Evidence in support of this view came from experiments in which calcium

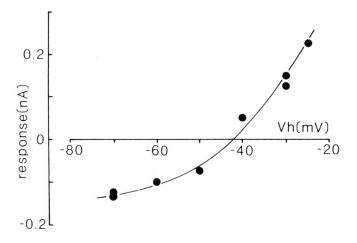

FIGURE 6. Relationship between FMRFamide response and membrane potential of a neuron in the visceral ganglion of *Lymnaea*. This response is very similar to that observed in the F2 neuron of *Helix*.

ions were injected into the neuron under voltage clamp and the peptide tested on the induced potassium current. In line with the above view, FMRFamide peptides reduced the calcium-induced potassium current. Confirmation of an action resulting in suppression of potassium channels has recently been obtained using patch-clamp techniques.[3] In these experiments, single-channel potassium currents (with a slope conductance of about 60 pS at a patch potential of $+20$ mV and an estimated P_K of 25×10^{-14} cm^3/sec) observed only at depolarized membrane potentials, were reversibly suppressed by application of peptide to a region of the membrane other than that to which the patch pipette was applied. In other words, the suppression resulting from peptide application must involve some secondary mechanism, presumably an intracellular second messenger.

COMBINATION OF RESPONSES

In many neurons, more than one of the above responses was observed. For example in the C1, or "giant serotonin neuron" of the cerebral ganglion of *Helix*, both the increase in potassium conductance and the decrease in potassium conductance were seen. Thus the response could be inhibitory at the resting membrane potential, but excitatory if for some reason the neuron was depolarized during application of the peptide. In some identified neuron of the visceral and right parietal ganglia, responses resulting from a combination of the increase in potassium conductance and the increase in sodium conductance were observed (see Figure 6). In such cells, the predominant effect of peptide application was excitatory. An interesting observation made on these neurons was that the analog YGGFMRFamide activated the increase in potassium permeability alone, and thus produced inhibition. This interesting observation also suggests that there is more than one receptor to the FMRFamide-peptides in *Helix* neurons and that whereas FMRFamide itself stimulates more than one receptor on these neurons YGGFMRFamide activates only one of them. (YGGFM and YGGMRF were without effect.)

Can the Different Types of Responses Observed in Various Muscle Preparations be Explained in Terms of Membrane Action of FMRFamide Involving (1) an Increased Potassium Permeability, (2) an Increased Sodium Permeability, and (3) a Decreased Potassium Permeability?

The effects of FMRFamide peptides on various muscle preparations have been studied and characterized (see above). By making some basic assumptions about the way in which these muscles contract, a program was written on a microcomputer in an attempt to determine whether the effects on the muscle could be explained if the peptides have a similar action on muscle cell membranes as they do on neuronal membranes, i.e., an increase in potassium conductance (gK), an increase in sodium conductance (gNa), and a decrease in potassium conductance (gK). Basically the contraction of the muscles depends on the level of intracellular free calcium, which in turn depends on membrane potential.

Membrane potential was calculated using

$$E_m = (E_{Na}g_{Na} + E_Kg_K + E_Lg_L)/(g_{Na} + g_K + g_L) \tag{1}$$

$$g_{Na} = nG_{Na} \tag{2}$$

$$g_K = kG_K \exp(1 + 10E_m) \tag{3}$$

where G_{Na} and G_K are the resting conductances of sodium and potassium, respectively. The constants n and k vary with the application of peptides and the conditions set (see below). The exponential term associated with g_K simulates the delayed rectification normally observed with K currents in neurons.

For the increase in conductance responses, a maximum value was assigned to n and k (n_{max} and k_{max}) and solving n and k by using simple dose-response relationships:

$$n = (n_{max}Cp)/(Cp + K_n) \tag{4}$$

$$k = (k_{max}Cp)/(Cp + K_k) \tag{5}$$

where Cp is the concentration of peptide and K_n and K_k are the dissociation constants for the receptors mediating an increase in g_{Na} and g_K, respectively. In some instances a desensitization of the increase in g_{Na} was simulated as this response in *Helix* neurons is rapidly desensitized.

The effect of FMRFamide on the radula protractor muscle of *Busycon* could readily be explained in terms of an increase in membrane permeability to sodium ions. The resulting depolarization of the membrane would cause an increase in intracellular calcium ions and the muscle would contract. One way in which "wane" of the contraction occurs could be dependent on the rate of desensitization of the receptors involved. Experimental data and computer simulations are compared in Figure 1.

The above example is relatively easy to explain in terms of the neuronal actions of FMRFamide. The *Helix* tentacle retractor muscle is more complicated, showing both excitatory and inhibitory effects to FMRFamide and YGGFMRFamide. The excitatory effect of YGGFMRFamide was assumed to be mediated by the voltage-dependent decrease in potassium permeability such as that observed in certain identified neurons to the peptide. An increase in sodium permeability activated by YGGFMRFamide has not been seen in any of the neurons studied so far.

The excitatory response in the tentacle muscle consists of induced contractions in addition to an increase in background tension. To explain these effects further assumptions were made:

1.　The muscle cells have a slow oscillating sodium conductance which leads to a sinusoidally varying membrane potential.

2.   The contractions observed occur when internal free calcium ion concentration reaches a threshold value. Equation 1 was therefore altered to

$$E_m = (E_{Na}g_{Na} + E_K g_K + E_L g_L + E_{Na}g_s)/(g_{Na} + g_K + g_L + g_s) \qquad (6)$$

The slow oscillating conductance was simulated by varying g_s:

$$g_s = 2f(t)g_{Na} \qquad (7)$$

and

$$f(t) = 1 + \sin t + 0.3(\sin(t - \pi/3) + 0.8)^5 \qquad (7a)$$

The decrease in g_K response is markedly voltage dependent and is described here using a sigmoidal relationship between membrane potential and decrease in g_K as follows (c f. Dingledine[5]):

$$D = (D_{max}Cp)/(Cp + K_d) \qquad (8)$$

$$d = (D/2)(1 + a/(1 + a^2)) \qquad (9)$$

$$a = m(E_m - mid) \qquad (10)$$

and finally

$$k = (1/d)(k_{max}Cp)/(Cp + K_k) \qquad (11)$$

In Equation 8, D is the maximum response possible at a peptide concentration Cp and K_d is the dissociation constant of the receptor mediating this response. Equations 9 and 10 simulate the voltage-dependent characteristic of the response. In Equation 10, m is a dimensionless number that determines the slope of the curve and mid is the membrane potential at which $d = D/2$.

The change of the response of the muscle from excitatory to inhibitory as the concentration of peptide increases could be easily simulated by assigning different values to the dissociation constants K_n, K_k, and K_d. Figures 2 and 3 compare the computer-simulated results with experimental data.

By modifying the equations slightly, we could simulate some of the results obtained from bivalve hearts also. The comparisons are shown in Figure 4.

DISCUSSION

The main conclusion we wish to draw is that the different mechanical responses produced by FMRFamide-peptides on molluscan muscles may be explained in terms of known actions of these peptides on ion channels in neuronal membranes. It is not necessary to suggest more complex effects involving for instance the release of other neurohormones within the tissues, although, of course, this does not mean that such effects do not occur in some cases.

The responses of some preparations vary considerably. Painter and Greenberg[8] observed that in some species (e.g., *Dosinia discus* and *Ligumia subrostrata*) individual isolated hearts may respond with excitation, inhibition, or a mixed response with low doses of FMRFamide. This interesting observation reminded us of voltage-clamp experiments on the *Helix* C1 and

F2 neurons.[3] On both of these neurons, at least two actions of FMRFamide are seen: on the C1 a reduction in gK and an increase in gK; on the F2 an increase in gNa and an increase in gK. The contribution made to the overall current response recorded in these neurons in C1s and F2s from different animals varied, considerably in some cases, reflecting presumably a difference in the proportions of the receptor-ion channel complexes of the cell in that animal. Perhaps similar variability in the receptor-channel complexes occurs in the hearts. In line with the considerations presented above, it is easy to imagine how such differences could underlie the variability observed by Painter and Greenberg.[8] If this view is correct, one wonders about the mechanism used to control the occurrence of the receptor-channel complexes and the physiological importance of their modulation.

ACKNOWLEDGMENT

We wish to thank the British M.R.C. for financial support.

REFERENCES

1. **Cottrell, G. A.,** Actions of "molluscan cardio-excitatory neuropeptide" on identified 5-hydroxytryptamine-containing neurones and their follower neurones in *Helix pomatia, J. Physiol.,* 284, 130, 1978.
2. **Cottrell, G. A.,** FMRFamide neuropeptides simultaneously increase and decrease K$^+$ currents in an identified neurone, *Nature (London),* 296, 5852, 1982.
3. **Cottrell, G. A., Davies, N. W., and Green, K. A.,** Multiple actions of a molluscan cardio-excitatory neuropeptide and related peptides on identified *Helix* neurones, *J. Physiol.,* 356, 315, 1984.
4. **Cottrell, G. A., Greenberg, M. J., and Price, D. A.,** Differential effects of the molluscan neuropeptide FMRFamide and the related Met-enkephalin derivative YGGFMRFamide on the *Helix* tentacle retractor muscle, *Comp. Biochem. Physiol.,* 75C, 373, 1983.
5. **Dingledine, R.,** *N*-methyl aspartate activates voltage-dependent calcium conductance in rat hippocampal pyramidal cells, *J. Physiol.,* 343, 385, 1983.
6. **Greenberg, M. J. and Price, D. A.,** Cardioregulatory peptides in molluscs, in *Peptides: Integrators of Cell and Tissue* Function, Bloom F. E., Ed., Raven Press, New York, 1980, 107.
7. **Nagle, G. T. and Greenberg, M. J.,** A highly sensitive microbioassay for the molluscan neuropeptide FMRFamide, *Comp. Biochem. Physiol.,* 71C, 101, 1982.
8. **Painter, S. D. and Greenberg, M. J.,** A survey of the responses of bivalve hearts to the molluscan neuropeptide FMRFamide and to 5-hydroxytryptamine, *Biol. Bull.,* 162, 311, 1982.
9. **Price, D. A. and Greenberg, M. J.,** Structure of a molluscan cardioexcitatory neuropeptide, *Science,* 197, 670, 1977.

NEUROPEPTIDES WITH CARDIOEXCITATORY AND OPIOID ACTIVITY IN OCTOPUS NERVES*

K.-H. Voigt and R. Martin

SUMMARY

In extracts of the brain and vena cava from the cephalopod *Octopus vulgaris* we found two cardioexcitatory peptides. One is identical with the molluscan cardioexcitatory peptide FMRFamide, first discovered by Price and Greenberg,[1] and the other is extended at the N-terminus by YGG, thus containing the principle opioid structure. The peptides were identified by applying region-specific radioimmunoassays, opiate receptor binding assays, and a bioassay of the cardioexcitatory potency. Different chromatographic procedures and enzymatic treatment confirmed the existence of both peptides.

The occurrence of YGGFMRFamide (corresponding to Met[5]-enkephalin-Arg[6]-Phe[7]amide) is of particular interest, since it represents a connection between the opioid precursor system with the neuropeptides belonging to the FMRFamide family. Furthermore, the colocalization of FMRFamide and YGGFMRFamide immunoreactivities together with α-MSH-like activity may point to an evolutionarily very old precursor system containing a number of biologically active peptides which were in vertebrates synthesized by different precursors.

INTRODUCTION

Since the isolation of the molluscan cardioexcitatory peptide FMRFamide by Price and Greenberg,[1] a number of structurally related peptides have been found in other invertebrates[2,3] and vertebrate species.[4] The functions of the tetrapeptide-amide include activity on the heart[5,6] and other muscles,[7,8] as well as on neural tissue.[9] Very recently, Leung and Stefano[10] found in ganglia of *Mytilus edulis* Met[5]- and Leu[5]-enkephalin, which are presently considered as the principle opioid structures in vertebrates.[11] Among the various opioid peptides, the C-terminal portion of proenkephalin A,[12] consisting in YGGFMRF-OH,[13] corresponds to the primary amino acid sequence of both, the opioid Met-enkephalin at the N-terminal and FMRF at its C-terminal part. However, only the peptides with an amidated C-terminus were active on the different muscle preparations.[5,6,8]

With immunocytochemical methods, Martin and co-workers[14,15] observed the coexistence of α-MSH-like activity, FMRFamide, and of enkephalin in a particular population of nerve endings of *Octopus vulgaris*. As enkephalin immunoreactivity appeared only after tryptic treatment, it was suggested that the opioid may be part of a longer peptide containing the C-terminal structure FMRFamide.

Beside the immunocytochemical observation of FMRFamide in *Octopus vulgaris,* there are reports on a strong cardioexcitatory activity of nerve tissue extracts from this mollusc species.[16,17]

Therefore we started identifying the cardioexcitatory peptides in brain extracts of the octopus with particular interest on the supposed heptapeptide amide YGGFMRFamide, which involves cardioexcitatory[6,18] and opioid activity.[13]

MATERIALS AND METHODS

The identification of FMRFamide[1] and the heptapeptide YGGFMRFamide was performed by applying a number of assay systems, which were specific for different portions of the

---

* Dedicated to Prof. H. Kewitz on the occasion of his 65th birthday.

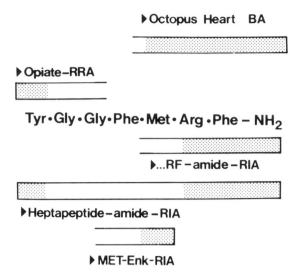

FIGURE 1.   Amino acid sequence of the *Octopus* heptapeptide (corresponding to YGGFMRF-NH$_2$ in single letter notation) and the assay systems used for identification of the peptide after gel chromatography and HPLC techniques. The regions of the peptide, which reacted specifically in a bioassay (BA), opiate assays (RRA) and radioimmunoassays (RIA), are marked by dotted columns.

molecules (Figure 1). The sensitive methods allowed the characterization of the peptides using only small amounts of material (about 30 brains from *Octopus vulgaris*.).

Extraction and Separation

Brain tissue from 30 medium-sized specimens (males and females, 300 to 1200 g) of *Octopus vulgaris* was homogenized with a minipotter on ice (one brain in 4 mℓ) in 1 *M* acetic acid and 1 m*M* phenyl-methan-sulfonyl-fluoride after heating in a boiling water bath for 20 min. The extracts were sonicated (1.5 min), centrifuged (30 min, 4°C, 32.000 × g), and the supernatants lyophilized. Dissected tissue of the anterior vena cava was extracted following the same protocol.[17] The lyophilized material was transported from the marine biology laboratories at Naples (Italy) and Banyuls (France) to Ulm and taken up in 1.67 *M* acetic acid for gel chromatography on Sephadex G-50-fine (column 0.9 × 96 cm) (Figure 5). Peptides were eluted with 1.67 *M* acetic acid and 1-mℓ fractions were collected and evaporated. The recovery of immunoreactivity was about 95%. Individual fractions were measured in a FMRFamide RIA (Figure 2A,B) or were combined for subsequent testing of their cardioexcitatory potency (Figure 3) and opioid receptor binding (Figure 4). HPLC separation was performed on an ultrasphere octadecasilica (ODS) column (Beckman; 1 × 25 cm; particle size 5 μm) in a gradient HPLC system (Kontron, Zurich, Switzerland): 50 m*M* KH$_2$PO$_4$, 0.1% phosphoric acid, 5% acetonitrile, pH 2.7. Peptides were eluted with an acetonitrile gradient (indicated by stippled lines in Figure 5) in a flow rate of 1 mℓ × min^{-1}. Before adding the prepurified material (Figure 5), the column was extensively washed with methanol (50%), and elution volumes with the buffer gradient only, preceding the addition of extracts, were measured in the RIA systems to exclude contaminations. Calibrations of the column (indicated by arrows in Figure 6a to f) were assured after each elution of extracts by chromatographing 5 ng of standards (FMRFamide and YGGFLRFamide) and measuring the fractions in RIA. Aliquots were desalted by gel chromatography on Sephadex G-25 (column 0.9 × 15 cm, elution with 1.67 *M* acetic acid), evaporated, and tested in the bioassay or in the opiate binding assay. Recovery of immunoreactivity was 85 to 90%. The peak fractions containing immunoreactivity were combined and subjected to a second HPLC system: ODS column (Beckman; 4.6 × 250 mm particle size 5 μm), buffer as described above, but adjusted to pH 7.5. Enzymatic treatment was applied for identification of the middle portion of the heptapeptide material. The combined fractions of the single rechromatographed peak were evaporated, desalted by Sephadex G-25-chromatography, and taken up in 300 μℓ 0.05 *M* Tris/HCl, pH 8.5. After addition of 10 μg trypsin (TPCK treated, from bovine pancreas; Sigma, Munich, Germany) the mixture was incubated at 37°C for 3 hr, followed by treatment with 0.5 μg carboxypeptidase-B (from porcine pancreas; Sigma) for 30 min at 37°C, and finally heated for 10 min at 96°C. The reaction mixture was evaporated and rechromatographed.

Radioimmunoassays (RIA)

Antisera against FMRFamide, YGGFMRFamide, and its homologue with a free C-terminus, and against Met-enkephalin were raised in rabbits by injection of the antigens coupled with 1-ethyl-3(3-dimethylaminopropyl)-carbodiimide to bovine serum albumin. Inhibition curves representing the specificity are shown in Figure 2A to D. The RIAs were conducted in 500 μℓ phosphate-buffered saline (0.14 *M* NaCl, 0.005 *M* phosphate buffer, pH 7.5, 0.5% bovine serum albumin) and incubated with the antisera for 24 hr at 4°C. Working antiserum dilutions were 1:100.000 (2A); 1:80.000 (2B); 1:30.000 (2C); and 1:16.000 (2D). Bound and free label were separated by addition of 500-μℓ suspension of dextran-coated charcoal in 0.05 *M* phosphate buffer, pH 7.5. Results are expressed as ratio of the bound label of the test tube (= B) to the initial bound label in the absence of competing peptides (= B_o).

The peptides in Figure 2A, B, and D were iodinated with ^{125}I using the Chloramin-T-method, and 10.000 cpm of tracer was added. The Met-enkephalin-RIA was performed with the (^3H)-labeled compound (47,7 μCi/m*M*; New England Nuclear (NEN), Chicago, U.S.). For the FMRF-a-RIA the parallelism of extracts and the standard preparation are shown for vena cava neuropil (v.c.) and brain tissue from *Octopus vulgaris* (Figure 2A). The FMRFamide radioimmunoassay detected equimolarily peptides with the C-terminal structure -RFamide supplemented by leucine or methionine. Crossreactivity against γ_1-MSH was about 1% but not evident for the desamidated peptides. The antiserum against the heptapeptide-amide was a gift from D. Price and its specificity has been described in Reference 20. Loss of the N-terminal portion YGG reduced the crossreactivity 50-fold thus enabling the differentiation between the tetra- and the heptapeptide. The antiserum against Met-enkephalin is described in Reference 9.

In Vitro Preparation of the Heart

Preparations of isolated systemic hearts of *Octopus vulgaris* were perfused with running seawater (16 to 20 mℓ min^{-1}) and the test substances were added for 1 min. Changes of the output pressure were amplified (SP 1400, Statham, Oxnard, U.S.A.). digitized (PCM 3K13, Johne & Reilhover, Munich, Germany), and stored online. Baseline values (f_o = 1) of the heartbeat were measured 5 min before infusion of the test substances and about 15 min thereafter. The data were further processed by a computer program to calculate the relative increase of frequency (Figure 3) and amplitude for any time interval. The C-terminal structure -RF-NH$_2$ is essential for cardioexcitatory potency and even the preceding amino acid may attribute to potency since analogs with neutral amino acids (we tested leucine and methionine), but not with aspartate as in case of γ-MSH, were active (Table 1). Extracts of different brain regions and the pituitary gland prepared from rats were ineffective (up to 100 μg protein). Naloxone did not antagonize the action at the heart, neither when added to synthetic opioid-amides (YGGFMRFamide and YGGFLRFamide), nor to purified material from brain and vena cava tissue of the octopus. The preparation responded significantly to 10^{-9} *M* of the active peptides.[6]

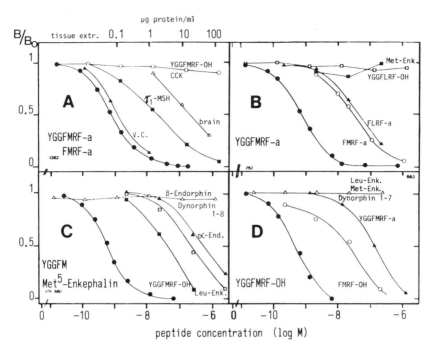

FIGURE 2.   Inhibition curves of radioimmunoassays (RIA) which are specific for the C-terminal -RFamide (A), the whole heptapeptide-amide structure (B), Met-enkephalin (C), and YGGFMRF-OH (D). In (A) the parallelism of extracts and the standard preparation are shown for vena cava neuropil (v.c.) and brain tissue from *Octopus vulgaris*.

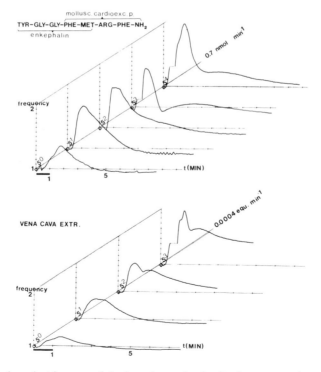

FIGURE 3.   Dose-dependent increase of the heart beat using in vitro heart preparations of *Octopus vulgaris*. Changes in frequency are given as differences from the baseline ($F_o = 1$) at the x-axis, concentrations of the test substances at the z-axis and, the time at the y-axis. Concentrations of extracts of the vena cava after gel chromatography on Sephadex G-50 are calculated as organ equivalents. The three-dimensional time-dose-response curves for the synthetic heptapeptide and the extracts are obviously very similar.

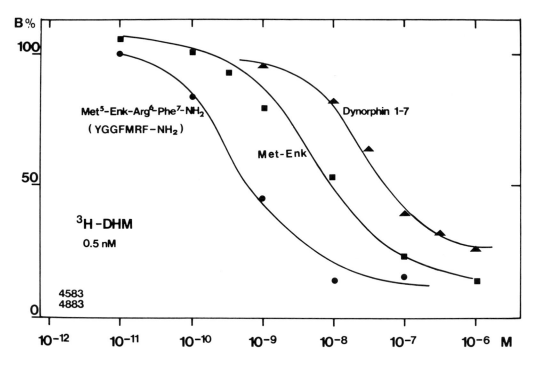

FIGURE 4.   Displacement curves of (³H)dihydromorphine from rat brain homogenates by addition of dynorphin 1-7, Met-enkephalin, and the *Octopus* heptapeptide YGGFMRFamide.

Opiate-Binding Assays

Occurrence of the C-terminal opioid structure Tyr-Gly . . . was tested with different binding assays. For measurement in opiate-binding assays, the separated material was desalted, evaporated, and taken up with 0.05 M Tris/HCl and adjusted to pH 7.4 with 1 M KOH. Binding assays were performed with membrane preparations of rat brain at 4°C for 120 min using(³H)-dihydromorphine (DHM; New England Nuclear) in a concentration of 0.5 nM as ligand (Figure 4). Incubation was terminated by rapid vacuum filtration (Whatman GF-B-glas fiber) or by centrifugation (13.000 × g for 2.5 min).

Contraction of the Locust Skeletal Muscle

The contraction of a muscle bundle of the jumping muscle from the male locust *Schistocerca gregaria* was tested in a preparation as described.[7] The peptides were added for 5 min and the influence on neurally evoked contractions and excitatory junction potentials (EJP) were measured (Figure 8).

RESULTS

The individual fractions after gel chromatography on Sephadex G-50 (Figure 6) were measured in the FMRFamide RIA (Figure 2A) and the mean peak of immunoreactive material was eluted with an apparent molecular weight of 400 to 1000 daltons. The tissue extracts and the separated products did not react in radioimmunoassays (RIA) for Met-enkephalin and Met⁵-enkephalin-Arg⁶-Phe⁷, thereby indicating that the heptapeptide in octopus nerve extracts appears only in its amidated form and that their opioid activity is not due to authentic Met-enkephalin, but probably within a larger molecule. In the opiate-receptor assay (Figure 4) only the combined fractions of this peak showed a displacement of (³H)-DHM, whereas in the bioassay also the higher molecular weight fractions exerted the same activity (Figure 5).

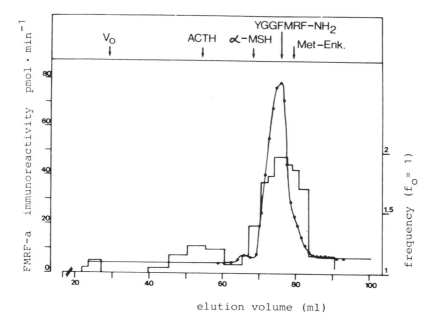

FIGURE 5.   Gel chromatography on Sephadex G-50-fine of octopus brain extracts. Individual fractions (●————●) were measured in a FMRFamide RIA (Figure 2A) or were combined for subsequent testing of their cardioexcitatory potency, shown for frequency at the right ordinate (Figure 4) and opioid receptor binding. At the top, the peak positions of iodinated (^{125}I) peptides are indicated by arrows for calibrations. The material of the single peak of FMRF-like activity, corresponding also to the principle region of bioactivity (i.e., fraction volume 70 to 82) was combined for further separation by two different HPLC systems (Figures 6 and 7).

The mean peak corresponded to about 80% of the immunoreactivity and bioactivity and was further separated at pH 2.7 into three regions of immunoactivity by a semipreparative high-performance liquid chromatography (HPLC) column (Figure 6). The peaks coeluted with FMRF-amide (b), its sulfoxydized form (a), and YGGFMRFamide (e). The occurrence of YGGFMRFamide was confirmed by a RIA system that was specific for the structure of the whole heptapeptide-amide (Figure 2B). Further separation of the immunoreactive material by a different HPLC system using an analytical column and elution at pH 7.5 supported this finding. The rechromatography of the FMRFamide peak resulted in a single region of immunoreactivity cochromatographing with the peptide standard (not shown). At Figure 7A, the FMRFamide immunoreactivity of the combined peak fractions (hatched bar in Figure 5) was eluted exclusively at the time of the standard preparation YGGFMRFamide.

For a characterization of the middle portion of the heptapeptide, the rechromatographed material (Figure 7A) was incubated with trypsin (33 μg mℓ$^{-1}$ for 3 hr at 37°C) followed by treatment with carboxypeptidase-B (1.5 μg mℓ$^{-1}$ for 30 min at 37°C). The resulting product was again separated by HPLC. A single peak of Met-enkephalin immunoreactivity appeared (Figure 7B), whereas activity in the FMRFamide RIA was totally abolished. This observation is similar to that made with immunocytochemical methods.[14]

Opiate-binding activity was detected only at the position of the octopus heptapeptide (hatched bar in Figure 5 and the rechromatographed peak in Figure 8A). Inhibition of total binding corresponded to 50 to 70 pmol YGGFMRFamide in 1 g brain tissue. The octopus heptapeptide exhibited strongest displacement in a predominantly μ-type specific binding system using (^{3}H)-dihydromorphine (NEN) in comparison to (^{3}H)-D-Ala^2, D-Leu^5)-enkephalin (Amersham, England) and (^{3}H)-ethylketazocine (NEN).

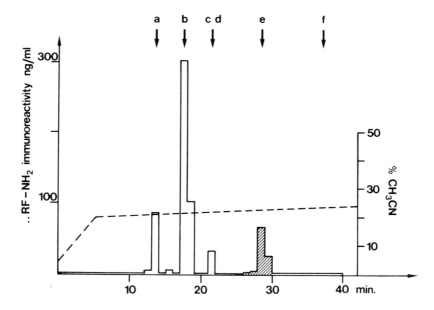

FIGURE 6.   HPLC separation of the immunoreactive (FMRFamide RIA) and cardioactive material after G-50-chromatography (Figure 5). The individual fractions were evaporated and measured in different RIA systems (here shown for FMRFamide). The fractions of the principle peaks, one appearing at the elution time of FMRFamide standard (b) and another (hatched bar) of YGGFMRFamide (f), were combined and subjected to a second HPLC system (Figure 7). Elution of standard peptides is indicated as follows: (a) sulfoxidized FMRFamide, (b) FMRFamide, (c) FLRFamide, (d) sulfoxidized YGGFMRFamide, (e) YGGFMRFamide, (f) YGGFLRFamide.

If one considers the high specificity of the used assay systems, then the results allow the assumption that the octopus nerves contain the heptapeptide YGGFMRFamide together with the molluscan cardioexcitatory tetrapeptide FMRFamide.[1] The approximate ratios were 70 pmol/g octopus brain for the heptapeptide and 200 pmol/g for the tetrapeptide.

In searching for a bioassay system with a specificity for nRFamide comparable with that of the isolated octopus heart,[6] we used a preparation of a locust jumping muscle[7] (Figure 8). Measuring the influence of FMRFamide-related peptides on neurally evoked contractions and the corresponding excitatory synaptic potentials, we observed that this model was also very useful for structure activity studies (Table 1). But in contrast to the equimolar activity of the tetrapeptide- and the heptapeptide-amide in the octopus heart preparation, the contractions of the locust muscle have been 100 times more responsible with the heptapeptide. Importantly, this activity was not antagonized by the simultaneous addition of 0.1 mM naloxone, and Met- and Leu-enkephalins were inactive.

DISCUSSION

In tissue extracts of brain and of the anterior vena cava from the cephalopod *Octopus vulgaris*, we identified two cardioexcitatory peptides: one is identical with the tetrapeptide FMRFamide, which first was found in a bivalve mollusc,[1] and the other neuropeptide has the apparent structure of YGGFMRFamide. Since immunoreactivities for the opioid portion of the latter, observed by applying antibodies against Met-enkephalin[14,15] and a monoclonal antibody, which reacts with the extreme N-terminal sequence YGG..,[21,22] occurred in identical nerve endings as FMRFamide, we suggest that both peptides may belong to the same precursor system. In contrast to the presence of separate enkephalin-containing nerve terminals[21]

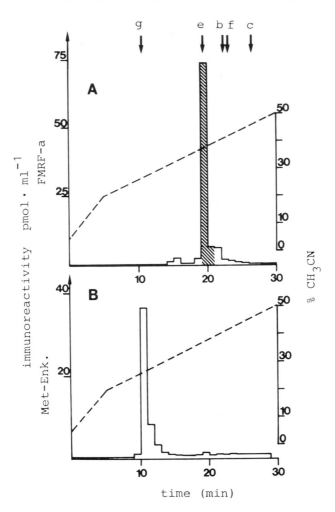

FIGURE 7.   HPLC rechromatography (at pH 7.5) of the heptapeptide
material (hatched bar in Figure 6) and measurement in the FMRFamide-
RIA (A). (B) the rechromatography of the material of the single peak
in (A) after enzymatic treatment (trypsin and carboxypeptidase b). The
immunoreactivity of the heptapeptide disappeared (not shown), and the
released Met-enkephalin was detected. Standard peptides (a) to (f) as
in Figure 6; (g) Met-enkephalin.

and the isolation of the pentapeptides in extracts of a mussel,[10] in octopus nerve extracts
we found opioid material only in fractions of the octopus heptapeptide. This was assured
by enzymatic cleavage at the position of the -R[6]-residue, which has destroyed the -RFamide
structure and liberated the N-terminal portion comprising Met-enkephalin. The conversion
of the heptapeptide to Met-enkephalin and also to FMRFamide by endopeptidases of purified
rat striatal membranes[23] may also indicate a functional role for the heptapeptide as a precursor
for two bioactive neuropeptides.

The octopus heptapeptide has two different functions mediated by distinct portions of the
molecule. The N-terminal part binds to opioid receptors of rat brain, preferably of the μ-
type, an affinity similar to that of another opioid amide isolated from bovine brain[24] The
C-terminal structure -RFamide supplemented by a neutral amino acid has been found to be
crucial for peptide action at the heart[5,6] and other muscles of molluscs[8] as well as for pre-
and postsynaptic effects on locust skeletal muscles.[7] As these functions were not antagonized
by naloxone,[6,7] opiate receptors seemed not to be involved.

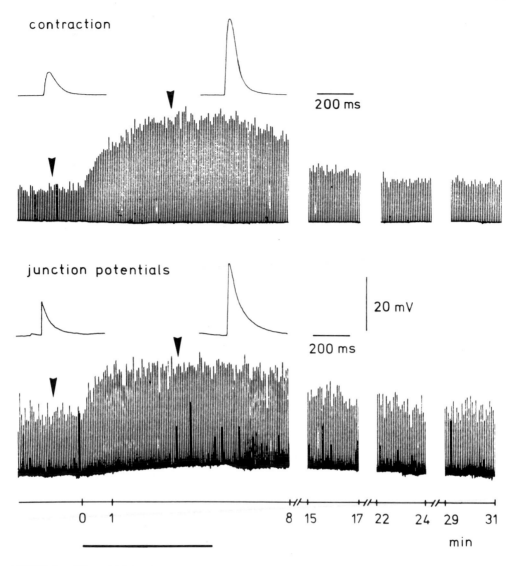

FIGURE 8.   Effect of YGGFMRFamide (10-8) on neurally evoked contractions and excitatory junction potentials. Synchronous recordings from one experiment on a locust jumping muscle. Application of the peptides is indicated by the bar at the bottom. The insets show individual responses, marked by the arrowheads and presented in an enlarged time scale. Initial resting membrane potential − 77 mV. (Taken from Walther, Chr., Schiebe, M., and Voigt, K. H., *Neurosci. Lett.*, 45, 99, 1984. With permission.)

However, some reports[7,25,26] point to a particular function of YGGFMRFamide, which is different from the molluscan tetrapeptide[1]: Cottrell and co-workers[25] found differential effects of both neuropeptides on the *Helix* tentacle retractor muscle. In a locust jumping muscle preparation,[7] the heptapeptide was about 100 times more active, whereas both peptides are equipotent when tested on molluscan hearts.[6,18] This potentiation was obviously not due to an additional interaction with opioid receptors as naloxone did not antagonize the action. Furthermore, a vasodilator effect in the rat caused by the heptapeptide was three times more potent than FMRFamide.[26]

The structure of the two cardioexcitatory peptides in the octopus is related to a variety of neuropeptides which belong to different precursor systems (Table 2), thus representing the connection between the FMRFamide family and the opioids.[28,29] The amidation of the C-

Table 1
STRUCTURE ACTIVITY
RELATIONSHIP FOR THE
CARDIOEXCITATORY ACTIVITY (IN
***OCTOPUS VULGARIS*) AND THE**
POTENTIATION OF MUSCLE
CONTRACTION (LOCUST JUMPING
MUSCLE)[6,7]

	Relative potency	
	---	---
Peptide	**Octopus heart**	**Locust muscle**
FMRF-NH$_2$	1	1
FLRF-NH$_2$	1	1
FMRF-OH	Ø	Ø
YGGFM	Ø	Ø
YGGFMRF-NH$_2$	1.5	100
YGGFLRF-NH$_2$	1.8	100
YGGFMRF-OH	<0.001	<0.001
...LITRORY-NH$_2$	Ø	Ø
...HFRWDRF-NH$_2$	Ø	<0.001

Table 2
LIST OF NATURALLY OCCURRING
NEUROPEPTIDES WITH STRUCTURAL
HOMOLOGIES TO THE OPIOID AND
CARDIOEXCITATORY PORTION OF THE OCTOPUS
HEPTAPEPTIDE (ONLY A SELECTION OF OPIOIDS
IS SHOWN)

Opioid-pre-cursor fragments	...HFRWD*RF-NH*$_2$	γ_1-MSH
	YGGFM	Met[5]-enkephalin
	*YGGFM*RRV-NH$_2$	Metorphamide
	YGGFMRF	C-terminal heptapeptide
	*YGGF*L*R*R	Dynorphin 1—8
Invertebrate cardioactive peptides	*FMRF-NH*$_2$	*Macrocallista nimbosa*
	...D*FIRF-NH*$_2$	*Helix aspersa*
	...*YLAF*P*RM-NH*$_2$	*Aplysia*
	YGGFMRF-NH$_2$	*Octopus vulgaris*
Vertebrate gut and brain peptides	D*YM*GW*MDF-NH*$_2$	CCK-octapeptide
	LP*LRF-NH*$_2$	Chicken brain
	...VVTRH*RY-NH*$_2$	Avian PP (29—36)
	...MLTRP*RY-NH*$_2$	Bovine
	...LVTRQ*RY-NH*$_2$	Porcine PP (29—36)
	...LITRO*RY-NH*$_2$	Porcine NPYY (29—36)
		Porcine brain NPY (29—36)

terminus would imply that glycine should follow in the hypothetical precursor molecule. This kind of amidation of the Arg-Phe-Gly residue exists already in the processing of proopiomelanocortin to produce γ_1-MSH.[30] It would be very interesting to know whether the vertebrate gut/brain peptides, characterized by an amidated C-terminus of a cyclic amino acid and a penultimate arginine (e.g., pancreatic polypeptide, neuropeptide Y) may belong to the class of the evolutionary very old FMRF-peptides.

ACKNOWLEDGMENTS

This work was supported by Deutsche Forschungsgemeinschaft (SFB 87/B2; Heisenberg-Stipendium for K. H. Voigt). We are grateful to Dr. E. Weber (San Francisco, U.S.A.) and Drs. D. A. Price and M. J. Greenberg (St. Augustine, U.S.A.) for the generous gift of some antisera. We thank the Stazione Zoologica di Napoli, Naples, Italy, and the Laboratoire Arago, Banyuls-sur-Mer, France, for hospitality.

REFERENCES

1. **Price, D. A. and Greenberg, M. J.,** Purification and characterization of a cardioexciatory neuropeptide from the central ganglia of a bivalve mollusc, *Prep. Biochem.,* 7, 261, 1977.
2. **Price, D. A.,** The FMRF-amide-like peptide of *Helix aspersa, Comp. Biochem. Physiol.,* 72c, 325, 1982.
3. **Morris, H. R., Panico, M., Karplus, A., Lloyd, P. E., and Riniker, B.,** Elucidation by FAB-MS of the structure of a new cardioactive peptide from *Aplysia, Nature (London),* 300, 643, 1982.
4. **Dockray, G. J., Reeve, J. R., Jr., Shively, J., Gayton, R. J., and Barnard, C. S.,** A novel active pentapeptide from chicken brain identified by antibodies to FMRFamide, *Nature (London),* 305, 328, 1983.
5. **Painter, S. D. and Greenberg, M. J.,** A survey of the responses of bivalve hearts to the molluscan neuropeptide FMRFamide and to 5-hydroxytryptamine, *Biol. Bull.,* 162, 311, 1982.
6. **Voigt, K. H., Kiehling, C., Frösch, D., Schiebe, M., and Martin, R.,** Enkephalin-related peptides: direct action on the octopus heart, *Neurosci. Lett.,* 27, 25, 1981.
7. **Walther, Chr., Schiebe, M., and Voigt, K. H.,** Synaptic and non-synaptic effects of molluscan cardioexcitatory neuropeptides on locust skeletal muscle, *Neurosci. Lett.,* 45, 99, 1984.
8. **Painter, S. D., Morley, J. S., and Price, D. A.,** Structure-activity relations of the molluscan neuropeptide FMRFamide on some molluscan muscles, *Life Sci.,* 31, 2472, 1982.
9. **Gayton, R. J.,** Mammalian neuronal actions of FMRFamide and the structurally related opioid Met-enkephalin-Arg[6]-Phe[7], *Nature (London),* 298, 275, 1982.
10. **Leung, M. K. and Stefano, G. B.,** Isolation and identification of enkephalins in pedal ganglia of *Mytilus edulis* (Mollusca), *Proc. Natl. Acad. Sci. U.S.A.,* 81, 955, 1984.
11. **Hughes, J., Kosterlitz, T. W., and Fothergill, L. A.,** Identification of two related pentapeptides from the brain with potent opiate agonist activity, *Nature (London),* 258, 577, 1975.
12. **Noda, M., Furutani, Y., Takahashi, H., Toyosato, M., Hirose, T., Inayama, S., Nakanishi, S., and Numa, S.,** Cloning and sequence analysis of cDNA for bovine adrenal preproenkephalin, *Nature (London),* 295, 202, 1982.
13. **Rossier, J., Audigier, Y., Ling, N., Cros, J., and Udenfriend, S.,** Met-enkephalin-Arg[6]-Phe[7], present in high amounts in brain of rat, cattle and man, is an opioid agonist, *Nature (London),* 288, 88, 1980.
14. **Martin, R., Frösch, D., Kiehling, C., and Voigt, K.-H.,** Molluscan neuropeptide-like and enkephalin-like material coexists in octopus nerves, *Neuropeptides,* 2, 141, 1981.
15. **Martin, R., Schäfer, M., and Voigt, K. H.,** Enzymatic cleavage prior to antibody incubation as a method for neuropeptide immunocytochemistry, *Histochemistry,* 74, 457, 1982.
16. **Berry, C. F. and Cottrell, G. A.,** Neurosecretion in the vena cava of the cephalopod *Eledone cirrhosa, Z. Zellforsch. Mikrosk. Anat.,* 104, 107, 1970.
17. **Voigt, K. H., Kiehling, C., Frösch, D., Bickel, U., Geis, R., and Martin, R.,** Identity and function of neuropeptides in the vena cava neuropil of octopus, in *Molluscan Neuro-Endocrinology,* Proc. Int. Minisymp. Molluscan Endocrinol., Lever, J. and Boer, H. H., Eds., North-Holland, Amsterdam, 1983.
18. **Greenberg, M. H., Painter, S. D., and Price, D. A.,** The amide of the naturally occurring opioid (Met)-enkephalin-Arg[6]-Phe[7] is a potent analog of the molluscan neuropeptide FMRF-amide, *Neuropeptides,* 1, 309, 1981.
19. **Voigt, K. H., Weber, E., and Martin, R.,** Neuropeptides: subcellular localization of ACTH and related peptides, in *Structure and Activity of Natural Peptides,* Voelter, W. and Weitzel, G., Eds., Walter de Gruyter, New York, 1981.
20. **Price, D. A.,** FMRFamide: assays and artefacts, in *Molluscan Neuro-Endocrinology,* Proc. Int. Minisymp. Molluscan Endocrinol., Lever, J. and Boer, H. H., Eds., North-Holland, Amsterdam, 1983.
21. **Martin, R., Haas, C., and Voigt, K. H.,** Opioid and related neuropeptides in molluscan neurons, in *Handbook of Comparative Opioid and Related Neuropeptide Mechanisms,* Vol. I, Stefano, G. B., Ed., CRC Press, Boca Raton, Fla., 1986, in press.

22. **Meo, T., Gramsch, Ch., Inan, R., Höllt, V., Weber, E., Herz, A., and Riethmüller, G.,** Monoclonal antibody to the message sequence Tyr-Gly-Gly-Phe of opioid peptides exhibits the specificity requirements of mammalian opioid receptors, *Proc. Natl. Acad. Sci. U.S.A.,* 80, 4084, 1983.

23. **Marks, N., Benuck, M., and Berg, M. J.,** Metabolism of a heptapeptide opioid by rat brain and cardiac tissue, *Life Sci.,* 31, 1845, 1982.

24. **Weber, E., Esch, F. S., Böhlen, P., Paterson, S., Corbett, A. D., McKnight, A. T., Kosterlitz, H. W., Barchas, J. D., and Evans, C. J.,** Metorphamide: isolation, structure and biologic activity of an amidated opioid octapeptide from bovine brain, *Proc. Natl. Acad. Sci. U.S.A.,* 80, 7362, 1983.

25. **Cottrell, G. A., Greenberg, M. J., and Price, D. A.,** Differential effects of the molluscan neuropeptide FMRFamide and the related Met-enkephalin derivative YGGFMRFamide on the *Helix* tentacle retractor muscle, *Comp. Biochem. Physiol.,* 75c, 373, 1983.

26. **Koo, A., Chan, W. S., Ng, W. H., and Greenberg, M. J.,** Microvascular vasodilator effect of FMRF-amide and Met-enkephalin-Arg[6]-Phe[7]-amide in the rat, *Microcirculation,* 2, 393, 1982 to 1983.

27. **Greenberg, M. J., Painter, S. D., Doble, K. E., Nagle, G. T., Price, D. A., and Lehman, H. K.,** The molluscan neurosecretory peptide FMRFamide: comparative pharmacology and relationship to the enkephalins, *Fed. Proc. Fed. Am. Soc. Exp. Biol.,* 42, 82, 1983.

28. **Dockray, G. J., Vaillant, C., Williams, R. G., Gayton, R. J., and Osborne, N. N.,** Vertebrate brain-gut peptides related to FMRFamide and Met-enkephalin-Arg[6]-Phe[7], *Peptides,* 2, 25, 1981.

29. **Böhlen, P., Esch, F., Shibasaki, T., Baird, A., Ling, N., and Guillemin, R.,** Isolation and characterization of a γ_1-melanotropin-like peptide from bovine neurointermediate pituitary, *FEBS Lett.,* 128, 67, 1981.

EVOLUTIONARY ASPECTS OF OPIATE INVOLVEMENT IN THERMOREGULATION

Martin Kavaliers and Maurice Hirst

SUMMARY

The broad spectrum of invertebrate and vertebrate species that produce opioid peptides raises the possibility that these simple peptides may be involved in the mediation of fundamental biologic functions. Thermoregulation is one basic function displayed by all animals. Their survival is dependent on the appropriate control of body temperature and avoidance of thermal extremes. These behavioral and physiological thermoregulatory responses are markedly sensitive to opiate substances. In invertebrates and lower ectothermic vertebrates, such as fish, opiate agonists and antagonists produce simple and opposite actions on environmental preferences. In contrast, in the higher ectothermic and endothermic vertebrates opiate mechanisms appear to regulate existing temperatures, indicating a hierarchical control of temperature regulation in which an opioid-based system is an component. This suggests an early evolutionary development and phylogenetic continuity of opiate sensitive thermoregulatory mechanisms.

INTRODUCTION

The discovery of endorphins and enkephalins, endogenous peptides with morphine-like properties, in the 1970s initiated an expanding interest in central neuropeptide function. These peptides were originally examined for analgesic qualities, but numerous investigations over the past decade have explored putative regulatory roles for these substances.[1-5]

While initial studies suggested that opioid peptides and their receptors existed within mammalian systems,[6] subsequent investigations have demonstrated that they have a much wider phylogenetic distribution.[7-9] The broad spectrum of invertebrate and vertebrate species that produce opioid peptides suggests that these simple peptides may be involved in the mediation of fundamental biologic functions.[10-12]

Thermoregulation is a basic feature of all animals, their survival being dependent on the appropriate control of body temperature.[13] Morphine and its congeners, administered peripherally or centrally, alter core temperature in mammals. Low doses of morphine produce hypothermia, while higher doses cause an initial phase of hypothermia followed by an elevation of body temperature.[5,14,15] Accordingly, one of the prime roles for endogenous opioids may be in the control of responses to environmental temperature.

The ability to respond to adverse thermal environments is demonstrated by all motile animals. In general, ectotherms regulate their body temperature by behaviorally selecting the most appropriate thermal regimes. This permits animals to maintain body temperature within defined limits ("set points"). These limits can vary with time of day, among other factors.[17]

Ectotherms exposed to excessively high temperatures will respond quickly by seeking out cooler locations. Similar responses occur to cold exposure when a warm region is available. In the absence of this, body temperature falls and motility decreases.[17]

Endotherms use both physiological and behavioral means to control body temperature.[13,18,20] To maintain relatively constant temperature, endotherms have developed several mechanisms of heat production and dispersion that operate when the animals are exposed to fluctuations in environmental temperature. In the face of more extreme changes, both behavioral and physiological mechanisms are employed to maintain constant body temperature. As in the ectotherms, such factors as time of day, season, and state of health can exert influences on the physiologically regulated temperatures of endotherms.[18-20]

The major direction of this review is to trace the phylogenetic development of thermo-regulatory responsiveness to opiates. A variety of other functions have also proved to be sensitive to opioids in invertebrates and in nonmammalian vertebrates.[11-13] Thus, the opioid peptides have significant influences on a variety of behavioral and neural processes, a number of which may be involved in thermoregulation and themselves be thermally dependent.

THE INFLUENCE OF OPIATE COMPOUNDS ON INVERTEBRATE THERMAL BEHAVIOR

Thermal Responses of the Terrestrial Snail, *Cepaea nemoralis*

Terrestrial snails are ectothermic, their body temperatures being dependent on the ambient temperatures. In the wild, *C. nemoralis* utilizes behavioral means to maintain body temperature in a selected range. This preferred range is influenced by numerous environmental factors and can vary with the morphological type of this polymorphic species.[21]

When placed on a thermally aversive surface, fully hydrated *Cepaea* display a characteristic elevation of the anterior portion of the foot[22,23] (Figure 1). This response has not been observed in snails exposed to temperatures normally present in their natural northern habitats, but becomes increasingly evident as the environmental temperature is experimentally raised toward 40°C.[23] The foot-lifting response suggestive of discomfort is influenced by low doses of opiate substances (Figures 2 and 3). Morphine increases the time taken for treated animals to demonstrate this stereotypic behavior. In contrast, narcotic antagonists such as naloxone reduce this latency. Further, narcotic antagonists such as naloxone can prevent the alteration of response latency shown by the analgesic compounds.[22,23] As noted above, *Cepaea* is genetically and to a lesser extent behaviorally polymorphic. This is visually evident in the different shell colors (pink or yellow) and banding patterns that are present.[21] This genetic diversity is seen also in their sensitivity to the opiate compounds (Figure 3).

The effect of morphine is produced by the benzomorphan, levorphanol, but not by the stereoisomer, dextrophan, suggesting that a receptor that interacts with opiates in generating this effect has stereospecific requirements.[24] Based upon behavioral responses to many synthetic narcotic analgesics, Martin and colleagues[25] suggested that there are several different types of opiate receptors. Accordingly, snails were injected with cyclazocine (10.0 μg) and phencyclidine (10.0 μg) and their responses to being placed on the heated surface monitored. These drugs did not alter significantly the foot-lifting latency, suggesting the probability that kappa and sigma forms of the receptor do not participate in this behavior.[26]

The increase in latency seen after narcotic and analgesic drugs was similarly observed in *Cepaea* treated with opiate peptides, β-endorphin (1 to 10 μg), Met-enkephalin (1 to 10 μg), and in a number of cases the molluscan cardioexcitatory "opioid"[27] peptide FMRFamide (1 to 10 μg). In these studies FMRF and Met-enkephalin were noticeably more potent than the larger peptide, β-endorphin.[28] Naloxone antagonized the effects of these peptides. It is of interest here that FMRFamide also has potent behavioral and physiological effects in mammals.[29]

In another series of experiments, fully hydrated *Cepaea* were placed on a circular thermal gradient ranging from 5 to 40°C. Over 2 hr they selected their preferred temperature of close to 28°C. Following this, animals were given doses of morphine or naloxone and returned to their pretreatment location on the surface.[30] Their selected temperatures were continuously monitored for the next 3 hr. Snails given morphine displayed a preference for higher temperature ranges in a dose-dependent fashion[30] (Figure 4). In contrast, *Cepaea* given naloxone exhibited a dose-dependent preference for lower temperatures. These altered thermal choices were evident within 15 min after drug administration and lasted for an hour or longer (Figure

FIGURE 1.   Characteristic elevation of the anterior portion of the extended foot of a fully hydrated *C. nemoralis* exposed to an aversive thermal stimulus of 40°C.

4). There was no indication at any time that snails were experiencing locomotor disturbances. Naloxone given with morphine obtunded the effect of morphine. In parallel to the above-described influences of the opiate peptides on the foot-lifting behavior, β-endorphin, FMRFamide, and Met-enkephalin caused treated snails to select higher temperatures.[28]

Thermal Responses of Larvae of the Mealworm, *Tenebrio molitor*

Fourth instar larvae of the mealworm, *Tenebrio molitor*, were injected with morphine (10 μg), Met-enkephalin (10 μg), β-endorphin (10 μg), or naloxone (100 μg). Their behavior and preference of temperatures were recorded following their return to pretreatment locations on a thermal gradient varying from 20 to 40°C. Prior to drug treatment, the larvae showed a preference for a temperature of about 32°C. After injection, animals given morphine or enkephalin moved towards higher temperature, whereas those receiving naloxone wriggled in the other direction (Figure 5). Larvae receiving β-endorphin displayed no consistent changes in thermal preference. The changes in preferred temperature that took place were evident for approximately 1 hr after which the animals returned to the pretreatment preference temperature (Figure 5).

THE INFLUENCE OF OPIATE COMPOUNDS ON VERTEBRATE THERMAL BEHAVIOR

Thermal Responses of the Goldfish, *Carassius auratus*

Teleost fish can regulate their body temperatures by behavioral exploitation of the thermal environment.[17] The absence of significant peripheral thermoregulatory adjustments, coupled with similarities in endothermic and ectothermic vertebrate central thermoregulatory controls, including locations of thermally sensitive nuclei and integration of thermal inputs,[18,31] indicates that fish can serve as a model for investigation of the mechanisms underlying central

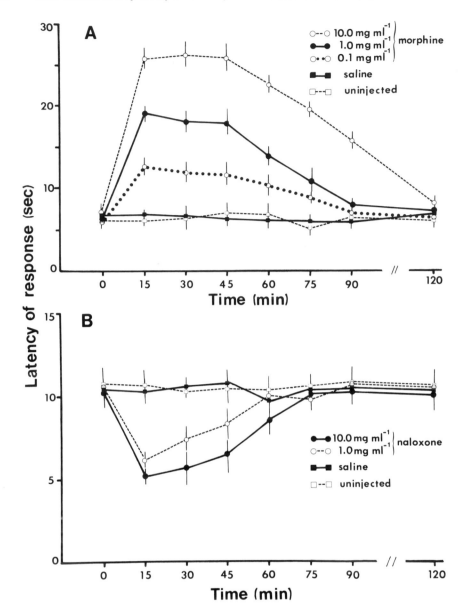

FIGURE 2.   Time course of the effects of (A) morphine sulfate (0.1, 1.0, and 10.0 μg) and (B) naloxone hydrochloride (1.0 and 10.0 μg) on the latency of the foot-lifting response to 40°C in two different morphs of *C. nemoralis*. Controls were given saline injections (1.0 μℓ) or no injections. Ten different individuals were used for each dose and treatment. Vertical lines denote two standard errors. (Modified from Kavaliers, M., Hirst, M., and Teskey, G. C., *Science*, 330, 99, 1983. With permission.)

control of body temperature in mammals. For this reason thermal responses of fish have been modified by a variety of experimental manipulations.[32,33]

Thermoregulatory behavior of individual goldfish was determined in horizontal thermal gradients of 10 to 30°C (Figure 6). Fine 32-gauge copper-constantan thermocouples used to record preferred and body temperatures in these small fish (5 to 15 g) were attached to the individual so as not to impede locomotor activity. Animals were injected intraventricularly with low doses of β-endorphin or Met-enkephalin after being removed form the preferred

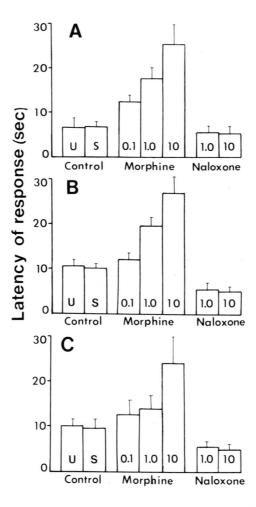

FIGURE 3.   Variations between morphological types of *C. nemoralis:* (A) pink unbanded, (B) yellow unbanded, (C) yellow banded in the latencies of their responses to a 40°C thermal stimulus after injection of either morphine sulfate (0.1, 1.0, and 10.0 μg) or naloxone hydrochloride (1.0 and 10.0 μg). U represents untreated individuals and S represents saline (1.0 μℓ)-injected individuals. Ten different individuals were used for each dose and treatment. Vertical lines denote two standard errors.

location. Injection was rapid and the fish was returned quickly to its original position in the gradient. Within the dose range of β-endorphin examined (0.5 to 15 pg/g), fish moved towards warmer or cooler regions. The lower doses (0.5, 1.0, 5.0 pg/g) caused a behavioral hyperthermia, whereas the highest dose (15 pg/g) produced behavioral hypothermia[34] (Figure 7). Slightly greater and more rapid responses were obtained with Met-enkephalin.[35] Similar types of responses were also observed with peripherally administered morphine (1.0 to 20 mg/kg). These alterations of preference were attained rapidly and continued for several hours. Within the observation period of 6 hr, the hypothermia resulting from the high doses of β-endorphin or Met-enkephalin changed towards the control thermal preference, there was no evidence of hypothermia changing to hyperthermia. A dose of naloxone (1 mg/kg) given peripherally produced behavioral hypothermia. This dose, given with β-endorphin, obliterated the opioid-induced changes in thermal preferences. Fish given β-endorphin moved to their selected region of the temperature gradient and did not show any deviation in

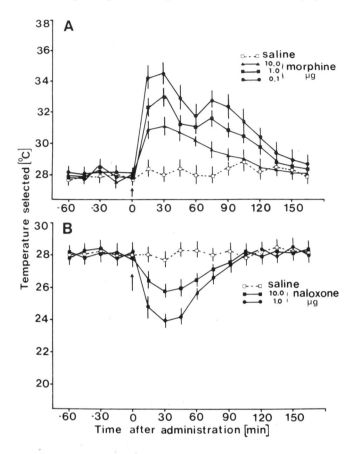

FIGURE 4.   Effects of administrations of (A) morphine sulfate (0.10, 1.0, and 10.0 μg) and (B) naloxone hydrochloride (1.0 and 10.0 μg) on the behaviorally selected temperatures of individual *C. nemoralis* (n = 10). Saline (1.0 μℓ)-injected individuals were used as controls. Arrows denote injection times (mid-photophase). Vertical lines denote two standard errors. (Modified from Kavaliers, M. and Hirst, M., *Neuropharmacology,* in press. With permission.)

locomotor levels from those seen in control, saline-injected animals.[34] In contrast, fish given the above-mentioned doses of β-endorphin displayed a dose-dependent increase in locomotor activity when held in a constant temperature of 22°C, the temperature preferred by control fish. Naloxone reduced locomotor activity after either intraventricular or peripheral injection.[34]

Thermal Responses of Tadpoles of the Bullfrog, *Rana catesbiana*

Two-year-old bullfrog tadpoles were placed in the horizontal temperature gradient (Figure 6). Their preferred temperatures were recorded continuously by means of attached copper-constantan thermocouples. Morphine (10 mg/kg), naloxone (10 mg/kg), or a mixture of these were injected intraperitoneally and the animals returned to their preinjection temperature locations. Their behaviorally selected temperatures were measured for the next 3 hr period. As seen in Figure 8, morphine produced a behavioral hyperthermia which was antagonized by naloxone. In this experimental animal, naloxone also had a slight but significant effect on thermal preferences.

Thermal Responses of the Curly-Tailed Lizard, *Leiocephalus carinatus*

The iguanid lizard, *Leiocephalus carinatus,* which is native to Haiti and Cuba, has been accidentally introduced into the southwestern U.S. where it experiences a more diverse range

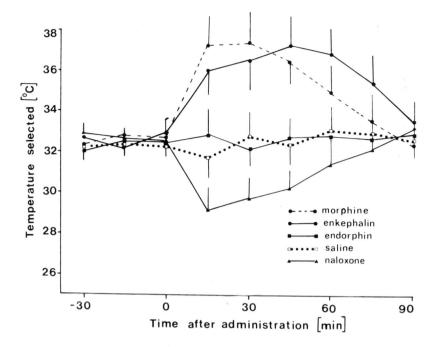

FIGURE 5. Effects of administration of morphine sulfate (10.0 μg), Met-enkephalin (10.0 μg), β-endorphin (10.0 μg), or naloxone (100 μg) on the behaviorally selected temperatures of larvae of *Tenebrio* (n = 10). Saline (1.0 μℓ)-injected animals were used as controls. Vertical lines denote two standard errors.

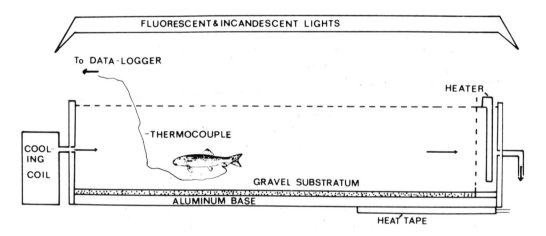

FIGURE 6. Thermal gradient used for investigations of behavioral thermoregulation in fish and bullfrog tadpoles.

of environmental temperatures.[36] The curly-tailed lizard along with other species of lizards displays a variety of thermally influenced behaviors, such as locomotor activity, feeding, and in its social interactions with its conspecifics.[37] Thus, for lizards the ability to regulate body temperature has both behavioral and physiological implications.

Following intraperitoneal administration of morphine (1.0, 10 mg/kg), lizards in a temperature gradient of 20 to 50°C displayed dose- and time-dependent changes in preferred temperatures.[38] As shown in Figure 9, the low doses produced hyperthermia, whereas the high dose initiated a more complex pattern — hyperthermia, hypothermia, and finally hyperthermia. Of particular interest was that the opiate-induced changes occurred with

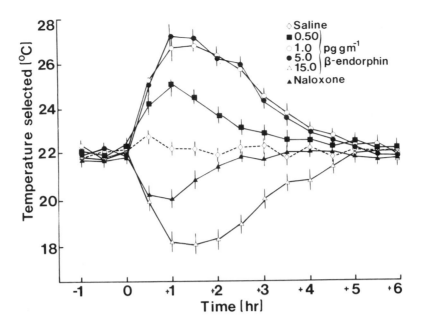

FIGURE 7.   Effects of central intraventricular injections of β-endorphin on the temperatures selected by individual goldfish in a horizontal thermal gradient. Saline (1.0 or 2.0 μℓ)-injected individuals were used as controls. Each point represents the mean of data with two standard errors for five fish replicated three times. (Modified from Kavaliers, M., *Peptides*, 3, 679, 1982. With permission.)

noticeable consistency across a broad range of initially selected temperatures which were from about 30 to 38°C.

In a similar manner, naloxone (1.0, 10 mg/kg) caused a dose-dependent hypothermia from the originally preferred temperature and could antagonize the changes in thermoregulation induced by morphine (Figure 9).

Thermal Responses of the Japanese Quail, *Coturnix coturnix japonica*

In contrast to the ectotherms, endotherms such as the Japanese quail can utilize both behavioral and physiological mechanisms to regulate body temperature.[18,19] As shown by Thornhill and West,[39] thermoregulation in the adult Pekin duck is influenced by peripheral injections of morphine. In this species morphine (10 mg/kg) elicited a hyperthermic response which could be antagonized by naloxone (10 mg/kg). In studies with the Japanese quail, morphine and naloxone were injected intraperitoneally during both the daytime and at night. The Japanese quail, while adopting both procedures for regulating body temperature, displays a significant day-night rhythm in this parameter with the highest temperatures occurring during the day. Following morphine administration during the day, the quail displayed a rapid hypothermia, with a reduced body temperature of 38 to 39°C being evident in approximately 30 min. This hypothermic response declined during the next hour (Figure 10A). At nighttime, morphine generated a more complex response of hyperthermia within 30 min and a hypothermia within the next hour. The hypothermic response observed during the day is consistent with that seen by Panksepp[40] and Nistico[41] who administered morphine and β-endorphin to young chickens. Naloxone given during the daytime also caused a decline in body temperature which disappeared by 90 min following injection (Figure 10B). There was no evidence of a change in body temperature in quail injected with naloxone during the dark cycle.

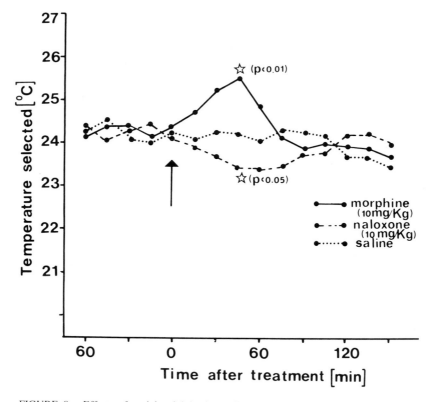

FIGURE 8. Effects of peripheral injections of morphine sulfate (10 μg) and naloxone hydrochloride (10 μg) on the temperatures selected by individual bullfrog tadpoles in a horizontal thermal gradient. Saline (0.10 μℓ)-injected individuals were used as controls. Vertical lines represent two standard errors of the mean. Stars denote significant changes in preferred temperatures.

DISCUSSION

The studies described above were undertaken to determine the universality of opioid modulation of body temperature and to obtain indications of the evolutionary development of opiate involvement in thermoregulation. The results of the experiments demonstrated a widespread responsiveness of species to opiates, for all organisms tested showed similar basic thermoregulatory responses after injection of opiate compounds. These observations coupled with the behavioral and physiological evidence from mammals of opioid peptide regulation of body temperature[5,14-16,42,43] suggest an early evolutionary development and phylogenetic continuity of opiate sensitive thermoregulatory mechanisms.

The species used in the investigations included both invertebrates and ectothermic vertebrates. It was considered that if endogenous opioids were functionally involved in mediating natural thermoregulatory processes, then opiate agonists and antagonists should produce opposite actions on environmental temperature preferences. Such actions were observed in the invertebrates and aquatic lower vertebrates species examined. In contrast, responses in the curly-tailed lizard and Japanese quail in which opiate-sensitive responses appeared to modulate existing regulated temperatures[1] suggested the development of hierarchical controls of temperature regulation in which an opioid-based system may be a component. This arrangement has been proposed by Satinoff[20] to be present in mammals. In all cases, opioid peptides produced alterations in an animal's behavior and physiology which facilitated regulated changes in body temperature.

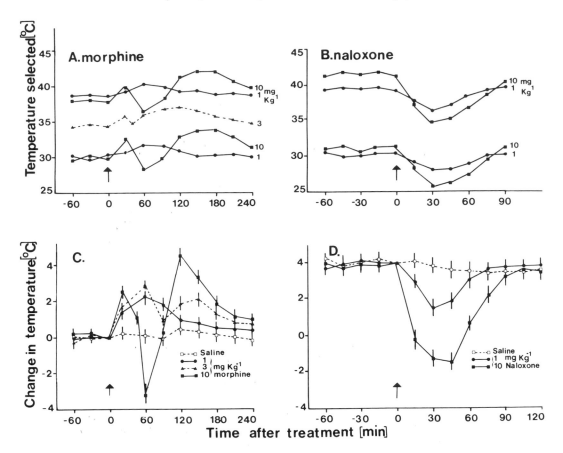

FIGURE 9.   Effects of intraperitoneal administrations of morphine and naloxone on the behaviorally selected body temperatures of curly-tailed lizards. Examples of different doses of (A) morphine sulfate (1.0, 3.0, and 10.0 mg/kg) and (B) naloxone (1.0 and 10.0 mg/kg) on individual animals of different initial body temperatures are shown. Mean representation of the effects of (C) morphine (n = 10, except n = 5 for 3.0 mg/kg) and (D) naloxone (n = 10) along with saline (10 mℓ/kg) controls are provided. Here each point represents the mean difference in absolute body temperature of lizards from just before injection time (mid-photophase) rather than their actual body temperatures. Vertical lines denote two standard errors. (Modified from Kavaliers, M., Courtenay, S., and Hirst, M., *Physiol. Behav.*, 32, 221, 1984. With permission.)

While the above-mentioned conclusions followed from the experimental results, it should be noted that in some experiments conflicting findings were observed when high doses of opiate agonists were administered. These latter results may reflect pharmacological-toxicological actions of the drugs in excess of any responses associated with physiological influences on thermoregulation. Similarly, in most experiments, doses of naloxone given were either 1.0 or 10 mg/kg. These doses were adapted from ones commonly used in mammalian studies[5] and may not be appropriate for all species. It has been suggested that high doses of naloxone may have nonspecific actions.[44]

As noted before, snails, the earliest evolutionary representative species examined, displayed a marked difference in thermoregulatory behavior when given levorphanol or its enantiomorph, dextrophan. This implies that the receptor reacting with the administered opiate has stereospecific requirements. Such diverse thermoregulatory effects to these stereochemically related isomers are found in rats in which levorphanol produces hyperthermia, whereas dextrophan is ineffective.[45] It would be of interest to determine whether the intermediate species show a continuity of this requirement for stereospecificity. In rats, cyclazocine can cause body temperature changes that resemble those produced by morphine.[46]

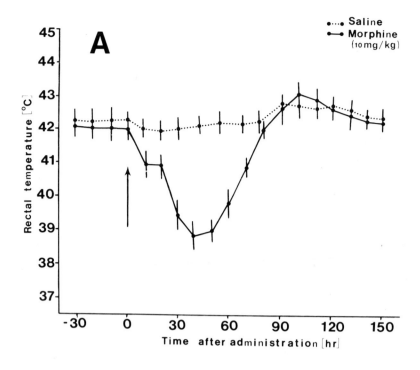

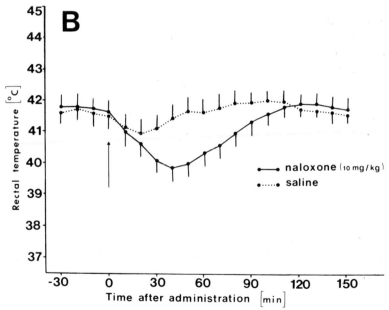

FIGURE 10.   Effects of peripheral injections of (A) morphine sulfate (10 mg/kg) and (B) naloxone hydrochloride (10 mg/kg) on the mid-photophase body temperatures of individual Japanese quail. Saline (0.50 mℓ)-injected birds were used as controls. Vertical lines denote two standard errors.

This was not found in the snail where cyclazocine had minimal effect. While this may indicate that kappa-opiate receptors may not be present or important in thermoregulatory processes in the snail, further investigations would be required before any definitive conclusions could be made.

A further finding of possible evolutionary significance resulted from experiments in which β-endorphin and Met-enkephalin were injected. In snails and insect larvae, Met-enkephalin was significantly more potent than the larger opioid peptide, β-endorphin. In the goldfish, both substances were effective in producing behavioral hyperthermia and were of relatively comparable potency. In mammals, Met-enkephalin can produce hyperthermic changes, as can β-endorphin,[14] but the former is much less potent. Met-enkephalin is metabolized rapidly by enzymes on neural tissue in the mammalian central nervous system.[47] It would be of interest to examine whether similar or altered degradation is present in invertebrates and lower vertebrates. It should also be noted that the majority of investigations of opioid influences on mammalian thermoregulation have been conducted with restrained animals held at a constant temperature.[14] The lack of ability to both behaviorally and physiologically thermoregulate has been shown to influence the responses obtained that are recorded.[14,42,48] This effect of experimental conditions and ambient temperature is borne out in the present studies with goldfish, where animals held at a constant temperature showed a dose-dependent increase in activity following β-endorphin administration, while fish tested in the thermal gradients showed no changes in activity after injection.

Various models have been proposed for mammalian thermoregulation.[20,49] The simplest may involve ionic — calcium and sodium — influences on thermal set point(s) and sensors.[49,50] It is known that morphine and levorphanol alter calcium binding to membranes isolated from rat brain,[51,52] which could be responsible for changes in calcium fluxes in vivo. In mammals, intracerebroventricular injection of calcium ions causes hypothermia.[50] The opposite action of the hyperthermic response is produced by low doses of opiate agonists.[14] While speculative, it is possible that the underlying process as through which endogenous opiate systems influence thermoregulation involve calcium-dependent changes in sensory and reactive systems. Calcium-ionic mechanisms display a very early evolutionary function and it is conceivable that they may form the basis of a thermoregulatory mechanism that has become more refined and intricate through the phylogenetic corridor.

REFERENCES

1. **Snyder, S. H.**, Brain peptides as neurotransmitters, *Science*, 209, 976, 1980.
2. **Henry, J. L.**, Circulating opioids: possible physiological roles in central nervous function, *Neurosci. Biobehav. Rev.*, 6, 229, 1982.
3. **Amir, S., Brown, Z. W., and Amit, A.**, The role of endorphins in stress: evidence and speculations, *Neurosci. Biobehav. Rev.*, 4, 77, 1980.
4. **Terenius, L.**, Endorphins and modulation of pain, *Adv. Neurol.*, 33, 59, 1982.
5. **Clark, W. G.**, Effects of opioid peptides on thermoregulation, *Fed. Proc. Fed. Am. Soc. Exp. Biol.*, 40, 2754, 1981.
6. **Pert, C. B., Aposhian, D., and Snyder, S.**, Phylogenetic distribution of opiate receptor binding, *Brain Res.*, 75, 356, 1974.
7. **Leung, M. K. and Stefano, G. B.**, Isolation and identification of enkephalins in pedal ganglia of *Mytilus edulis* (Mollusca), *Proc. Natl. Acad. Sci. U.S.A.*, 81, 955, 1984.
8. **Stefano, G. B., Scharrer, B., and Assanah, P.**, Demonstration, characterization and localization of opioid binding sites in the midgut of the insect *Leucophaea maderae* (Blattaria), *Brain Res.*, 253, 205, 1982.
9. **Roth, J., LeRoith, D., Shiloach, J., Rosenzweig, J. L., Lesniak, M. A., and Hawankova, A.**, The evolutionary origins of hormones, neurotransmitters, and other extracellular chemical messengers, *N. Engl. J. Med.*, 306, 523, 1982.

10. **Josefsson, J.-O. and Johansson, P.,** Naloxone-reversible effect of opioids on pinocytosis in *Amoeba proteus, Nature (London),* 282, 78, 1979.
11. **Stefano, G. B.,** Comparative aspects of opioid-dopamine interaction, *Cell Mol. Neurobiol.,* 2, 167, 1982.
12. **Kavaliers, M., Hirst, M., and Teskey, G. C.,** Opioid-induced feeding in the slug, *Limax maximus, Physiol. Behav.,* 33, 765, 1984.
13. **Crawshaw, L. I., Moffitt, B. P., Lemons, D. E., and Downey, J. A.,** The evolutionary development of vertebrate thermoregulation, *Am. Sci.,* 69, 543, 1981.
14. **Clark, W. G. and Lipton, J. M.,** Brain and pituitary peptides in thermoregulation, *Pharmacol. Ther.,* 22, 249, 1983.
15. **Thornhill, J. A., Hirst, M., and Gowdey, C. W.,** Changes in the hypothermic response of rats to daily injections of morphine and the antagonism of the acute response by naloxone, *Can. J. Physiol. Pharmacol.,* 56, 483, 1978.
16. **Yehuda, S. and Kastin, A. J.,** Peptides and thermoregulation, *Neurosci. Biobehav. Rev.,* 4, 459, 1980.
17. **Reynolds, W. W. and Casterlin, M. E.,** Behavioral thermoregulation and the ''final preferendum'' paradigm, *Am. Zool.,* 19, 211, 1979.
18. **Crawshaw, L. I.,** Temperature regulation in vertebrates, *Ann. Rev. Physiol.,* 42, 473, 1980.
19. **Hammel, H. T.,** Phylogeny of regulatory mechanisms in temperature regulation, *J. Therm. Biol.,* 8, 37, 1983.
20. **Satinoff, E.,** Neural organization and evolution of thermal regulation in mammals, *Science,* 201, 16, 1978.
21. **Jones, J. S., Leith, B. H., and Rawlings, P.,** Polymorphism in *Cepaea:* a problem with too many solutions?, *Annu. Rev. Ecol. Syst.,* 8, 109, 1977.
22. **Kavaliers, M., Hirst, M., and Teskey, G. C.,** A functional role for an opiate system in snail thermal behavior, *Science,* 330, 99, 1983.
23. **Kavaliers, M. and Hirst, M.,** Tolerance to the morphine-influenced thermal response in the terrestrial snail, *Cepaea nemoralis, Neuropharmacology,* 22, 1321, 1983.
24. **Hirst, M. and Kavaliers, M.,** in preparation.
25. **Martin, W. R., Eadea, C. G., Thompson, J. A., Huppler, R. E., and Gilbert, P. E.,** The effects of morphine- and nalorphine-like drugs in the non-dependent chronic spinal dog, *J. Pharmacol. Exp. Ther.,* 197, 517, 1976.
26. **Hirst, M. and Kavaliers, M.,** in preparation.
27. **Greenberg, M. J., Painter, S. D., Doble, K. E., Nagle, G. T., Price, D. A., and Lehman, H. K.,** The molluscan neurosecretory peptide FMRDamide: comparative pharmacology and relationship to the enkephalins, *Fed. Proc. Fed. Am. Soc. Exp. Biol.,* 42, 82, 1983.
28. **Kavaliers, M., Hirst, M., and Teskey, G. C.,** The effects of opioid and FMRFamide peptides on thermal behavior in the snail, *Neuropharmacology,* 24, 621, 1985.
29. **Gayton, R. J.,** Mammalian neuronal actions of FMRPamide and the structurally related opioid Met-enkephalin-Arg[6]-Phe[7], *Nature (London),* 298, 275, 1982.
30. **Kavaliers, M. and Hirst, M.,** The presence of an opioid system mediating behavioral thermoregulation in the terrestrial snail, *Cepaea nemoralis, Neuropharmacology,* in press.
31. **Nelson, D. O. and Prosser, C. L.,** Temperature-sensitive neurons in the preoptic region of sunfish, *Am. J. Physiol.,* 241, R259, 1981.
32. **Green, M. D. and Lomax, M.,** Behavioral thermoregulation and neuroamines in fish, *Chromus chromus, J. Therm. Biol.,* 1, 237, 1976.
33. **Kavaliers, M. and Hawkins, M.,** Bombesin alters behavioral thermoregulation in fish, *Neuropharmacology,* 28, 1361, 1981.
34. **Kavaliers, M.,** Pineal mediation of the thermoregulatory and behavioral activating effects of β-endorphin, *Peptides,* 3, 679, 1982.
35. **Kavaliers, M.,** Peptides, the pineal gland and thermoregulation, in *The Pineal and Its Hormones,* Vol. 92, Reiter, R. J., Ed., Alan R. Liss, New York, 1982, 207.
36. **Behler, J. L. and King, F. W.,** *The Audubon Society Field Guide to North American Reptiles and Amphibians,* A. A. Knop, New York, 1979.
37. **Huey, R. B. and Slatkin, M.,** Cost and benefits of lizard thermoregulation, *Q. Rev. Biol.,* 51, 363, 1976.
38. **Kavaliers, M., Courtenay, S., and Hirst, M.,** Opiates influence behavioral thermoregulation in the curly-tailed lizard, *Leiocephalus carinatus, Physiol. Behav.,* 32, 221, 1984.
39. **Thornhill, J. A. and West, N. H.,** Opiate modulation of thermoregulation in adult Pekin ducks, *Can. J. Physiol. Pharmacol.,* 62, 288, 1984.
40. **Panksepp, J., Vilbert, T., Nean, N. J., Coy, H., and Kastin, A. J.,** Reduction of distress vocalization in chicks by opiate-like peptides, *Brain Res. Bull.,* 3, 663, 1978.
41. **Nistico, G., Rotiroti, D., Naccari, F., DeSarro, G. B., and Marho, E.,** Effects of intraventricular β-endorphin and D-Ala-[2]-methionine enkephalinamide on behavior, spectrum power of electrocortical activity and body temperature in chicks, *Res. Commun. Chem. Pathol. Pharmacol.,* 28, 295, 1980.

42. **Rezvani, A. H., Gordon, C. J., and Heath, J. E.,** Action of preoptic injections of β-endorphin on temperature regulation in rabbits, *Am. J. Physiol.,* 243, R104, 1982.

43. **Rezvani, A. H. and Heath, J. E.,** Reduced thermal sensitivity in the rabbit by β-endorphin injection into the preoptic/anterior hypothalamus, *Brain Res.,* 292, 297, 1984.

44. **Sawynok, J., Pinsky, C., and Labella, F. S.,** Mini review on the specificity of naloxone as an opiate antagonist, *Life Sci.,* 25, 1621, 1979.

45. **Thronhill, J. A., Hirst, M., and Gowdey, C. W.,** Changes in core temperature and feeding in rats by levorphanol and dextrophan, *Can. J. Physiol. Pharmacol.,* 57, 1028, 1979.

46. **Thornhill, J. A.,** Disruption of Core Temperature and Feeding Responses in Rats by Narcotic Analgesics and Other Centrally Acting Drugs, Ph.D. Thesis, University of Western Ontario, Canada, 1978.

47. **Schwartz, J. C. and Malfroy, B.,** Biological inactivation of enkephalins and the role of enkephalin-dipeptidyl-carboxypeptidase (''enkephalinase'') as neuropeptides, *Life Sci.,* 29, 17(5), 1981.

48. **McDougall, J. N., Marques, P. R., and Burks, F. I.,** Restraint alters the thermic response to morphine by postural interference, *Pharmacol. Biochem. Behav.,* 18, 495, 1983.

49. **Myers, R. D.,** Hypothalamic control of thermoregulation — neurochemical mechanisms, in *Handbook of the Hypothalamus,* Morgane, P. J. and Panksepp, J., Eds., Marcel Dekker, New York, 1980, 83.

50. **Myers, R. D. and Veale, W. L.,** The role of sodium and calcium ions in the hypothalamus in the control of body temperature of the unanaesthetized cat, *J. Physiol.,* 212, 411, 1971.

51. **Ross, D. H., Medina, M. A., and Cardenus, H. L.,** Morphine and ethanol: selective depletion of regional brain calcium, *Science,* 186, 63, 1974.

52. **Werz, M. A. and Macdonald, R. L.,** Opioid peptides selective for mu- and delta-opiate receptors induce calcium-dependent action potential duration by increasing potassium conductance, *Neurosci. Lett.,* 42, 173, 1983.

Immunocytochemistry

IMMUNOHISTOLOGICAL DETECTION OF MET-ENKEPHALIN-LIKE NEUROPEPTIDE IN THE BRAINS OF THE NEMERTEANS (TRIPLOBLASTIC ACOELOMATE INVERTEBRATES)

Christian Rémy and Danièle Brossard

SUMMARY

This immunohistological study was performed with an anti-Met-enkephalin antiserum in two different species of nemerteans: *Lineus viridis* and *Lineus ruber*. The cephalic nervous system of these two worms is made up of two dorsal and two ventral ganglia. In each dorsal ganglion, from 80 to 100 immunoreactive cells were located in the rostral half, and a group of about 35 immunoreactive perikarya form a cell cluster in the caudal half. In the ventral ganglia, the cells synthetizing Met-enkephalin-like material are rather scarce. All these immunoreactive cells are paraldehyde fuchsin negative. Yet, they are located in the immediate vicinity of type A and type B fuchsinophil neurosecretory cells. We were not able, however, to demonstrate interrelations between these two cell categories as clearly as in the insect *Locusta*. Since control tests were conclusive, we were thus able to demonstrate that substances related to morphinomimetic polypeptides, previously found in the central nervous system of higher invertebrates, were also present in the brains of lower invertebrates. In the latter, which often exhibit a remarkable ability to regenerate, such substances are synthesized in a much greater number of cells than in higher invertebrates which are not capable of regeneration such as insects, molluscs, and achaetous annelids. To our knowledge, nemerteans are the most primitive group of invertebrates in which a peptide related to Met-enkephalin could be detected. Morphinomimetic neuropeptides, which were at first thought to be characteristic of vertebrates, seem therefore to be largely distributed in the animal kingdom.

INTRODUCTION

The presence of invertebrate neurosecretory products related to vertebrate neuropeptides was first demonstrated in 1975 with radioimmunoassays and in 1977 with immunohistological techniques in the central nervous systems of various gastropods,[1] of an oligochaetous annelid,[2] and of an insect.[3] Since these pioneer works, similar relationships concerning practically all the vertebrate neuropeptides known to date have been observed in many invertebrate classes, including the lowest ones such as coelenterates.[4,5] The presence of materials similar to corticotropin and β-endorphin was even detected in unicellular organisms.[6]

Vertebrate morphinomimetic neuropeptide-like substanced have been identified mainly in the central nervous systems of the higher invertebrates: insects,[7-12] molluscs,[13-16] and annelids.[17-20] The presence of endogenous opiates in invertebrates has been clearly established. In the triploblastic acoelomate invertebrates, i.e., platyhelminths, nemathelminths, and nemerteans, immunological relationships of neuropeptides have not been much studied. The only results obtained concern on the one hand neurosecretory products related to somatostatin and corticotropin and on the other hand an endorphin-like neuropeptide, detected respectively in planarians[21] and in a parasitic nematode.[22]

The neurosecretory system in nemerteans has only been studied with classical staining methods[23,25] and no immunohistological investigation had been undertaken. We therefore carried out an immunohistological study of the cephalic nervous systems of two heteronemerteans: *Lineus viridis* and *Lineus ruber* using an anti-methionine (anti-Met-) enkephalin antiserum. The lowest invertebrates in which a Met-enkephalin-like neuropeptide had been detected were the oligochaetous[18] and the achaetous[19,20] annelids. We therefore thought it relevant to try and find a similar neuropeptide in worms which are more primitive animals than annelids.

MATERIALS AND METHODS

Animal Material

Adult *Lineus* were collected in August on the beach at the marine institute of Roscoff (France) and bred in separate tanks in a dark laboratory and kept at the temperature of 12°C. Seven *L. viridis* and six *L. ruber* were anesthetized (MS 222) and the anterior ends of the worms were cut and fixed for 8 hr in Bouin Hollande fixative containing no acetic acid to which was added a saturated mercuric chloride solution (10% v/v). These anterior ends, including brains, were washed overnight in running tap water, dehydrated, and embedded in paraffin. Serial sections 7 μm thick were cut in transverse or longitudinal planes.

Antisera

The anti-Met-enkephalin and anti-α-endorphin antisera that we used for our study had already enabled us to obtain positive immunohistological results in the central nervous systems of insects[7,8] and annelids.[17] These two antisera were the kind gifts of Dr. M. P. Dubois, Directeur de Recherches, INRA, Station de Physiologie de la Reproduction, Nouzilly 37380, France. They were obtained by immunizing rabbits with synthetic Met-enkephalin or α-endorphin conjugated to human serum albumin (HSA). The details of the preparation of the immunogens, the different stages of immunization, and the characteristics of the antisera have been described elsewhere.[8,26]

Immunoreaction

The indirect immunofluorescence technique was used.[27] Anti-Met-enkephalin and anti-α-endorphin antisera diluted (1:100) in veronal buffer + 2‰ HSA were exposed to dehydrated sections for 12 hr in moist atmosphere at 6°C. After washing, fluorescein isothiocyanate-conjugated sheep antirabbit immunoglobulins were used to determine the fixation of each antiserum on sites containing the specific antigen. Evans blue (0.1%) was used as a counterstain. Slides were scanned using a fluorescence microscope (Leitz Ortholux®, lamp HBO 200, primary filters BG12, secondary filter K 530) and fluorescent cells photographed by Ektachrome® film.

Control tests included omission of fluorescent sheep antirabbit immunoglobulins, replacement of the specific antiserum by normal serum, and the use of antigen-saturated antiserum (500 μg of specific antigen per milliliter of undiluted antiserum).

Staining

Some photographed sections with immunoreactive cells were washed and stained with paraldehyde fuchsin after oxidation.

RESULTS

Our investigations were merely carried out in the cephalic nervous system or brain. In *Lineus* the brain is made up of four gnaglia: on each side, a dorsal and a ventral ganglia partially fused (the dorsal ganglion being bigger than the ventral one.) The two dorsal ganglia and the two ventral ganglia are, respectively, united by a dorsal commissure and a ventral commissure. The brain thus forms a ring round the rhynchocoel (Figure 1).

Lineus viridis

Met-enkephalin immunoreactivity was found in a great number of cell bodies (about 80 on each side) mainly located in the dorsal ganglia. In the rostral half of each dorsal ganglion,

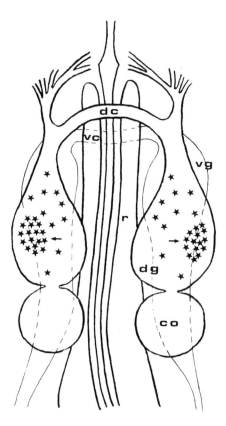

FIGURE 1. Stereogram of *Lineus* brain showing immunoreactive perikarya (★) in dorsal ganglia: disseminated cell bodies in the rostral halves, cell clusters (arrows) in the caudal halves. co, Cerebral organ; dc, dorsal commissure; dg, dorsal ganglion; r, rhynchocoel; vc, ventral commissure; vg, ventral ganglion.

approximately 40 immunoreactive perikarya are disseminated throughout the cellular area surrounding the neuropil (from 1 to 4 perikarya werc visible on each side in some of the transversal sections). These perikarya are fusiform (7 × 3 μm) or round-shaped (5 μm). In two specific section planes located in the posterior part of this rostral half, we were able to observe in the two dorsal ganglia of several *L. viridis* couples of perfectly symmetrical pear-shaped cell bodies of fairly large size (10 × 5 and 20 × 10 μm) (Figures 2 and 3). The caudal half of each dorsal ganglion is characterized by the presence, in its anterior part, of a group of fusiform (7 × 3 μm) or round (5 μm) immunoreactive perikarya forming a clear-cut highly fluorescent cell cluster (Figure 4). The right and left clusters are symmetrical, positioned laterally, close to the neuropils, and made up of about 35 cell bodies each. Behind the clusters, the anti-Met-enkephalin antiserum crossreacts with about ten fusiform or round perikarya dissemianted as far as the back sections adjacent to the cerebral organs.

In the ventral ganglia, cell bodies with Met-enkephalin-like material are sparse and disseminated in the anterior part of the ganglion. Their number does not exceed ten on each side.

Lineus ruber

The results obtained were quite comparable to those described in *L. viridis*. Immunoreactive cells were found in the dorsal ganglia mainly. Their number, however, is more important than in *L. viridis*, since there are about 100 of them in each ganglion. This

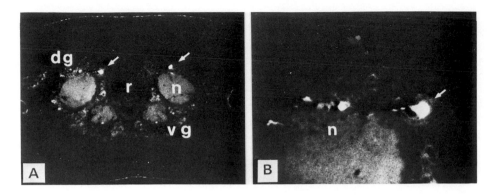

FIGURE 2.   *Lineus viridis.* (A) Transverse section in the rostral half of the brain. Met-enkephalin-like material in some cell bodies and in a couple of symmetrical perikarya (arrows) close to the neuropils (n) or dorsal ganglia (dg). r, Rhynchocoel; vg, ventral ganglion. (Magnification × 65.) (B) High magnification of (A) (left side). (Magnification × 270.)

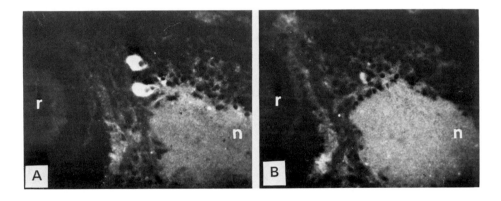

FIGURE 3.   *Lineus viridis.* Transverse sections in the rostral half of a dorsal ganglion. n, Neuropil; r, rhynchocoel. (A) Two large pear-shaped Met-enkephalin immunoreactive peikarya. (Magnification × 270.) (B) Adjacent section to (A) treated with antigen-inactiveted Met-enkephalin antiserum. No reaction at level of perikarya. (Magnification × 270.)

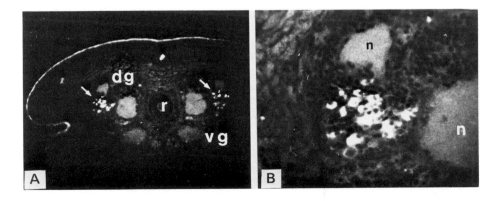

FIGURE 4..   *Lineus viridis.* (A) Transverse section in the caudal half of the brain. Left and right immunoreactive cell clusters (arrows) are clearly visible in dorsal ganglia (dg). r, Rhynchocoel; vg, ventral ganglion. (Magnification × 65.) (B) High magnification of (A) (left side). n, Neuropil. (Magnification × 270.)

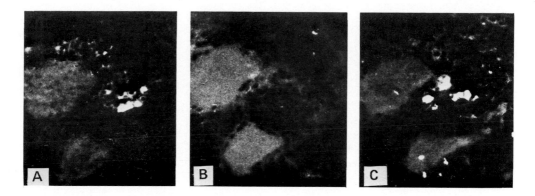

FIGURE 5. *Lineus ruber.* Three adjacent transverse sections of same immunoreactive cell cluster. (A,C) First and last sections treated with Met-enkephalin-antiserum. (Magnification × 270.) (B) Middle section treated with antigen-inactivated Met-enkephalin antiserum. No reaction at level of perikarya. (Magnification × 270.)

difference is mainly due to the fact that a greater number of isolated cells are distributed throughout the rostral half of both dorsal ganglia (approximately 60 perikarya). These cell bodies are quite comparable to those described in *L. viridis,* their size and shape being similar. Yet, we were not able to observe in the six animals studied, the large symmetrical pear-shaped cells characteristic of *L. viridis.* In the caudal half, two clusters of highly immunoreactive perikarya are also present (Figure 5). They are rigorously similar to the clusters described in *L. viridis,* their identical cell bodies have similar location, shapes, sizes, and numbers. Some isolated cells are visible behind the clusters, where cerebral organs start. The ventral ganglia are poor in immunoreactive perikarya.

In *L. viridis* and *L. ruber* numerous processes in the neuropil contained immunoreactive material. In a few favorable cases, we were able to observe in the same section, fluorescent perikarya and their axons penetrating into the neuropil. In the various immunoreactive cells of the anterior part of the brain, fluorescence intensity varies from one cell to the next. In the cell clusters, however, all the cells seem to contain significant amount of Met-enkephalin-like material since they are all highly fluorescent (Figures 4B, 5A, 5C, 6A, and 5C).

No evidence was found for the existence of any neuropeptide related to Met-enkephalin in the cerebral organs.

In all of the animals examined, control tests carried out in the isolated cells and the cellular clusters were all conclusive. In particular, the Met-enkephalin immunoreactive material in cell bodies and processes was not visible in adjacent sections treated with antigen-inactivated antiserum (Figure 5). The control tests were easily performed in cell clusters (visible in five or six adjacent sections) and in the bigger isolated cells of *L. viridis* (each perikaryon being visible in two or three consecutive sections). We have also completed these tests by using, on every other section, anti-Met-enkephalin antiserum and anti-α-endorphin antiserum, mainly at the level of the cell clusters. None of the Met-enkephalin immunoreactive perikarya described above showed immunoreactivity against the anti-α-endorphin antiserum (Figure 6).

After being photographed, many brain sections containing Met-enkephalin immunoreactive cell bodies were washed and stained with paraldehyde fuchsin (PF). These immunoreactive cells are PF negative. However, the use of this stain has enabled us to observe that the perikarya or processes synthetizing Met-enkephalin-like neuropeptide can always be found very close to PF-positive neurosecretory perikarya or processes (Figures 7 and 8).

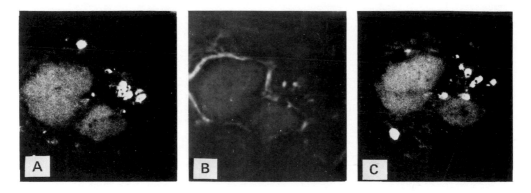

FIGURE 6.   *Lineus viridis.* Three adjacent transverse sections of same immunoreactive cell cluster. (A,C) First and last sections treated with Met-enkephalin-antiserum. (Magnification × 270.) (B) Middle section treated with anti-α-endorphin antiserum. No reaction at level of perikarya. (Magnification × 270.)

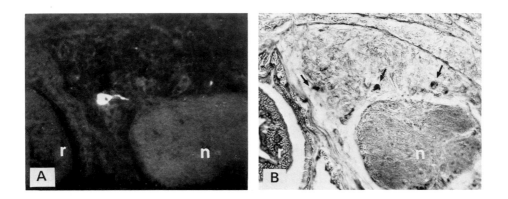

FIGURE 7.   *Lineus viridis.* (A) Transverse section in the rostral half of a dorsal ganglion. Met-enkephalin immunoreactive perikaryon close to the neuropil (n). (Magnification × 270.) (B) Same section as in (A), rinsed after examination of fluorescence and stained with paraldehyde fuchsin (PF). The immunoreactive perikaryon is PF negative. Note the presence of three PF-positive perikarya (arrows) in the immediate vicinity. n, Neuropil. (Magnification × 270.)

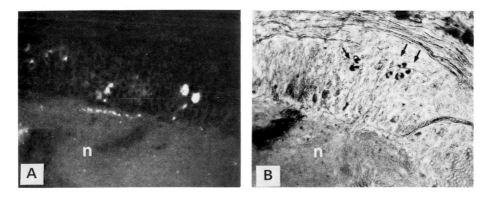

FIGURE 8.   *Lineus viridis.* (A) Longitudinal section in the rostral half of a dorsal ganglion. Met-enkephalin immunoreactive cell bodies and processes close to the neuropil (n). (Magnification × 270.) (B) Same section as in (A), rinsed and stained with PF. Immunoreactive cell bodies and processes are PF negative. Sic PF-positive perikarya (arrows) are visible in the immediate vicinity. n, Neuropil. (Magnification × 270.)

DISCUSSION

Unlike what Simantov et al.[28] asserted, enkephalins do not appear to be typical vertebrate opioid peptides. The numerous results which have been obtained since 1978[7] provide sufficient evidence that morphinomimetic peptides, and more particularly enkephalins, are present in the nervous systems of invertebrates.

Enkephalin-like neuropeptides have been mainly detected in insects, molluscs, and annelids. The immunohistological investigations that were carried out in *Lineus* with anti-Met-enkephalin antiserum show that similar neuropeptides are also present in the central nervous systems of lower invertebrates, since nemerteans are phylogenetically more primitive animals than annelids. Such results can be compared with those obtained in a nemathelminth,[22] where a polypeptide related to α-endorphin was detected in the digestive and genital tracts.

Met-enkephalin immunoreactive cells of *Lineus* are mainly located in the immediate vicinity of type A and type B fuchsinophil neurosecretory cells which have been studied for years.[23-25] This is reminiscent of the neurosecretory pattern observed in the brains of migratory locusts. In the brain of *Locusta*, four perikarya synthetize a Met-enkephalin-like material and give rise to a network of ramifications surrounding the main areas throught which positive PF neurosecretory products pass.[8] In *Locusta*, these four cells could act as neuromodulators of the brain neurosecretory activity. In *Lineus*, a similar activity could also be considered. Yet, the interrelations between both categories of cells are less clearly visible than in locusts.

In the brains of *L. viridis* and *L. ruber*, anti-Met-enkephalin antiserum allows one to detect a great number of immunoreactive cells. The results are clearly different from those obtained with the same antiserum in the higher invertebrates in which the immunoreactive cells are much scarcer: 4 cells[8] or 4 groups from 2 to 9 cells[12] in insects; 2 groups from 10 to 15 cells[14] in gastropods; 1 cell in each ganglion[19] or sparse cells[20] in achaetous annelids. Oligochaetous annelids are the only higher invertebrates to have numerous neurons with Met-enkephalin-like material. Leaving aside polychaetes about which nothing has been published so far, oligochaetes are the only ones among annelids, molluscs, or insects to exhibit a remarkable ability for regeneration. Just like oligochaetes, Lineidae are also able to regenerate. It is at present too soon to draw reliable conclusions concerning this field since, to our knowledge the presence of endogenous opiates has not yet been detected in invertebrates capable of regeneration such as diploblastics or echinoderms. In a recent paper on vertebrates, however, mention is made of a higher plasma level of endogenous opiates (β-endorphin) in animals capable of limb regeneration such as newts.[29]

In nemerteans, neurohormones are to date still not very well known. Only the existence of a brain neurosecretory product that controls reproduction (inhibiting action) has been demonstrated in these worms.[30-33] However, the type of neurosecretory cells synthetizing this gonad-inhibiting hormone has not been identified. Our results provide further evidence that these animals are capable of secreting neuropeptides. thanks to immunohistology, neurons containing Met-enkephalin-like material can be located in a very precise way. We will therefore be able to implement a research program in order to study more specifically the regeneration process of the central nervous system and the possible part played by endogenous opiates in this regeneration process. It has been observed that *Lineus* exhibitsd an ability rather unusual in metozoa to regenerate a complete brain following excision of either the right or left half of this organ. If regeneration of neurosecretory cells occurs, and more particularly that of the cells synthetizing Met-enkephalin-like neuropeptide, we will be able to observe the different stages of their differentiation. It will also be of interest to study the evolution of these cells in the remaining part of the brain after unilateral excision, or in transplants after grafting additional brains. From the phylogenetic point of view, our immunohistological investigations will be furthered with other antineuropeptide antisera in *Lineus* (Anopla) and extended to Enopla, the other class of the phylum.

ACKNOWLEDGMENT

We want to thank Dr. M. P. Dubois for kind gifts of Met-enkephalin and α-endorphin antisera.

REFERENCES

1. **Grimm-Jorgensen, Y., McKelvy, J., and Jackson, I. M. D.**, Immunoreactive thyrotrophin releasing factor in gastropid circumoesophageal ganglia, *Nature (London)*, 254, 620, 1975.
2. **Sundler, F., Håkanson, R., Alumets, J., and Walles, B.**, Neuronal localization of pancreatic polypeptide (PP) and vasoactive intestinal peptide (PP) and vasoactive intestinal peptide (VIP) immunoreactivity in the earthworm *(Lumbricus terrestris)*, *Brain Res. Bull.*, 2, 61, 1977.
3. **Rémy, C., Girardie, J., and Dubois, M. P.**, exploration immunocytologique des ganglions cérébroïdes et sous oesophagiens du phasme *Clitumnus extradentatus*: existence d'une neurosécrétion apparentée à la vasopressine-neurophysine, *C.R. Acad. Sci. Paris*, 285, 1495, 1977.
4. **Grimmelikhuijzen, C. J. P., Dockray, G. J., and Yanaihara, N.**, Bombesin-like immunoreactivity in the nervous system of *Hydra*, *Histochemistry*, 73, 171, 1981.
5. **Grimmelikhuijzen, C. J. P.**, Coexistence of neuropeptides in *Hydra*, *Neuroscience*, 9, 837, 1983.
6. **Le Roith, D., Liotta, A. S., Roth, J., Shiloach, J. Lewis, M. E., Pert, C. B., and Krieger, D. T.**, Corticotropin and β-endorphin-like materials are native to unicellular organisms, *Proc. Natl. Acad. Sci. U.S.A.*, 79, 2086, 1982.
7. **Rémy, C., Girardie, J., and Dubois, M. P.**, Présence dans le ganglion sous-oesophagien de la chenille processionnaire du pin *(Thaumetopoae pityocampa)* de cellules révélées en immunofluorescnece par un anticorps anti-α-endorphine, *C. R. Acad. Sci. Paris*, 286, 651, 1978.
8. **Rémy, C. and Dubois, M. P.**, Immunohistological evidence of methionine enkephalin-like material in the brain of the migratory locust, *Cell Tissue Res.*, 218, 271, 1981.
9. **Stefano, G. B. and Scharrer, B.**, High affinity binding of an enkephalin analog in the cerebral ganglion of the insect *Leucophaea maderae* (Blattaria), *Brain Res.*, 225, 107, 1981.
10. **Hansen, B. L., Hansen, G. N., and Scharrer, B.**, Immunoreactive material resembling vertebrate neuropeptides in the corpus cardiacum and corpus allatum of the insect *Leucophaea maderae*, *Cell Tissue Res.*, 225, 319, 1982.
11. **Duve, H. and Thorpe, A.**, Immunocytochemical identification of α-endorphin-like materials in neurons of the brain and corpus cardiacum of the blowfly, *Calliphora vomitoria* (Diptera), *Cell Tissue Res.*, 233, 415, 1983.
12. **El Salhy, M., Falkmer, S., Kramer, K. J., and Speirs, R. D.**, Immunohistochemical investigations of neuropeptides in the brain, corpora cariaca, and corpora allata of an adult lepidopteran insect, *Manduca sexta*, *Cell Tissue Res.*, 232, 295, 1983.
13. **Martin, R., Frosch, D., Weber, E., and Voigt, K.-H.**, Met-enkephalin-like immunoreactivity in a cephalopod neurohemal organ, *Neurosc. Lett.*, 15, 253, 1979.
14. **Schot, L. P. C., Boer, H. H., Swaab, D. F., and Van Noorden, S.**, Immunocytochemical demonstration of peptidergic neurons in the central nervous system of the pond snail *Lymnaea stagnalis* with antisera raised to biologically active peptides of vertebrates, *Cell Tissue Res.*, 216, 273, 1981.
15. **Stefano, G. B. and Leung, M.**, Purification of opioid peptides from molluscan ganglia, *Cell Mol. Neurobiol.*, 2, 347, 1982.
16. **Stefano, G. B. and Martin, R.**, Enkephalin-like immunoreactivity in the pedal ganglion of *Mytilus edulis* (Bivalvia) and its proximity to dopamine-containing structures, *Cell Tissue Res.*, 230, 147, 1983.
17. **Rémy, C. and Dubois, M. P.**, Localisation par immunofluorescence de peptides analogues à l'α-endorphine dans les ganglions infra-oesophagiens du lombricidé *Dendrobaena subrubicunda* Eisen, *Experientia*, 35, 137, 1979.
18. **Alumets, J., Håkanson, R., Sundler, F., and Thorell, J.**, Neuronal localisation of immunoreactive enkephalin and α-endorphin in the earthworm, *Nature (London)*, 279, 805, 1979.
19. **Zipser, B.**, Identification of specific leech neurones immunoreactive to enkephalin, *Nature (London)*, 283, 857, 1980.
20. **Osborne, N. N., Patel, S., and Dockray, G.**, Immunohistochemical demonstration of peptides, serotonin and dopamine-β-hydroxylase-like material in the nervous system of the leech *Hirudo medicinalis*, *Histochemistry*, 75, 573, 1982.

21. **Schilt, J., Richoux, J. P., and Dubois, M. P.,** Demonstration of peptides immunologically related to vertebrate neurohormones in *Dugesia lugubris* (Turbellaria, Tricladida), *Gen. Comp. Endocrinol.*, 43, 331, 1981.
22. **Kerboeuf, D. and Dubois, M. P.,** Mise en évidence chez un Nématode parasite de Vertébrés *(Heligmosomoides polygyrus)* de cellules révélées en immunofluorescence par un anticorps anti-α endorphine, *C. R. Acad. Sci. Paris*, 293, 675, 1981.
23. **Lechenault, H.,** Sur l'existence de cellules neurosécrétrices dans les ganglions cérébroïdes des Lineidae (Hétéronémertes), *C. R. Acad. Sci. Paris*, 255, 194, 1962.
24. **Servettaz, F. and Gontcharoff, M.,** Etude cytochimique des cellules présumées neurosécrétrices dans le système nerveux central des Hétéronémertes Lineidae, *Gen. Comp. Endocrinol.*, 30, 285, 1976.
25. **Ferraris, J. D.,** Neurosecretion in selected Nemertina. A histological study, *Zoomorphology*, 91, 275, 1978.
26. **Rémy, C., Girardie, J., and Dubois, M. P.,** Vertebrate neuropeptide-like substances in the suboesophageal ganglion of two insects: *Locusta migratoria* (Orthoptera) and *Bombyx mori* (Lepidoptera). Immunocytological investigation, *Gen. Comp. Endocrinol.*, 37, 93, 1979.
27. **Coons, A. H., Leduc, E. J., and Connoly, J. M.,** Studies on antibody production. I. A method for the histochemical demonstration of specific antibody and its application to the study of the hyperimmune rabbit, *J. Exp. Med.*, 102, 49, 1955.
28. **Simantov, R., Goodman, R., Aposhian, D., and Snyder, S. h.,** Phylogenetic distribution of a morphine-like peptide "enkephalin", *Brain Res.*, 111, 204, 1979.
29. **Vethamany-Globus, S., Globus, M., and Milton, G.,** A comparison of β-endorphin levels in regenerating and nonregenerating vertebrates, *J. Exp. Zool.*, 227, 475, 1983.
30. **Bierne, J.,** Maturation sexuelle anticipée par décapitation de la femelle chez l'Hétéronémerte *Lineus ruber* Müller, *C.R. Acad. Sci., Paris*, 259, 4841, 1964.
31. **Bierne, J.,** Localisation dans les ganglions cérébroïdes du centre régulateur de la maturation sexuelle chez la femelle de *Lineus ruber* Müller (Hétéronémertes), *C.R. Acad. Sci. Paris*, 262, 1572, 1966.
32. **Gierne, J.,** Recherches sur la différenciation sexuelle au cours de l'ontogenèse et la régénération chez le Némertien *Lineus ruber* Müller, *Ann. Sci. Nat. Zool. Biol. Anim.*, 12, 181, 1970.
33. **Rué, G. and Bierne, J.,** Synthèse de RNAs et de protéines dans les oocytes d'Hoplonémertes soumis et soustraits à l'influence de la gonadostatine (GIH), *J. Physiol. Paris*, 78, 579, 1982.

OPIOID-PEPTIDE AND SUBSTANCE P IMMUNOREACTIVITY IN CYTOLOGICAL PREPARATIONS AND TISSUE HOMOGENATES OF THE LEECH

Tom Flanagan and Birgit Zipser

SUMMARY

Evidence compiled in this monograph supports the belief that some neuropeptide families are broadly distributed across vertebrate and invertebrate species. The interest in exploiting an invertebrate preparation to model the biology of families of phylogenetically homologous peptides lies in the potential to provide precise analysis of cellular interactions. Such precision may be uniquely available in invertebrate preparations. The central nervous system (CNS) of the leech is particularly well suited to allow identification of classes of homologous neurons and to allow physiological characterizations of interactions between sets of identifiable cells. A brief outline of the organization of the leech nervous system highlights its utility as a model system.

THE LEECH AS A MODEL SYSTEM

Segmentally Repeated Neuronal Networks

The leech CNS is composed of 32 embryonically homologous segmental ganglia; 4 fuse to form a cephalic "brain" (the subesophageal ganglion), 7 fuse to form a tail "brain", and 21 persist as unfused segmental ganglia.[1] The embryonic homology of ganglia facilitates identifying homologous populations of cells in segmental ganglia along the ventral nerve cord. Leech neurons are relatively large (~30 to 70 μm), occupy characteristic positions within each segmental ganglia, and offer excellent electrophysiological accessibility (thus, facilitating identification of neuron types). Each segmental ganglia contains a relatively small and numerically constant population of neurons (typically 400),[2] and most leech neurons have been individually identified using physiological, electrophysiological, or anatomical methods. Neurons in the segmental ganglia are usually bilaterally paired, effect reflexive circuits within the body segment, and constitute a core population common to all segmental ganglia. Neuronal networks between members of cells within the core population are represented in each segmental ganglion, and the resolution of synaptic connectivity has been clear enough to allow for quantal analysis of chemically transmitting synapses.[3] The cellular organization of the segmental ganglion of the leech readily allows for identification of individual cells, mapping of neuronal networks, analysis of synaptic transmission, and comparisons among homologous cell types.

Specialized Cells and Transmitter Systems

In addition to a homologous core population, selected segmental ganglia contain sets of specialized neurons.[4] These neurons can be viewed as higher-order neurons which typically play a role in effecting neuronal output coordinated across a number of segmental ganglia. Several function to regulate heartbeat,[5] swimming,[6,7] and feeding behavior programs.[8] Some of these specialized cells may qualify as command neurons which modulate circuits. Understanding the biology of rare cell types benefits greatly from knowing the transmitters that they use. Leech neurons which synthesize, contain, or sequester acetylcholine,[9,10] dopamine and serotonin,[11,12] and gamma aminobutyric acid[13] have been identified. Leech neurons which do not use these transmitters may use peptidergic neurotransmitters. For this specific reason, several laboratories are energetically searching for potentially peptidergic cells in the leech CNS. In a broader context, it is of interest to identify any features shared among higher-order leech neurons which may allow one to identify relationships between rare cell sets. Rare leech neurons differentiate into sets of cells with mutually exclusive effects upon behavior (i.e., leech do not swim and mate simultaneously), and rules which govern their differentiation and interactions regulate the highest level of organization within the CNS. The origins and lineages of leech neurons can be mapped with embryonic markers,[14] such that studies of the development and differentiation of antigenically homologous cell sets may be done in the leech.[15,16]

Our interests in identifying leech neurons which potentially contain phylogenically conserved peptide transmitters have two objectives: (1) to identify families of neurons which use similar peptidergic transmitters, and (2) to characterize the cellular effects of peptidergic inputs at identified synapses. The leech CNS offers both a privileged cellular perspective for identifying families of cells and excellent electrophysiological accessibility for pharmacological and electrophysiological comparisons of synaptic effects across these cell families. In this report, we will summarize the results of our studies using antisera raised against two vertebrate neuropeptides (leucine enkephalin and substance P) and our colleagues' studies using antisera raised against two invertebrate neuropeptides (molluscan FMRFamide and insect proctolin).

ENKEPHALIN-LIKE IMMUNOREACTIVITY IN THE LEECH

Immunoreactive Soma
Unpaired Ventral Cells

As previously reported,[17] a set of unpaired, 30-μm neurons containing an enkephalin-like antigen has been immunocytologically identified in whole mounts of segmental ganglia of the mud leech *Haemopis marmorata*. This cell set was identified as enkephalin-containing on the basis of two criteria: (1) selective labeling with 4 different polyclonal antisera raised against Leu enkephalin, and (2) blocking of staining with antisera preabsorbed overnight with 20 to 500 μg mℓ$^{-1}$ Leu-enkephalin. These unpaired enkephalin-immunoreactive cells lie in the posteriomedial region of the ventral surface of the segmental ganglia, and represent the most regularly staining enkephalin-like cell type seen in *Haemopis*. Bilateral pairs of concurrently immunoreactive cells were occasionally seen in the subesophageal ganglia and segmental ganglia 5, 6, and 7 of some individual leech, but this infrequent immunoreactivity has not been examined in detail and will not be considered in this report. The unpaired ventromedial cells are confined to segmental ganglia 8 through 21, and their distribution is complementary with unpaired ventromedial serotonin-containing cells (which lie in segmental ganglia 1 through 7).

Dorsolateral Cells

Continued studies of enkephalin-like immunoreactive cells in *Haemopis* have extended earlier findings to include other cell sets lying within the segmental ganglionic series. A summary of immunoreactive cells stained with six different antisera raised against Leu-enkephalin is reported in Table 1. One of three antisera used in initial studies occasionally labeled a pair of dorsolateral cells which appeared in alternating (even-numbered) segmental ganglia. In subsequent studies, two additional antisera also label dorsolateral pairs of enkephalin immunoreactive cells which appear in alternating segmental ganglia (Figure 1A). One of these antisera (Immunonuclear Lot #3538101) strongly labeled only these paired cells. The increased resolution of the paired dorsolateral cells is at least in part due to a change in tissue preparation (i.e., the removal of both the dorsal and ventral connective capsules and extended incubation in concentrated primary antisera); however, the cell-specific expression of enkephalin-like antigens is also probably a reflection of genuine differences between individual leech and between different laboratory populations of leech (see below).

Table 1
FAMILIES OF LEECH NEURONS
IMMUNOCYTOLOGICALLY STAINED WITH A PANEL OF
ENKEPHALIN-DIRECTED ANTIBODIES

Antisera to Leu-enkephalin	Unpaired, ventromedial	Paired, dorsolateral	Paired, ventrolateral
Immunonuclear lot #1	+ + +	(−)	(−)
Immunonuclear lot #2	(−)	+	(−)
Miller/chang #1	(+)	(−)	(−)
Miller/chang #2	(+)	(−)	(−)
Immunonuclear lot #3	(−)	+ + +	(−)
Immunonuclear lot #4	+ + +	(+)	+ + +

Note: Immunoreactive intensity: soma and neurites, + + +; soma and 1° neurites, +; weakly reactive soma, (+); no reaction (−). Whenever any soma are labeled, we see immunoreactive varicosities in the neuropil.

The Nine-Cell Set

Studies in progress are using an antisera which labels as many as nine cells per segmental ganglia (Immunonuclear Lot #8324010). This staining pattern includes cells which appear to correspond with both the unpaired ventromedial cells (prevously seen in ganglia 8 through 21) and with the paired dorsolateral cells (previously seen in alternating ganglia), and additionally stains two ventrolateral and one anterior ventromedial pair of cells. Panels of enkephalin-directed antisera which label overlapping sets of leech neurons suggest that leech contain families of antigenically related, enkephalin-like cells. The most frequently labeled cell sets are interpreted as containing antigens most closely related to enkephalin itself. Less frequently labeled cell sets may contain an enkephalin-like sequence within a larger molecule (possibly a opioid-peptide precursor) not recognized by all antisera. Selective preabsorption studies (using Met-enkephalin, Leu-enkephalin, and extended forms of Leu-enkephalin), or enzymatic treatment of cytological preparations, may allow us to identify subfamilies within this larger cell set.

Variable Immunoreactivity

Immunocytological studies of enkephalin-like antigens in *Haemopis* are confounded by interindividual variations in the extent (although not in the pattern) of somatic immunoreactivity. Immunoreactive variability frequently yields acceptable staining in only one out of four leeches examined. This variability is in contrast with the consistent cytological staining we see with antisera raised against serotonin and with a variety of monoclonal antibodies raised against leech nerve cord homogenates.[18] For this reason, we feel that variable enkephalin-like immunoreactivity reflects differences in the accumulation of antigen rather than differences in antigen accessibility among individual leech. We have not carefully examined changes in enkephalin-like immunoreactivity on an annual basis, but we suspect that some of the variability we have seen has a seasonal basis (see below). Alternately, interindividual variations in enkephalin-like immunoreactivity could reflect physiological differentiation of hermaphroditic leech into male or female morphs.[19,20] A third possibility is that altered enkephalin-like immunoreactivity in the leech is one indicator of the general health of the leech.

Immunoreactive Varicosities

Neuropil Associations

All 6 antisera raised against Leu-enkephalin and screened against whole mount preparations

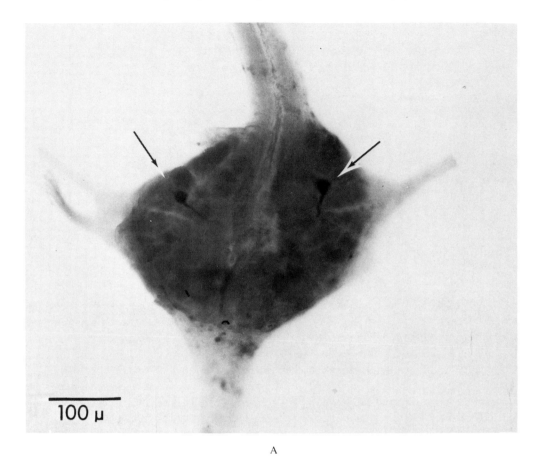

A

FIGURE 1.   Dorsolateral cells in a midbody ganglion of *Haemopis marmorata* stained with antisera raised against Leu-enkephalin (A). At higher magnification, immunoreactive fibers (long arrow) and neuronal varicosities (short arrows) are apparent (B).

of *Haemopis* stain neuronal fibers and varicosities which are 1 to 3 μm in diameter (Figure 1B). Most leech neurons give rise to neuronal varicosities, and synaptic transmission occurs across such varicosities in the leech neuropil.[21] Immunocytological staining of enkephalin-like varicosities is only fully blocked when antisera are absorbed with high concentrations of peptide, such that enkephalin-like immunoreactivity is most strongly expressed at these putative secretory sites (a distribution expected of a secretory product). In *Haemopis*, these immunoreactive varicosities lie in a circumscribed planar domain,[18] and thus appear to act within a selected region of the neuropil. Each leech neuron type projects to a characteristic portion of the neuropil, and neurons which project into similar regions of the neuropil may interact synaptically. Comparing the distribution of immunoreactive varicosities in the neuropil with the morphologies of identified leech neurons may allow identification of cells responsive to enkephalin-like antigens. For example, touch responsive mechanosensory neurons project to a circumscribed planar compartment which topographically resembles the neuropil region containing enkephalin-immunoreactive varicosities; thus, touch responsive mechanosensory neurons represent a set of cells which may interact with enkephalin-like projections.

Neurohemal and Visceral Associations

Enkephalin immunoreactive fibers and varicosities are found throughout the leech CNS. Immunoreactive fibers project within interganglionic connectives, but have never been seen

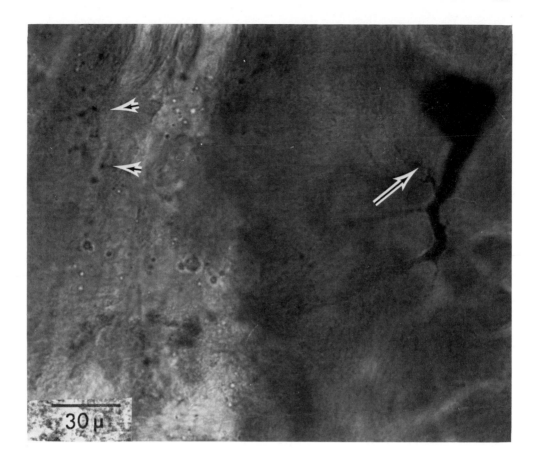

FIGURE 1B

within peripherally projecting ganglionic roots. Antisera raised against serotonin, however, stain interganglionic and peripherally projecting serotonin-containing fibers with equivalent facility. This suggests that enkephalin-like neurons within the CNS are interganglionic interneurons. While immunoreactive varicosities within segmental ganglia are confined to a narrow compartment in the neuropil, similar varicosities are found broadly distributed throughout the supraesophageal ganglion (a structure well known for its rich supply of neurosecretory cells[22-24]). Enkephalin-like agents may therefore play a role in modulating activites of cephalic neurosecretory cells, or, alternatively, may act as neuroendocrine agents themselves. Our recent studies with antisera #8324010 indicates that the leech foregut contains enkephalin-immunoreactive fibers (cell bodies giving rise to these fibers were not seen).

Correlative Studies

Immunoreactivity in Nerve Cord Homogenates

Acetic acid extracts of homogenized leech nerve cord contain enkephalin-like antigens. In collaboration with John Farah and Chris Molineaux (USUHS, Bethesda, Md.), we screened leech nerve cord homogenates with a panel of opioid-peptide RIAs (Table 2). The results of this screening study indicate the presence of dynorphin-like and Leu-enkephalin-like antigens that vary reciprocally on a seasonal basis. In an earlier RIA, we found 2 to 3 fM of Met-enkephalin-like immunoreactivity per nerve cord; however, Met-enkephalin immunoreactivity was not detected in ''spring'' and ''winter'' leech, and it remains to be seen

Table 2
RIA ESTIMATES OF OPIOID-
PEPTIDE CONTENT IN
HOMOGENATES OF LEECH NERVE
CORDS

Season sampled	Leu-enkephalin	Dynorphin
Winter	4.2 fmol/cord	12.0 fmol/cord
Spring	6.3 fmol/cord	9.0 fmol/cord

whether Met-enkephalin can be consistently detected during intermediate seasons. The low levels, and the multiple species of opioid peptide detected with RIA are consistent with the immunocytological evidence for the existence of sets of sparsely distributed cells which contain a family of enkephalin-like peptides. A more thorough biochemical analysis of these antigens (requiring a pooled sample of 200 isolated nerve cords) is in progress.

Opioid Pharmacology In Vivo

Leech respond behaviorally to externally applied morphine and to morphine injections. Kaiser[25] reported that morphine produced a biphasic hyperkinetic-quiescent locomotory response in free-swimming *Hirudo medicinalis*. We have found a similar biphasic response in *Haemopis* which were injected with 1.5 μmol of morphine sulfate. Control animals injected with 100 μℓ H_2) were initially hyperactive after injection, but rapidly reattached themselves to the walls of a holding tank and remained highly responsive to subsequent tactile stimulation. Morphine-injected leech were also initially hyperactive, but within 5 to 10 min they fell to the bottom of the tank and remained unresponsive to all but the most extreme tactile stimulation for a period of 4 days. These animals fully recovered by day 5.

Opioid Pharmacology In Vitro

Morphine appears to inhibit synaptic transmission in leech. In collaboration with Jorgen Johansen and Anne Kleinhause (Yale University, New Haven, Conn.), we have begun to examine the effects of millimolar morphine perfusions on synaptic transmission between a pair of identified leech neurons.[26] In our preliminary studies, we have seen blocking of the 5–mV excitatory postsynaptic potential between the medial nociceptive dermal mechano-sensory cell onto the "nut" cell (a cell which contains as much acetylcholinesterase as identified leech motor neurons, but whose function is yet unknown). This synaptic inhibition was reversed during a 10-min wash period, and was not accompanied with a blockage of the action potential in the presynaptic cell. In an earlier study, Gardner and Walker[27] reported inhibitory effects of morphine and apomorphine on Retzius cells in *Hirudo*. Morphine applied in this fashion does not appear to act as a general anesthetic.

Opioid Receptor-Binding Studies

Physiological evidence for the existence of opioid receptors in the leech is presently being compared with biochemical evidence for opioid binding. In collaboration with George B. Stefano (SUNY, Old Westbury, N.Y.), preliminary studies of homogenates of leech nerve cord display saturable binding of ^{3}H-(D-Ala^2)-Met-enkephalin. Opioid receptor densities very seasonally in concert with enkephalin titers in the bivalve mollusk *Mytilus edulis*,[28] and, for this reason, opioid receptor binding may be expected to vary among individual leech (in parallel with enkephalin immunoreactivity).

Comparisons Across Leech Species

The cellular organization of segmental ganglia is strongly conserved among leech species within the Hirudinidae family,[29] and, therefore, it is reasonable to suspect that cells which contain phylogenetically conserved antigens can be identified across these species. A single study of cross-species enkephalin-like immunoreactivity has not yet been done; however, our studies of whole mounts of *Haemopis* ganglia can be compared with whole mounts of *Hirudo* ganglia studied by Lucy Leake and collaborators (Portsmouth Polytechnic, Hants, U.K.). Leake's group has found cells which are labeled with antisera directed against somatostatin, vasoactive intestinal polypeptide, substance P, and Met-enkephalin.[30] In their study, three pairs of cells are labeled with enkephalin-directed antisera. The extent to which these cells correspond with cells in *Haemopis* is not yet clear, but our collective results support the opinion that enkephalin-like cells are sparsely distributed in both leech species. Differences in staining patterns could potentially result from four factors: (1) leech species actually contain different populations of enkephalin-like cells, (2) leeches are not being examined under comparable physiological states, (3) enkephalin immunoreactivities are differentially affected by immunocytochemical methodologies, or (4) the antisera used in these studies recognize different enkephalins (Leake's group used antisera raised against Met-enkephalin, and we used antisera raised against Leu-enkephalin). The latter suggestion seems unlikely because the antisera we have used (though raised against Leu-enkephalin) crossreacts with both Met- and Leu-enkephalin. A comprehensive analysis of leech enkephalins will require a direct comparison of immunoreactivities across neighboring species.

SUBSTANCE P-LIKE IMMUNOREACTIVITY IN THE LEECH CNS

The Tachykinin Peptide Family

Tachykinins constitute a family of peptides found in the gut and CNS of vertebrates, the skin of amphibians, and the salivary glands of an octopus.[31,32] Substance P, as the archetypic tachykinin, is an 11-amino acid peptide which contains the conserved tachykinin C-terminus sequence (Phe-Phe-Gly-Leu-Met-NH$_2$). Antisera which recognize this conserved C-terminus have been used immunocytologically to identify substance P-like immunoreactivity in a variety of invertebrate species: hydra,[33] ascidians,[34] molluscs,[35,36] earthworms,[37] chelicerates,[38] crustaceans,[39] and insects.[40-42]

Immunoreactive Soma

In our studies, six antisera raised against substance P were screened against whole mounts of leech segmental ganglia.[43] One of these antisera, Leeman RD2-C2, consistently labeled four neurons, and their primary neurites and varicosities. These immunoreactive cells consist of two pairs of 40-μm neurons located ventromedially in two adjacent anterior ganglionic compartments: one cell pair lies in the fourth ganglionic compartment of the fused subesophageal ganglion, and one cell pair lies in the first free segmental ganglion (Figure 2A). RD2-C2 immunoreactivity is extremely consistent between successively sampled leech, and is fully blocked when antisera is absorbed overnight with 100 μg mℓ$^{-1}$ substance P.

Immunoreactive Fibers and Varicosities

Interganglionic projections from cells stained with RD2-C2 can be traced anteriorly into the supraesophageal ganglion and posteriorly into the sixth segmental ganglion (Figure 2B). No immunoreactive processes were detected in peripherally projecting ganglionic roots. The anterior interganglionic projections remain uniformly immunoreactive, while posterior projections are less immunoreactive in each successive segmental ganglia. Immunoreactive varicosities labeled with antisera RD2-C2 are widely distributed throughout the neuropil of segmental ganglia (in contrast with the circumscribed distribution of enkephalin-like im-

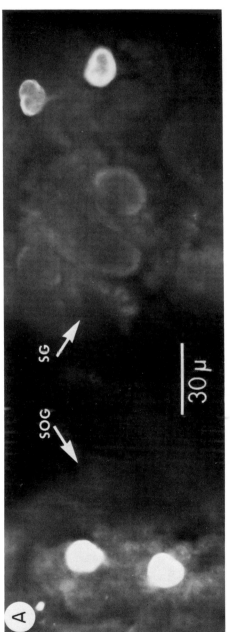
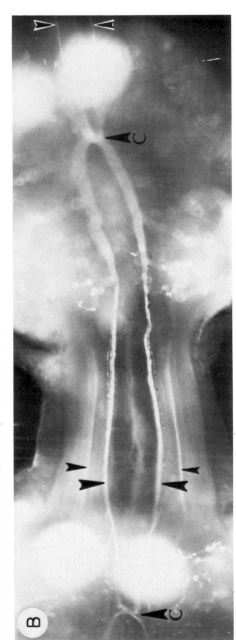

FIGURE 2. Ventromedial cells in the fourth subesophageal ganglion (SOG) and the first segmental ganglion (SG) of *Haemopis marmorata* stained with antisera RD2-C2 raised against substance P (A). At a different focal plain, immunoreactive fibers and neuronal varicosities are apparent (B). Interganglionic fibers (arrows) and ganglionic commissures (C) facilitate identifying the primary morphology of these neurons.

munoreactive varicosities). Substance P-like cells lie in selected anterior ganglia and potentially interact with a large population of cells. Strongly immunoreactive projections within the supraesophageal ganglion suggest that these cells may have neuroendocrine functions. The four neurons labeled with RD2-C2 are interganglionic interneurons whose functions remain to be identified. Their general morphology parallels those of second-order visual interneurons studied by Peterson.[44] We are presently "double labeling" the RD2-C2 immunoreactive cells (with intracellular lucifer injections, followed by rhodamine-tagged immunocytological analysis to confirm the identity of RD2-C2 dye-filled cells) to identify electrophysiological features of these cells.

Immunoreactivity in Nerve Cord Homogenates

Acid extracts of homogenates of leech nerve cords contain substance P-like antigens. In colloboration with Erik Floor and Susan Leeman (University of Massachusettes Medical School, Worcester, Mass.), we have assayed homogenates with two RIAs which used different substance P-directed antisera. Both RIAs indicate that extracts contain a substance P-like antigen, but they differ fivefold in their estimates of the titer of antigen present. The RD2-C2, which stains cells immunocytologically, provides the higher estimate of substance P-like antigen (~0.2 fmol per nerve cord). The differences in estimates of substance P-like antigen observed across our RIAs argue that the leech antigen is not identical with substance P. We suspect that the leech may contain a related substance P-like tachykinin, and we are now assaying homogenates of anterior and posterior portions of the leech nerve cord to compare the distribution of our immunocytologically detected antigen with antigen measured by RIA.

Comparisons Across Leech Species

Antisera RD2-C2 exclusively labels a set of four cells with corresponding primary morphologies in whole mounts of three leech species: *Haemopis marmorata, Hirudo medicinalis,* and *Macrobdella decorata*. Thus, both a substance P-like antigen and an unusual cell type containing that antigen is phylogenically conserved within the Hirudinidae family of leeches. Our whole mount studies reveal an extremely selective distribution of substance P-immunoreactive processes. This result is consistent with earlier studies in *Hirudo;*[45] however, more recently, substance P-immunoreactive cells have been immunocytologically detected in each segmental ganglion of *Hirudo*.[30] Leake[30] feels that one pair of these cells potentially correspond with leech Lydig cells, a cell type thought to serve some neuroendocrine function.[46] During out initial screening for substance P-like immunoreactivity in whole mounts of *Haemopis* ganglia, we stained a pair of cells with the Cuello monoclonal antibody which may correspond with the cells Leake's group has seen in *Hirudo*. In *Haemopis*, however, the Cuello antibody did not stain fibers or varicosities at that time, and we were unable to stain cells with this antisera in subsequent studies. Perhaps a substance P-like antigen seen by the Cuello antibody is expressed conditionally. We may not have been examining animals in appropriate physiological states. Five additional substance P-directed antisera uniformly failed to label cells in our studies of *Haemopis* (but we caution that positive control tissues were not concurrently tested to demonstrate that these antisera retained anti-substance P immunoreactivity during our screening). The very distinct distributions of substance P-immunoreactive cell populations in the leech makes it possible for us to compare homogenates of regions of the leech nerve cord to correlate extractable antigen titers (detected with RIA) with immunocytologically identified cell types.

FMRFamide-LIKE IMMUNOREACTIVITY IN THE LEECH

Molluscan FMRFamide

FMRFamide is a tetrapeptide (Phe,Met,Arg,Phe-NH$_2$) initially isolated from molluscs[47] and subsequently immunocytologically or biochemically reported in hydra, insects, and vertebrates.[48-50] At nanomolar doses FMRFamide increases the frequency of cardiac contraction and induces contraction of the radular protractor and the anterior byssus retractor muscles in selected molluscan species.[51,52]

Immunoreactive Soma

Using 2 antisera raised against synthetic FMRFamide, Kuhlman et al.[53] have identified about 50 immunoreactive neurons in whole mounts of segmental ganglia of *Hirudo medicinalis*. This set of cells (9 ventral pair and 18 dorsal pair) represents about one eighth of the cells in a segmental ganglion. It includes identified motor neurons (i.e., annulus erector neurons) and interneurons (i.e., swim-initiating interneuron #204), but appears to lack identified sensory neurons. Selected ganglia contain additional FMRFamide-immunoreactive cells (i.e., two pairs of reproductive motor neurons unique to ganglion 6, and a pair of heart modulator neurons common to ganglia 5 and 6). Immunoreactivity was fully blocked when antisera were preabsorbed with 100 μg FMRFamide mℓ^{-1}.

Comparisons Across Leech Species

In our studies with *Haemopis*, we find a similarly large set of FMRFamide-like immunoreactive neurons. The extent of FMRFamide immunoreactivity varies little among consecutively sampled leech (unlike enkephalin-like immunoreactivity where interindividual variations can be substantial), and is regularly seen in nerve fibers in both interganglionic and peripherally projecting nerve trunks. FMRFamide is not restricted to the leech CNS. Khulman[54] has mapped projections from ganglionic neurons onto heart muscle, and we have additionally seen FMRFamide immunoreactivity in cells and fibers within the wall of the foregut (Figure 3).

Pharmacological Activity of Synthetic FMRFamide

Pharmacological studies indicate that synthetic FMRFamide has both peripheral and central effects within the leech. At nanomolar doses, FMRFamide stimulates peripheral myogenic cardiac contractions and independently accelerates the central neurogenic heartbeat pattern generator.[55] FMRFamide-like antigens might be generally associated with rhythm-generating systems in the leech.

PROCTOLIN-LIKE IMMUNOREACTIVITY IN THE LEECH

Insect Proctolin

Proctolin is a pentapeptide (Arg,Tyr,Leu,Pro,Thr) initially detected in stomatodeal and proctodeal nerves of the cockroach.[56,57] At nanomolar doses, it evokes graded contraction of arthropod muscle[58,59] and increases the force[60-62] or frequency[61] of cardiac contractions.

Immunoreactive Soma

Using 3 antisera raised against synthetic proctolin, Li and Calabrese[63] have immunocytologically identified 2 populations of regularly immunoreactive neurons in segmental ganglia of *Hirudo*: a segmentally repeated set consisting of 14 cells (4 dorsal pair and 3 ventral pair) and a set of 2 to 6 cells distributed among selected anterior segmental ganglia (2 lie in ganglia 1, 3, and 4; 3 lie in ganglia 5, 6, and 7; and 6 lie in ganglion 2). This immunoreactivity is fully blocked when antisera are preadsorbed with 100 to 500 μg proctolin mℓ^{-1}.

Bioassay of Nerve Cord Homogenates

Homogenates of *Hirudo* nerve cord were bioassayed with the assay of O'Shea and

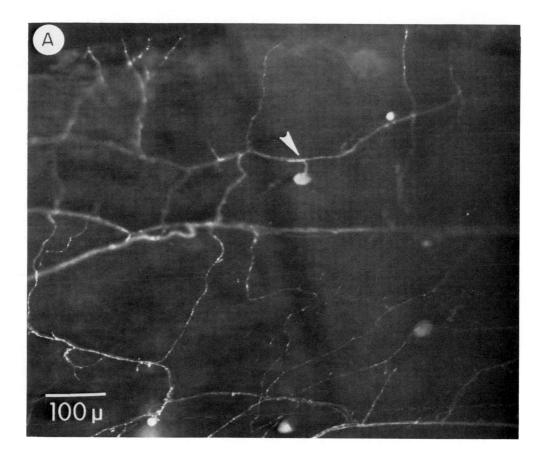

FIGURE 3.   Peripheral cells (arrow) and fiber tracts in the foregut of *Haemopis marmorata* stained with antisera raised aganist FMRFamide (A). At higher magnification, varicosities are seen along these fine fibers (B).

Adams[64] and ganglia were estimated to contain 0.27 fM of proctolin-like activity,[63] Li and Calabrese suspect, however, that this activity is associated with a molecular species which contains a proctolin sequence, but which is biochemically distinct from insect proctolin.

SUMMARY OF NEUROPEPTIDE-LIKE ANTIGENS IN THE LEECH

Relative Abundance of Peptide-Like Cells

Evidence is being accumulated to indicate that leeches contain four families of antigens which resemble previously described vertebrate or invertebrate neuropeptides (Table 3). These antigens are detectable in the femtamolar range in homogenates of leech nerve cord, and are immunocytologically distributed in a pattern consistent with the suggestion that they serve as neuronal secretory products. Immunoreactivity is found in both segmentally repeated and in selectively distributed cell types such that it is presently premature to conclude that higher-order leech neurons are generally differentiated from segmentally repeated leech neurons with respect to the transmitters they use.

Variability in Immunocytological Reactivity

Our studies of enkephalin-like immunoreactivity in *Haemopis* suggest that antigen titers vary on both a seasonal and an individual basis. Serotonin, the RD2-C2 substance P-like antigen, and a variety of antigens labeled with monoclonal antibodies raised against ho-

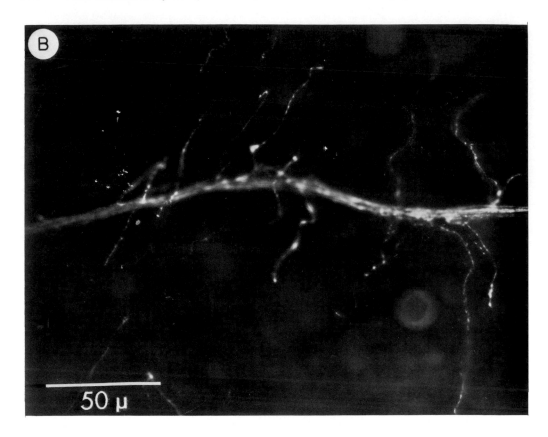

FIGURE 3B

Table 3
NEUROTRANSMITTER AND PEPTIDE-LIKE ANTIGEN
DISTRIBUTION IN LEECH MIDBODY SEGMENTAL GANGLIA

Transmitter (enzyme)	Percent of the ganglionic neuronal population	Estimated molar content per leech nerve cord
ACh	30[66]	ND
GABA	8.8[13]	ND
5HT	1.8[73]	860 pmol[74]
DA	0.5[11]	56 pmol[12]
OA	ND	650 pmol[67]
FMRFamide	13[53]	ND
Proctolin	3.5[63]	~9.3 fmol[63]
Opioids	2.3[75]	~15 fmol[76]
Substance P	0.5[30]	~0.2 fmol[77]

Note: ND, not determined.

mogenates of the leech cord remain uniformly detectable with our immunocytological methods.[18] Efforts to alter peptide-like antigen titers (by transecting distal processes from immunoreactive somata *in situ*, injecting animals with colchicine, incubating ganglia in a variety of secretagogues, or culturing ganglia with mixtures of forskolin, demecolcin, and insulin) have not produced a profound change in the expression of a variety of peptide-like immunocytological reactivities. Such results are consistent with the suggestion that these antigens are synthesized and released gradually, perhaps in parallel with changes in behavioral states.

Co-localization of Multiple Transmitter Types

Five nonpeptide neurotransmitter substances have been identified biochemically or cytologically within the leech CNS: serotonin and dopamine,[11,65] acetylcholine[9] and acetylcholinesterase,[10,66] gamma aminobutyric acid,[13] and octopamine.[67] In leech, the co-localization of peptide-like immunoreactivity in cells which carry these nonpeptidergic transmitters is apparently low. Peptide-like immunoreactivities have not been associated with identified serotonin- or dopamine-containing neurons, and octopamine-containing cells have not yet been identified such that their potential association with peptide-like antigens is unknown. Several identified peripherally projecting neurons do appear to contain co-localized peptide and nonpeptide transmitters. Several acetylcholinesterase-containing neurons are labeled with antisera directed against FMRFamide,[53] and one GABA-sequestering cell type appears to also contain a proctolin-like antigen.[63] These few cells, however, represent only a small portion of the total set of cholinergic and gabaminergic cells in the leech. Co-localization of transmitter types has not yet been reported for any leech interneuron, leaving open the possibility that peptide-containing leech interneurons may be purely peptidergic (and thus attractive cell types for subsequent physiological study).

A Need for Biochemical Analysis

The data reviewed in this report argue that identifiable leech neurons contain substances which resemble previously identified neuropeptides; however, the exact nature of this resemblance is not clear. Phyletically conserved peptide immunoreactivities do not, in themselves, demonstrate close structural or evolutionary relationships between crossreacting antigens.[68] To strengthen arguments for evolutionary relationships, immunocytological studies in the leech have been corroborated with radioimmunoassay, with biological activity assays, or with pharmacological studies, and biochemical studies currently in progress. These biochemical studies are likely to reveal two general features: (1) that immunocytologically identified peptides exist as families of molecules with related sequences, and (2) that only portions of these peptide sequences will truly be conserved across metazoan species. Differential immunocytological labeling of sets of leech neuron with different antisera to either enkephalin or to substance P suggest that leech contain multiple enkephalin- and substance P-like antigens. Different antisera raised to FMRFamide label a uniform population of leech neurons; however, because these antisera recognize only the C-terminal portion of FMRFamide (-Arg-Phe-NH$_2$),[69] and because they label very large sets of neurons, FMRFamide-directed antisera could be labeling cells which contain a variety of FMRFamide-like antigens. The C-terminus of gastrin and cholecystokinin (-Trp-Met-Asp-Phe-NH$_2$), for example, resembles the immunocytochemically recognized C-terminus of FMRFamide, and antisera directed against cholecystokinin also label many leech neurons.[70] FMRFamide-like antigens biochemically distinct from FMRFamide have been isolated from *Helix*[71] and *Limulus*,[72] and corresponding molecules probably also exist within leech. Neuropeptide-like antigens isolated from invertebrate species which differ from their analogs in vertebrate species represent a potential source of pharmacological analogs which could contribute to a general understanding of neuropeptide evolution and function.

ACKNOWLEDGMENTS

Unpublished observations of substance P-, FMRFamide-, and proctolin-like immunoreactivities in various leech species were generously provided by Drs. Leake, Calabrese,

Khulman, and Li. The spirit of cooperation among leech neurobiologists promises to contribute substantially to a coherent picture of neuropeptide biology in future studies of this invertebrate group.

REFERENCES

1. **Muller, K. J., Nicholls, J. G., and Stent, G. S.,** *Neurobiology of the Leech,* Cold Spring Harbor Laboratory, Cold Spring Harbor, N.Y., 1981.
2. **Macagno, E. R.,** Number and distribution of neurons in leech segmental ganglia, *J. Comp. Neurol.,* 190, 283, 1980.
3. **Nicholls, J. G. and Wallace, B. G.,** Quantal analysis of transmitter release at an inhibitory synapse in the C.N.S. of the leech, *J. Physiol.,* 281, 171, 1978.
4. **Zipser, B.,** Complete distribution patterns of neurons with characteristic antigens in the leech central nervous system, *J. Neurosci.,* 2, 1453, 1982.
5. **Stent, G. S., Thompson, W. J., and Calabrese, R. L.,** Neural control of heartbeat in the leech and in some other invertebrates, *Physiol. Rev.,* 59, 101, 1979.
6. **Kristan, W. B., Jr. and Calabrese, R. L.,** Rhythmic swimming activity in neurons of the isolated nerve cord of the leech, *J. Exp. Biol.,* 65, 643, 1976.
7. **Stent, G. S. and Kristan, W. B., Jr.,** Neural circuits generating rhythmic movements, in *Neurobiology of the Leech,* Muller, K. J., Nicholls, J. G., and Stent, G. S., Eds., Cold Spring Harbor Laboratory, Cold Spring Harbor, N.Y., 1981, 113.
8. **Lent, C. M. and Dickinson, M. H.,** Serotonin integrates the feeding behavior of the medicinal leech, *J. Comp. Physiol.,* 154A, 457, 1984.
9. **Sargent, P. B.,** Synthesis of acetylcholine by excitatory motorneurons in central nervous system of the leech, *J. Neurophysiol.,* 40, 453, 1970.
10. **Wallace, B. G.,** Distribution of AChE in cholinergic and non-cholinergic neurons, *Brain Res.,* 219, 190, 1981.
11. **Lent, C. M.,** Morphology of neurons containing monoamines within leech segmental ganglia, *J. Exp. Zool.,* 216, 311, 1981.
12. **Lent, C. M., Muller, R. L., and Haycock, D. A.,** Chromatographic and histochemical identification of dopamine within an identified neuron in the leech nervous system, *J. Neurochem.,* 41, 481, 1983.
13. **Cline, H. T.,** ³H-GABA uptake selectively labels identifiable neurons in the leech central nervous system, *J. Comp. Neurol.,* 215, 351, 1983.
14. **Weisblat, D. A., Harper, G., Stent, G. S., and Sawyer, R. T.,** Embryonic cell lineages in the nervous system of the glossiphonid leech *Helobdella triserialis, Dev. Biol.,* 76, 58, 1980.
15. **Macagno, E. R., Stewart, R. R., and Zipser, B.,** The expression of antigens by embryonic neurons and glia in segmental ganglia of the leech *Haemopis marmorata, J. Neurosci.,* 3, 1746, 1983.
16. **McKay, R. D. G., Hockfield, S., Johansen, J., Thompson, I., and Fredericksen, K.,** Surface molecules identifying groups of growing axons, *Science,* 222, 788, 1983.
17. **Zipser, B.,** Identification of specific leech neurons immunoreactive to enkephalin, *Nature (London),* 283, 857, 1980.
18. **Flanagan, T. R. J. and Zipser, B.,** Immunoreactive varicosities define compartments within the leech neuropile, *J. Neurocytol.,* in press.
19. **Lukowiak, K.,** Peptide modulation of neuronal activity and behavior in *Aplysia,* in *Handbook of Comparative Opioid and Related Neuropeptide Mechanisms,* Vol. 2, Stefano, G. B., Ed., CRC Press, Boca Raton, Fla., in press.
20. **Stefano, G. B. and Scharrer, B.,** High affinity binding of an enkephalin analog in the cerebral ganglion of the insect *Leucophaea maderae* (Blattaria), *Brain Res.,* 225, 107, 1981.
21. **Muller, K. J.,** Synapses between neurones in the central nervous system of the leech, *Biol. Rev.,* 54, 99, 1979.
22. **Hagadorn, I. R.,** The histochemistry of the neurosecretory system in *Hirudo medicinalis, Gen. Comp. Endocrinol.,* 6, 288, 1966.
23. **Orchard, I. and Webb, R. A.,** The projections of neurosecretory cells in the brain of the North-American medicinal leech, *Macrobdella decora,* using intracellular injection of horseradish peroxidase, *J. Neurobiol.,* 3, 229, 1980.
24. **Yagi, K., Bern, H. A., and Hagadorn, I. R.,** Action potentials of neurosecretory neurons in the leech *Theromyzon rude, Gen. Comp. Endocrinol.,* 3, 490, 1963.

25. **Kaiser, F.,** Beitrage zur Bewegungsphysiologie der Hirudineen, *Zool. Jahrb. Abt. Allg. Zool. Physiol.,* 65, 59, 1954.
26. **Johansen, J. and Kleinhause, A. L.,** A monosynpatic connection between the medial nociceptive and the nut cell in leech ganglia, *J. Comp. Physiol.,* A156, 65, 1985.
27. **Gardner, C. R. and Walker, R. J.,** The roles of putative neurotransmitters and neuromodulators in annelids and related invertebrates, *Prog. Neurobiol.,* 18, 81, 1982.
28. **Stefano, G. B. and Leung, M. K.,** Opioid aging and seasonal variations in invertebrate ganglia: evidence for an opioid compensatory mechanism, in *Handbook of Comparative Opioid and Related Neuropeptide Mechanisms,* Vol. 2, Stefano, G. B., Ed., CRC Press, Boca Raton, Fla., in press.
29. **Keyser, K. T. and Lent, C. M.,** On neuronal homologies within the central nervous system of leeches, *Comp. Biochem. Physiol.,* 58A, 285, 1977.
30. **Leake, L.,** personal communication, 1984.
31. **Erspamer, V.,** The tachykinin peptide family, *TINS,* 4, 267, 1981.
32. **Harmar, A. J.,** Three tachykinins in mammalian brain, *TINS,* 7, 57, 1984.
33. **Taban, C. H. and Cathieni, M.,** Localization of substance P-like immunoreactivity in *Hydra, Experientia,* 35, 811, 1979.
34. **Fritsch, H. A. R., VanNoorden, S., and Pearce, A. G. E.,** Localization of somatostatin-, substance P-, and calcitonin-like immunoreactivity in the neural ganglion of *Ciona intestinalis* L. (Ascidiaceae), *Cell Tissue Res.,* 202, 263, 1979.
35. **Schot, L. P. C., Boer, H. H., Swabb, D. F., and VanNoorden, S.,** Immunocytochemical demonstration of peptidergic neurons in the central nervous system of the pond snail *Lymnaea stagnalis* with antisera raised to biologically active peptides of vertebrates, *Cell Tissue Res.,* 216, 273, 1981.
36. **VanNoorden, S., Fritsch, H. A. R., Grillo, T. A. I., Polak, J. M., and Pearse, A. G. E.,** Immuno-cytochemical staining for vertebrate peptides in the nervous system of a gastropod mollusc, *Gen. Comp. Endocrinol.,* 40, 375, 1980.
37. **Aros, B., Wenger, T., Vigh, B., and Vigh-Teichmann, I.,** Immunohistochemical localization of substance P and ACTH-like activity in the central nervous system of the earthworm *Lumbricus terrestris* L., *Acta Histochem.,* 66, 262, 1980.
38. **Mancillas, J. R. and Brown, M. R.,** Neuropeptide modulation of photosensitivity. I. Presence, distribution, and characterization of a substance P-like peptide in the lateral eye of *Limulus, J. Neurosci.,* 4, 832, 1983.
39. **Mancillis, J. R., McGinty, J. F., Selverston, A. I., Karten, H., and Bloom, F. E.,** Immunocytochemical localization of enkephalin and substance P in retina and eyestalk neurons of lobster, *Nature (London),* 293, 576, 1981.
40. **Benedeczky, I., Kiss, J. Z., and Somogy, P.,** Light and electron microscopic localization of substance P-like immunoreactivity in the cerebral ganglion of locust with a monoclonal antibody, *Histochemistry,* 75, 123, 1982.
41. **El-Salhy, M., Falkmer, S., Kramer, K. J., and Speirs, R. D.,** Immunohistochemical investigations of neuropeptides in the brain, corpora cardiaca, and corpora allata of an adult lepidopteran insect, *Manduca sexta* (L.), *Cell Tissue Res.,* 232, 295, 1983.
42. **Hansen, B. L., Hansen, G. N., and Scharrer, B.,** Immunoreactive material resembling vertebrate neu-ropeptides in the corpus cardiacum and corpus allatum of the insect *Leucophaea maderae, Cell Tissue Res.,* 225, 319, 1982.
43. **Flanagan, T.,** A set of 4 leech neurons stained with an antisera against substance P, *Leech Neurobiol. Newsl.,* 6, 16, 1984.
44. **Peterson, E. L.,** Visual processing in the leech central nervous system, *Nature (London),* 30, 240, 1983.
45. **Osborne, N. N., Cuello, A. C., and Dockray, G. J.,** Substance P and cholecystokinin-like peptides in *Helix* neurons and cholecystokinin and serotonin in a giant neuron, *Science,* 216, 409, 1982.
46. **Keyser, K. T., Frazer, B. M., and Lent, C. M.,** Physiological and anatomical properties of Lydig cells in the segmental nervous system of the leech, *J. Comp. Physiol.,* 146A, 379, 1982.
47. **Price, D. A. and Greenberg, M. J.,** Structure of a molluscan cardioexcitatory neuropeptide, *Science,* 197, 670, 1977.
48. **Boer, H. H., Schot, L. P. C., Veenstra, J. A., and Reichelt, D.,** Immunocytochemical identification of neural elements in the central nervous system of a snail, a fish, and a mammal with an antiserum to the molluscan cardio-excitatory tetrapeptide FMRFamide, *Cell Tissue Res.,* 213, 21, 1980.
49. **Cottrell, G. A., Price, D. A., and Greenberg, M. J.,** FMRFamide-like activity in the ganglia and in a single identified neurone of *Helix aspera, Comp. Biochem. Physiol.,* 70C, 103, 1981.
50. **Grimmelikhuijzen, C. J. P., Dockray, G. J., and Schot, L. P. C.,** FMRFamide-like immunoreactivity in the nervous system of *Hydra, Histochemistry,* 73, 499, 1982.
51. **Greenberg, M. J. and Price, D. A.,** FMRFamide, a cardio-excitatory neuropeptide in molluscs: an agent in search of a mission, *Am. Zool.,* 19, 163, 1979.
52. **Painter, S. D.,** FMRFamide catch contractures of a molluscan smooth muscle: pharmacology, ionic de-pendence and cyclic nucleotides, *J. Comp. Physiol.,* 148A, 491, 1982.

53. **Khulman, J. R., Li, C., and Calabrese, R. L.,** FMRFamide-like substances in the leech: I. Immuno-cytochemical localization, *J. Neurosci.,* in press.

54. **Khulman, J. R.,** The Neuropeptide FMRFamide in the Leech: Immunocytochemical Localization and Bioactivity, B.A. Honors thesis, Harvard University, Cambridge, 1984.

55. **Khulman, J. R., Li, C., and Calabrese, R. L.,** FMRFamide-like substances in the leech: bioactivity in the leech heartbeat system, *J. Neurosci.,* in press.

56. **Brown, B. E.,** Proctolin: a peptide transmitter candidate in insects, *Life Sci.,* 17, 1241, 1975.

57. **Starratt, A. N. and Brown, B. E.,** Structure of the pentapeptide proctolin, a proposed neurotransmitter in insects, *Life Sci.,* 17, 1253, 1975.

58. **Adams, M. E. and O'Shea, M.,** Peptide cotransmitter at a neuromuscular junction, *Science,* 221, 286, 1983.

59. **Schwarz, T. L., Harris-Warrick, R. M., Glusman, S., and Kravitz, E. A.,** A peptide action in a lobster neuromuscular preparation, *J. Neurobiol.,* 11, 623, 1980.

60. **Benson, J. A., Sullivan, R. E., Watson, W. H., and Augustine, G. J.,** The neuropeptide proctolin acts directly on *Limulus* cardiac muscle to increase the amplitude of contraction, *Brain Res.,* 213, 449, 1981.

61. **Miller, T.,** Nervous versus neurohormonal control of insect heartbeat, *Am. Zool.,* 19, 77, 1979.

62. **Sullivan, R. E.,** Proctolin-like peptide in crab pericardial organs, *J. Exp. Zool.,* 210, 543, 1979.

63. **Li, C. and Calabrese, R. L.,** Evidence for proctolin-like substances in the central nervous system of the leech, *Hirudo medicinalis, J. Comp. Neurol.,* 232, 414, 1985.

64. **O'Shea, M. and Adams, M. E.,** Pentapeptide (proctolin) associated with an identified neuron, *Science,* 213, 567, 1981.

65. **McAdoo, D. J. and Coggshall, R. E.,** Gas chromatographic-mass spectrometric analysis of biogenic amines in identified neurons and tissues of *Hirudo medicinalis, J. Neurochem.,* 26, 163, 1976.

66. **Wallace, B. G. and Gillon, J. W.,** Characterization of acetylcholinesterase in individual neurons in the leech central nervous system, *J. Neurosci.,* 2, 1108, 1982.

67. **Webb, R. A. and Orchard, I.,** Octopamine in leeches. I. Distribution of octopamine in *Macrobdella decora* and *Erpobdella octoculata, Comp. Biochem. Physiol.,* 67C, 135, 1980.

68. **Jornvall, H., Persson, M., and Ekman, R.,** Structural comparisons of leukocyte interferon and pro-opiomelanocortin correlated with immunological similarities, *FEBS Lett.,* 137, 153, 1982.

69. **Weber, E., Evans, C. J., Samuelsson, S. J., and Borchas, J. D.,** Novel peptide neuronal system in rat brain and pituitary, *Science,* 214, 1248, 1981.

70. **Osborne, N. N., Patel, S., and Dockray, G.,** Immunohistochemical demonstration of peptides, serotonin and dopamine-β-hydroxylase-like material in the nervous system of the leech *Hirudo medicinalis, Histochemistry,* 75, 573, 1982.

71. **Price, D. A.,** The FMRFamide-like peptide of *Heliz aspera, Comp. Biochem. Physiol.,* 72C, 325, 1982.

72. **Watson, W. H., Groome, J. R., Chronwall, B. M., Bishop, J., and O'Donohue, T. L.,** Presence and distribution of immunoreactive and bioactive FMRFamide-like peptides in the nervous system of the horse-shoe crab, *Limulus polyphemus, Peptides,* 5, 585, 1984.

73. **Lent, C. M.,** Serotonin-containing neurons within the segmental nervous system of the leech, in *Biology of Serotonergic Transmission,* Osborne, N. N., Ed., John Wiley & Sons, New York, 1982, 431.

74. **Lent, C. M.,** Quantitative effects of a neurotoxin on serotonin levels in tissue compartments of the leech, *J. Neurobiol.,* 15, 309, 1984.

75. **Flanagan, T. and Zipser, B.,** unpublished observations.

76. **Flanagan, T., Farah, J., Molineaux, C., and Zipser, B.,** unpublished observations.

77. **Flanagan, T., Floor, E., Leeman, S., and Zipser, B.,** unpublished observations.

NEUROPEPTIDES IN CRUSTACEANS WITH SPECIAL REFERENCE TO OPIOID-LIKE PEPTIDES

Peter P. Jaros

SUMMARY

The invertebrate nervous system may be used as a model to investigate the mode of action of peptide hormones in regard to neuroendocrine control. In crustacea, as in all invertebrate phyla, neuropeptides have been found which are native to crustaceans and those which are identified mostly as immunoreactive vertebrate hormone-like material.[1] Two recently published reviews comprise the functional and biochemical aspects of native neurohormones produced in the cortices of the optic ganglia and in the "root cells" as well as in the anterior and posterior pericardial organ cells. These neuropeptides are stored in the sinus glands (SG), the pericardial organs (PO), and in the postcommissural organs (PCO), respectively.[2,3]

POSTCOMMISSURAL ORGANS

Since the discovery of neuropeptides in the PCO by Knowles,[4] and the study of Maynard and Maynard,[5] little work has been done on these neurohemal structures which store neuropeptides with chromatophorotropic activity. These neuropeptides originate from the tritocerebral commissure which connects the circumoesophageal connectives caudally.

PERICARDIAL ORGANS

The PO were first identified as neurohemal organs by Alexandrowicz and Carlisle[6] and are situated in the lateral pericardial cavity, forming a nerve plexus near the openings of the brachiocardiac veins supplied by axons from the thoracic ganglia. Neurosecretory material is stored in seven types of terminals as classified by an ultrastructural investigation.[2] Cardioexcitatory active biogenic amines and peptides were isolated in extracts from PO.[7]

Sullivan[8] and Livingstone et al.[9] established the presence of 5-HT in the hemolymph of the lobster. In addition, it is suggested that octopamine and dopamine are stored and released from the PO.[10] DL-Octopamine excites isolated perfused hearts and increases cyclic AMP in heart muscle and ganglia[11] of crabs and lobsters and increases spontaneous contractions of the hindgut of *Procambarus*.[12] Dopamine has effects similar to those of octopamine in respect to excitation of isolated hearts in lobster and crab, contraction of the hindgut, and increasing cyclic AMP levels.[2] Tryptic digestion reduces the cardioexcitatory activity of PO extracts; as a result, Maynard and Welsh[13] postulated the existence of additional peptidergic hormones with cardioexcitatory effects, Separation of PO extracts revealed two cardioexcitatory factors, a trypsin-sensitive and a trypsin-insensitive peptide (for review, see Keller[3]). Recently, Sullivan[14] demonstrated the occurrence of the insect putative-neurotransmitter proctolin in PO extracts of the crab, *Cardisoma carnifex*. Sullivan demonstrated that the trypsin-insensitive peptide and the synthetic pentapeptide proctolin were identical.[14]

In addition to the cardioexcitatory biogenic amines and peptides mentioned above, we demonstrated the presence of the crustacean hyperglycemic hormone (CHH) in the PO of *Carcinus maenas* by radioimmunoassay. One pair of PO contains approximately 4 ng CHH, which is less than 1% of the amount stored in the SG.[15]

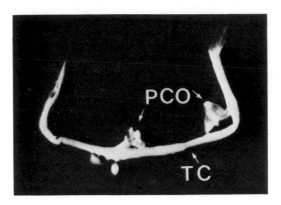

FIGURE 1.   Dissected postcommissural organs (PCO) from the shrimp, *Lysmata unicornis*, protruding from the tritocerebral commissure (TC). (Magnification × 200.) (From Chaigneau, J., *Neurohemal Organs of Arthropods*, Gupta, A. P., Ed., Charles C Thomas, Springfield, Ill., 1983, 53. With permission.)

SINUS GLAND

The sinus gland is the most thoroughly investigated neurohemal organ, situated dorsal or dorsolateral in the eyestalks of podophthalmian crustacean. In some species the location can differ; e.g., the SG is found attached to the medulla externa (e.g., Euphausiacea) or the medulla terminalis (e.g., *Cardisoma carnifex*). In edriophthalmian Crustacea like Isopoda, the SG protrudes from the optic ganglia in the head capsula (for review, see Chaigneau[16]). The SG is composed of axon terminals of neurosecretory cells located in the eyestalk ganglia (Figure 1). The main source of neural elements forming the sinus gland originates from the mtXo (medulla terminalis ganglionic X-organ) as demonstrated by cobalt chloride iontophoresis (Figure 2).[17,18] The large neurosecretory perikarya of the mtXo produces the crustacean hyperglycemic hormone[19-21] (Figure 3 and 4). Approximately 20% of the total protein content of the SG consists of CHH. The neurosecretory cells located in ganglia other than the medulla terminalis also innervate the SG (for reference, see Andrew[23]). From the 18 postulated neurosecretory factors attributed to the SG and the SG-mtXo system, only 2 are known by their analysis of the primary structure, the red-pigment concentrating hormone (RPCH),[24] an octapeptide, closely related to insect adipokinetic hormone,[25] and the distal-retinal pigment hormone (DRPH),[26] a peptide with 18 amino acid residues. The amino acid composition for the CHH of three different species is known.[27,28] CHH is a N-terminal blocked polypeptide with two intrachain disulfide bridges and a molecular weight of 5900 to 6800 daltons. In contrast to the aforementioned peptides, CHH is species or (at least to a limited extent) group specific.[29] It is released by the SG into the hemolymph where it circulates at a level of 0.2 pmol/mℓ as determined by RIA.[30] Injections of less than 3 pmol per animal of the isolated extract containing this hormone gave a clear hyperglycemic response.[3]

An amino acid composition has been reported by Huberman et al.[31] for the neurodepressing hormone (NDH)) isolated from eyestalks of *Procambarus* and *Penaeus*. It is said to be a nona- or decapeptide with N- and C-terminal presumably blocked. The material possesses no net electrical charge, and it is trypsin sensitive, heat stable, and dialyzable. It is active in the nanomolar range and depresses the electrical acitivty of motor and sensory neurons.[2]

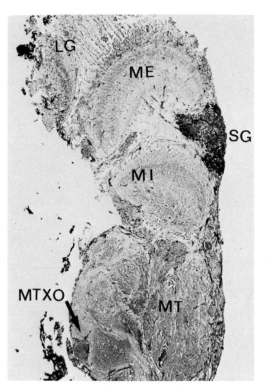

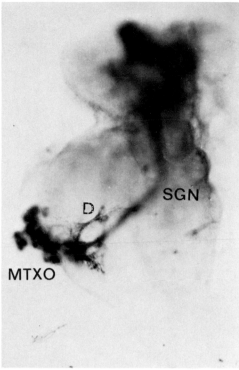

FIGURE 2. Section of eyestalk ganglia of *Carcinus maenas*. LG, lamina ganglionaris; ME, medullar externa; MI, medulla interna; MT, medulla terminalis; MTXO, medulla terminalis ganglionic X-organ; SG, sinus gland. Azan stained. (Magnification × 63.) (From Mangerich, S., unpublished. With permission.)

FIGURE 3. Pathway of the sinus gland nerve (SGN) arises from the neurosecretory cells of the medulla terminalis ganglionic X-organ (MTXO, traced by iontophoresis with cobalt dye in the crayfish *Orconectes limosus*. D, dendritic arborizations. (Magnification × 50.)

OPIOID-LIKE NEUROPEPTIDES IN CRUSTACEANS

Enkephalin-like material has been reported in the octopus,[32] the leech,[33] and the earthworm,[34] and by radioimmunoassay in the brain of locust.[35] Analysis of the primary structure of enkephalins in the mussel, *Mytilus edulis*, revealed the presence of Met- and Leu-enkephalin as well as Met-enkephalin-Arg^6-Phe.[36] The occurrence of opioid-like material in Crustacea has been reported. Mancillas et al.[37] demonstrated the presence of Leu-enkephalin-like material in the spiny lobster, *Panulirus interruptus* and *Procambarus* sp. Enkephalin immunoreaction was found in all retinular cells of the ommatidia and in axons which are situated between the medulla interna and the medullar terminalis. They join a tractus which comes from the MT and enters into the MI where the anons "terminate in a plexus of fine varicose fibers".[37]

Enkephalin-like material stored in axon terminals of the sinus gland has been shown by Jaros and Keller[38] in the crab, *Carcinus maenas,* and the crayfish, *Orconectes limosus*. Recently, the presence of FMRFamide-like material in the eyestalk of the prawn, *Palaemon serratus,* was demonstrated.[39]

The photoreceptor zone of Crustacea is composed of many subunits, the ommatidia. Each ommatidium presents a separate optical unit containing eight retinular cells in the group of decapods.[40] The retinular cells appear to contain enkephalin-like material. At the base of the receptor zone bundles of axons penetrate the fenestrated lamella, often called basement membrane. The axons pass the area of vascular sinuses which occupy the space proximal to the fenestrated lamella and enter the lamina ganglionaris (LG). The LG enkephalin-

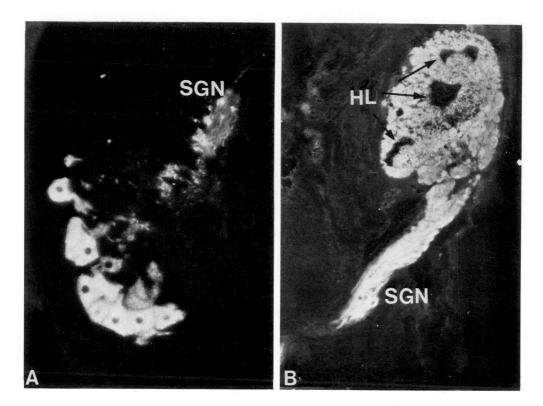

FIGURE 4.   Neurosecretory cells and neurohemal organ of *Carcinus maenas*. (A) Immunofluorescence of hyperglycemic hormone-producing perikarya in the medulla terminalis ganglionic X-organ (MTXO). (Magnification × 500.) (B) From the MTXO-cells the hyperglycemic hormone is transported by the sinus gland nerve (SGN) to the neurohemal organ. HL, hemolymph lacune. Immunofluorescence. (Magnification × 200.)

positive cells are found in the distal glial layer and in the external plexiform layer along with small varicose fibers running in the direction of the first chiasm, which connects the LG with the ME. Axons containing presumably neurosecretory granules are present in the LG (Figure 5) as observed previously.[41] Multiple points of synaptoid contact can be observed between monopolar cells and the retinular cell axons within the optic cartridges of the lamina.[40] No axonal connection between the enkephalin-like material containing axons in the LG and the SG has ever been observed.

Sinus Gland Enkephalin-Like Material

The sinus gland is closely packed with axon terminals of neurosecretory cells of the eyestalks. They are grouped around a large central hemolymph lacuna. The bulk of terminals is formed by axons derived from the medullar terminalis ganglionic X-organ (Figure 6), the site of the hyperglycemic hormone-producing cells.[20] Besides the mass of terminals with accumulated hyperglycemia hormone, a limited number of axons store enkephalin-like material (Figure 7). These terminals are filled with electron-dense granules, often oval or elongated, having a diameter of 82 ± 23 nm (n = 558) (Figure 8). Axon-axonic synapses have not been found in *Carcinus;* however, synaptoid contacts between axon terminals in the prawn, *Palaemon paucipes,* have been demonstrated.[42] Less than 5% of all terminals appear to contain enkephalin-like material. The origin of the perikarya from where the axons arise is unknown. Intrinsic cells[43] and glial cells do not react with anti-Leu-enkephalin serum.

Extracts from sinus glands of *Carcinus* and *Orconectes* displaced (3-iodotyrosyl-^{125}I) enkephalin (5-L-leucine) from the antiserum in a RIA. The overall amount of stored enkephalin-like material was about 0.6 pmol/mg total sinus gland protein (1 SG: 5 μg protein).

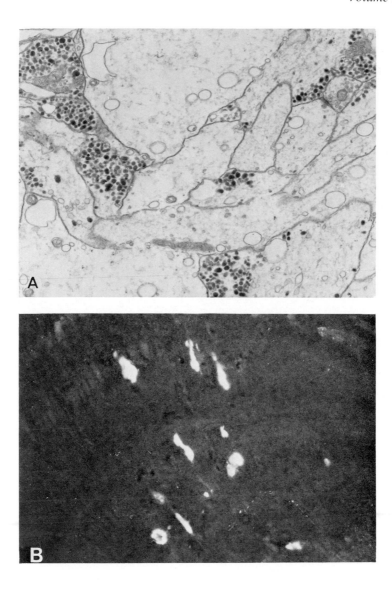

FIGURE 5. Lamina ganglionaris (LG). (A) Ultrastructure of axons with neurose-cretory granules in the LG of *Orconectes limosus*. (Magnification × 18,000.) (B) Immunoreactive neurons and axons in the LG of *Carcinus maenas*. (Magnification × 200.)

No cross reactivity with the hyperglycemic hormone was detected by means of the radioimmunoassay.

Enkephalin-Like Material in the Optic Ganglia

In all eyestalk ganglia, enkephalin-like material was demonstrated in axons and in small neurons. The neurons are situated in the lamina, the ME, MI, and MT, but never in groups of cells. In the medulla externa, these cells are in the neuropil of the ganglion. The same arrangement was found in the medulla interna. Between the MI and the MT, several axons were immunopositive, but no tractus was observed to enter the sinus gland. In the distal portion of the MT, situated dorsolaterally, immunoreactive enkephalin was found in a multilamellar structure, the organ of Bellonci. Attached to mtXo-sinus gland-tract multiple enkephalin-positive axons were observed. No immunoreactive-enkephalin fibers were found

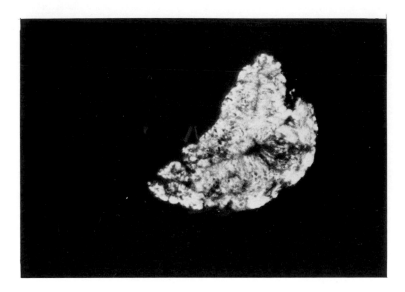

FIGURE 6.   Most of the axon terminals of the sinus gland accumulate the crustacean hyperglycemic hormone, demonstrated by a bright and distinct immunofluorescence in *C. maenas*. (Magnification × 200.)

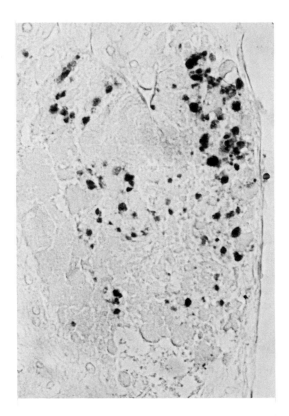

FIGURE 7.   Immunoreactive enkephalin in the sinus gland of *C. maenas*. Note the spotted pattern of distribution of the immunopositive axon terminals. (Magnification × 480.) (From Mangerich, S., unpublished. With permission.)

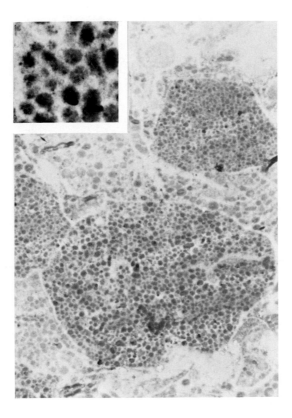

FIGURE 8. Electron micrograph showing the immunocyto-chemical localization of enkephalin-like material in axon terminals and neurosecretory granules (inset) of the sinus gland of *C. maenas*. (Magnification × 11,750; inset, magnification × 45,000.) (From Dircksen, H., unpublished. With permission.)

within the tractus (Figure 9). In the ventromedian portion of the MT, numerous axons with immunoreactive enkephalin and a small number of neurons were found in the close vicinity of the large CHH-producing neurosecretory cells of the mtXo. Dendritic arborizations of the CHH somata in this area have been reported.[17,18] Glantz et al.[44] occasionally observed neurons in this branching area which were inhibited by illumination of the cornea. The period of latency was 20 to 40 msec, depending on the intensity of illumination. The neurosecretory cells of the mtXo, however, seemed to be insensitive to the illumination.

Enkephalin Immunoreactivity in the Pericardial Organs

Immunohistological observations indicate the presence of enkephalin-like material in the pericardial organ of the crab, *Carcinus maenas*. The trunks of single axons showed a distinct immunoreaction. They are embedded in loosely packed connective tissue and hemolymph spaces. The axons terminate more diffusely at the margin of the trunks. Terminating fields with enkephalin-like material were in close contact with the pericardial blood sinus.

Inhibition of Hyperglycemia by Synthetic Leu-Enkephalin In Vivo and In Vitro

Injections of 1.5 nM snythetic Leu-enkephalin in the fiddler crab, *Uca pugilator*, decreases the blood glucose level by 50% after 90 min.[52] When injected together with extracts of 0.1 sinus gland equivalents which are known to raise the blood sugar levels dramatically,[45] only about 65% of the glucose level was obtained compared to injections of the same dose of sinus gland extracts without Leu-enkephalin. In vitro experiments to localize the site of action of the synthetic enkephalin were then performed. Isolated eyestalks of the crayfish,

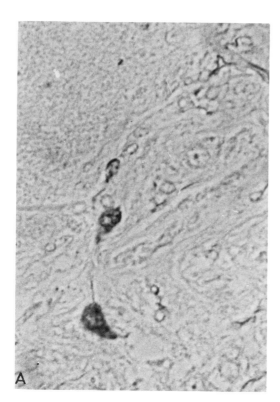 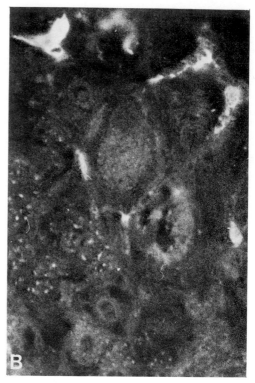

FIGURE 9.   Immunoreactive enkephalin in neurons and axons of the eyestalk of ganglia of *C. maenas*. (A) Small neurons with 8 to 12 μm diameter are found in the lamina terminalis. (Magnification × 760.) (B) Axons with enkephalin-like material in close proximity to the hyperglycemic hormone-producing cells of the medulla terminalis ganglionic X-organ. (Magnification × 430.)

Orconectes limosus, dissected from the cuticle were perfused with saline, supplemented either with serotonin, which increases the release of the crustacean hyperglycemic hormone (CHH), or with Leu-enkephalin, which appears to decrease it. As a control, perfusion was performed with both serotonin and enkephalin or with saline exclusively (Figure 10). The present results demonstrate that the release of CHH from sinus glands is inhibited by Leu-enkephalin. Since no release of neurosecretory material from the X-organ cells has been observed, it is tempting to speculate that the sinus gland is the site of the inhibitory action of injected enkephalin.

Isolation of Enkephalin-Like Peptides from the Sinus Gland

Acetic acid extracts of sinus glands were subjected to high-performance liquid chromatography. Separation was achieved by a μ-Bondapak® phenyl reversed-phase column and a solvent system consisting of trifluoroacetic acid and acetonitrile. Enkephalin-like peptides were detected in fractions with the same retention times as synthetic Met- and Leu-enkephalin and Met-enkephalin-Arg^6-Phe^7. The relevant fractions were identified in a radioimmunoassay by the use of an antiserum Leu-enkephalin (N1551; Amersham) which shows crossreactivity with Met-enkephalin and Met-enkephalin-Arg^6-Phe^7. Approximately 0.33 pmol Leu-enkephalin was found in 150 lyophilized sinus glands, which contain a total of 675 μg protein. Corrected in regard to their crossreactivity to the anti-Leu-enkephalin, 45 fmol Met-enkephalin and 150 fmol Met-enkephalin-Arg^6-Phe^7 were calculated for the same batch. The total amount of immunoreactive enkephalin in the eyestalk ganglia has not been measured at this time.

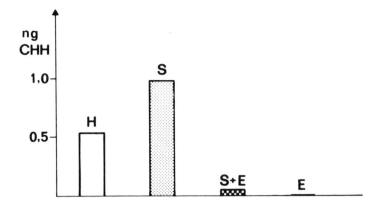

FIGURE 10.   Crustacean hyperglycemic hormone (CHH) release from sinus glands of isolated eyestalk ganglia from *Orconectes limosus*. Ganglia were perfused with serotonin (S), Leu-enkephalin (E), or serotonin and Leu-enkephalin (S + E). Harreveld saline was used as control. CHH was determined by the use of a radioimmunoassay. (From Resch, G., unpublished. With permission.)

Immunoreactive Molluscan Cardioexcitatory Tetrapeptide

Since the molluscan cardioexcitatory tetrapeptide (FMFRamide: Phe-Met-Arg-Phe-NH$_2$) has been characterized[46] in the CNS of the clam *Macrocallista nimbosa*, several reports have supported the idea that biologically active peptides may be distributed in many phyla of the animal kingdom as phylogenetic precursors.[47] Boer et al.[48] identified immunoreactive FMRFamide in insects, a fish, and in mouse brain. Dockray et al.[49] reported the existence of FMRF-like material in the brain of rats and cows and in the gastrointestinal system of rats, dogs, and chickens. FMRFamide is of particular interest in comparative studies, because it presents the C-terminal tetrapeptide of the heptapeptide Met-enkephalin-Arg[6]-Phe[7] and is related to the C-terminal amino acid sequence of cholecystokinin and gastrin.

FMRFamide-Like Material in the Eyestalk Ganglia of Crustaceans

The localization of FMRFamide-like material in the eyestalk ganglia of the prawn, *Palaemon serratus* has been demonstrated.[39] The authors described about 25 to 30 FMRFamide-positive cells in the lamina ganglionaris and the medulla terminalis. In the MT, five types of neurosecretory cells were found, situated near the organ of Bellonci and in two neurosecretory areas called mtXo$_1$ and mtXo$_2$. A network of FMRFamide-positive axons emerge from the neuropil of the MT. In the dorsolateral portion of the MT a tractus is formed and their individual axons split up in close contact with a distinct unstained axon bundle (Figure 11).

By using the same antiserum against FMRFamide in the crab, *C. maenas*, FMRFamide-like material has been localized in cells of the MT and the MI. Clusters of immunopositive neurons were found between the MT and the MI (unpublished observations of our laboratory; Figure 12). In accordance with Jacobs and Van Herp,[39] no immunoreactive axon terminals were detected in the sinus gland (Figure 13). In the vicinity of the CHH-producing perikarya of the mtXo multiple axons, a FMRFamide-like distribution pattern was observed (Figure 14). Attached to the mtXo gland tract various immunoreactive fibers were traced, but no axons were found running along with the tractus. The FMRFamide/Met-enkephalin-Arg[6]-Phe[7]-like immunoreactive axons seem to terminate on other neural elements; as a result the substance might have a neurotransmitter or neuromodulatory function.[39] Although immunoreactivity was completely obliterated after preabsorption of the antiserum with synthetic FMRFamide, the identification and isolation of the FMRFamide-like material have to be

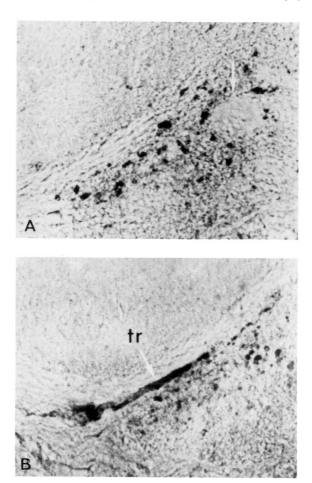

FIGURE 11.   FMRFamide-like material in axons of the medulla terminalis (MT) of *Palaemon serratus*. (A) Multiple immunopositive axons attached to an unstained nerve tract (arrow) in the dorsolateral
portion of the MT. (Magnification × 640.) (B) Dorsomedially situated
axons in the MT. Note the individual axons in contact to the unstained
nerve. tr, Tractus. (Magnification × 640.) (From Jacobs, A. A. C.
and Van Herp, F., *Cell Tissue Res.*, 235, 601, 1984. With permission.)

carried out. Studies on the crossreactivity of the antiserum with the Met-enkephalin heptapeptide are under progress.

PUTATIVE FUNCTIONS OF OPIOID-LIKE PEPTIDES IN CRUSTACEANS

On the basis of in vivo and in vitro experiments, it is suggested that at least Leu-enkephalin
might inhibit the release of CHH from sinus glands of the fiddler crab, *U. pugilator*. This
is supported by anatomical findings which demonstrate distribution of axonal pathways in
the optic ganglia and by electrophysiological results reported by Glantz et al.[44] In the mtXo
of *Pacifastacus* the authors recorded occasionally light-induced inhibitory responses in the
electrical activity of neurons. In the CHH-producing neurosecretory cells of the mtXo of
Orconectes limosus protein synthesis is significantly stimulated at the beginning of the dark
phase as revealed by ^{35}S-cysteine incorporation.[50] In the crayfish, the highest blood glucose
level occurred during the night.[51] These physiological responses seemed to be triggered by

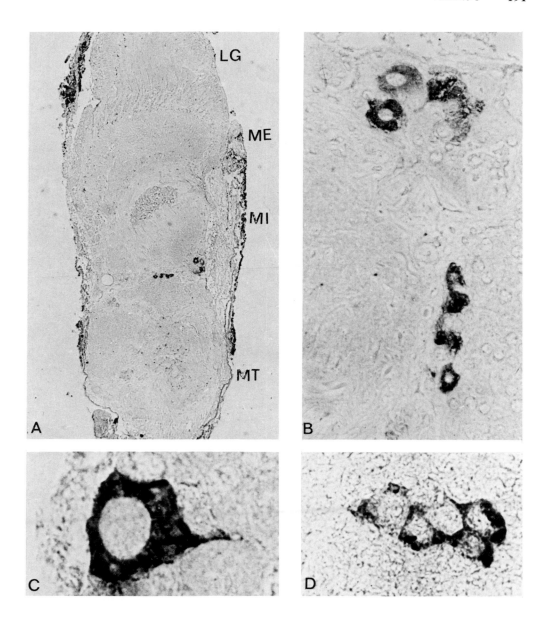

FIGURE 12.   Clusters of immunoreactive FMRFamide cells in the eyestalk ganglia of the crab, *Carcinus maenas*, and the prawn, *Palaemon serratus*. (A) Two types of neurons at the distal margin of the MT of *C. maenas*. (Magnification × 76.) (B) Neurons of the same section as shown in (A) with a distinct immunoreaction. The diameter of the upper cells is approximately 20 to 22 μm, of the lower cells 11 to 13 μm. (Magnification × 480.) (C) In *P. serratus* approximately 10 cells were found lateroexternally to the MTXO₁, situated in the dorsolateral portion of the MT, with a diameter of 28 μm. (Magnification × 1100.) (D) Group of FMRFamide-positive cells in the MT of *P. serratus*. (Magnification × 690.) (E) Longitudinal section of cells containing FMRF-like material in the MT of P. serratus. (Magnification × 576.) (F) Adjacent section of (E) incubated with anti-FMRFamide serum previously absorbed with synthetic FMRFamide. (Magnification × 576.) For abbreviations, see Figure 2. [(A) and (B) from Mangerich, S., unpublished; (C) to (F) from Jacobs and Van Herp.³⁹]

illumination conditions; light inhibits the CHH synthesis in the mtXo via neurons from the cornea. The occurrence of enkephalin-like material in the photoreceptor cells may suggest the involvement of opioid-like peptides in the modulation of light stimuli on blood sugar levels. Full characterization of the immunoreactive material and its physiological interactions

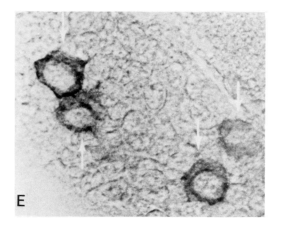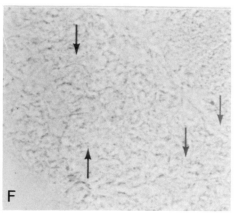

FIGURES 12E and 12F

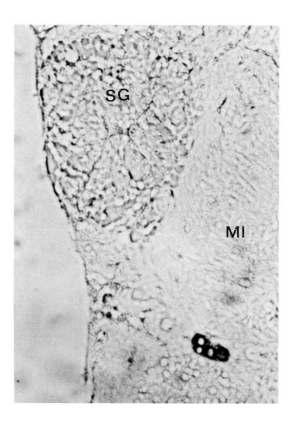

FIGURE 13.   In the sinus gland of *C. maenas*, FMRFamide immunoreaction has been demonstrated. Small immunopositive neurons are
situated in the medulla interna (MI) near the neurohemal organ. (Magnification × 190.) (From Mangerich, S., unpublished. With permission.)

must await its biochemical isolation. On the ultrastructural level synaptoid contacts between
axons containing opioid-like material and CHH neurons have to be *verified*, even though
they are extremely close to each other. Work in this direction is in progress in several
laboratories.

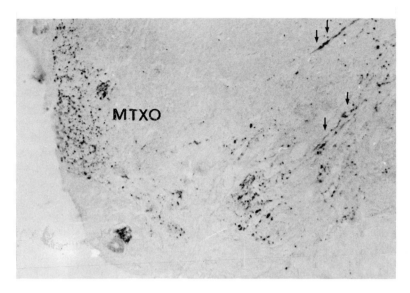

FIGURE 14.   FMRFamide-immunopositive axons in the medulla terminalis. Axons split-
ting up in the area of the medulla terminalis ganglionic X-organ (MTXO) of *C. maenas.*
A diffuse tractus is present in the central neuropil (arrows). Note immunopositive cells
proximal to the MTXO. (Magnification × 300.) (From Mangerich, S., unpublished.
With permission.)

ACKNOWLEDGMENT

Supported by MWF/NRW Grant FA 8996.

REFERENCES

1. **Greenberg, M. J. and Price, D. A.,** Invertebrate neuropeptides: native and naturalized, *Ann. Rev. Physiol.,*
   45, 271, 1983.
2. **Cooke, I. M. and Sullivan, R. E.,** Hormones and neurosecretion, in *The Biology of Crustacea,* Bliss, D.
   E., Ed., Vol. 3, Adwood, H. L. and Sandeman, D. C., Eds., Academic Press, New York, 1982, 205.
3. **Keller, R.,** Biochemistry and specificity of the neurohemal hormones in Crustacea, in *Neurohemal Organs
   of Arthropods,* Gupta, A. P., Ed., Charles C Thomas, Springfield, Ill., 1983, 118.
4. **Knowles, F. G.,** Hormone production with the nervous system of a crustacean, *Nature (London),* 176,
   564, 1951.
5. **Maynard, D. M. and Maynard, E.,** Thoracic neurosecretory structures in Brachyura. III. Microanatomy
   of peripheral structures, *Gen. Comp. Endocrinol.,* 2, 12, 1962.
6. **Alexandrowicz, J. S. and Carlisle, D. B.,** Some experiments on the function of the pericardial organs in
   Crustacea, *J. Mar. Biol. Assoc. U.K.,* 32, 175, 1953.
7. **Florey, E. and Florey, E.,** Uber die mogliche Bedeutung von Enteramin (5-Oxy-Tryptamin) als nervöse
   Aktionssubstanz bei Cephalopoden und dekapoden Crustaceen, *Z. Naturforsch.,* 9b, 58, 1954.
8. **Sullivan, R. E.,** Stimulus-coupled ³H-serotonin release from identified neurosecretory fibres in the spiny
   lobster, *Panulirus interruptus, Life Sci.,* 22, 1429, 1978.
9. **Livingstone, M. S., Schaeffer, S. F., and Kravitz, E. A.,** Biochemistry and ultrastructure of serotonergic
   nerve endings in the lobster: serotonin and octopamine are contained in different nerve endings, *J. Neurobiol.,*
   12, 27, 1981.
10. **Evans, P. D., Kravitz, E. A., and Talamo, B. R.,** Octopamine release at two points along lobster nerve
    trunks, *J. Physiol. (London),* 262, 71, 1976.
11. **Sullivan, R. E. and Barker, D. L.,** Octopamine increases cyclic AMP content of crustacean ganglia and
    cardiac muscle, *Neurosci. Abstr.,* 1, 394, 1975.
12. **Sullivan, R. E.,** Pharmacological investigation of crayfish hindgut: the actions of biogenic amines, ace-
    tylcholine and peptide hormone, *Physiologist,* 23, 16, 1980.

13. **Maynard, D. M. and Welsh, J. H.,** Neurohormones in the pericardial organs of brachyuran Crustacea, *J. Physiol. (London),* 149, 215, 1959.
14. **Sullivan, R. E.,** A proctolin-like peptide in crab pericardial organs, *J. Exp. Zool.,* 210, 543, 1979.
15. **Keller, R. and Jaros, P. P.,** Immunological studies on crustacean hyperglycemic hormones, presented at 9th Int. Symp. Comparative Endocrinol., Hong Kong, Dec. 7 to 11, 1981.
16. **Chaigneau, J.,** Neurohemal organs in Crustacea, in *Neurohemal Organs of Arthropods,* Gupta, A. P., Ed., Charles C Thomas, Springfield, Ill., 1983, 53.
17. **Andrew, R. D., Orchard, I., and Saleuddin, A S.,** Structural re-evaluation of the neurosecretory system in the crayfish eyestalk, *Cell Tissue Res.,* 190, 235, 1978.
18. **Jaros, P. P.,** Tracing of neurosecretory neurons in crayfish optic ganglia by cobalt iontophoresis, *Cell Tissue Res.,* 194, 297, 1978.
19. **Jaros, P. P.,** Immunocytochemical demonstration of the neurosecretory X-organ complex in the eyestalk of the crab *Carcinus maenas, Histochemistry,* 63, 303, 1979.
20. **Jaros, P. P. and Keller, R.,** Immunocytochemical identification of hyperglycemic hormone-producing cells in the eyestalk of *Carcinus maenas, Cell Tissue Res.,* 204, 379, 1979.
21. **Gorgels-Kallen, J. L. and Van Herp, F.,** Localization of crustacean hyperglycemic hormone (CHH) in the X-organ sinus gland complex in the eyestalk of the crayfish, *Astacus leptodactylus* (Nordmann, 1842), *J. Morphol.,* 170, 347, 1981.
22. **Jaros, P. P. and Keller, R.,** Radioimmunoassay of an invertebrate peptide hormone — the crustacean hyperglycemic hormone, *Experientia,* 35, 1252, 1979.
23. **Andrew, R. D.,** Neurosecretory pathways supplying the neurohemal organs in crustacea, in *Neurohemal Organs of Arthropods,* Gupta, A. P., Ed., Charles C Thomas, Springfield, Ill., 1983, 90.
24. **Fernlund, P. and Josefsson, L.,** Crustacean color-change hormone: amino acid sequence and chemical synthesis, *Science,* 1977, 173, 1972.
25. **Stone, J. V., Mordue, W., Batley, K. E., and Morris, H. R.,** Structure of locust adipokinetic hormone, a neurohormone that regulates lipid utilisation during flight, *Nature (London),* 263, 207, 1976.
26. **Fernlund, P.,** Structure of a light-adapting peptide hormone from the shrimp, *Pandalus* borealis, *Biochim. Biophys. Acta,* 439, 17, 1976.
27. **Keller, R.,** Purification and amino acid composition of the hyperglycemic neurohormone from the sinus gland of *Orconectes limosus* and comparison with the hormone from *Carcinus maenas, J. Comp. Physiol.,* 141, 445, 1981.
28. **Martin, G., Keller, R., Kegel, G., Besse, G., and Jaros, P. P.,** The hyperglycemic neuropeptide of the terrestrial isopod, *Porcellio dilatatus.* I. Isolation and characterization, *Gen. Comp. Endocrinol.,* 50, 208, 1984.
29. **Leuven, R. S. E. W., Jaros, P. P., Van Herp, F., and Keller, R.,** Species-or group-specificlty in biological and immunological studies of crustacean hyperglycemic hormone, *Gen. Comp. Endocrinol.,* 46, 288, 1982.
30. **Jaros, P. P. and Keller, R.,** Improvement and first application of a RIA for an invertebrate neurosecretory peptide hormone, in *Neurosecretion,* Farner, D. S. and Lederis, K., Eds., Plenum Press, New York, 1981, 517.
31. **Huberman, A., Aréchiga, H., Cimet, A., de la Rosa, J., and Aramburo, C.,** Isolation and purification of a neurodepressing hormone from the eyestalk of *Procambarus bouvieri* (Ortmann), *Eur. J. Biochem.,* 99, 203, 1979.
32. **Martin, R., Frösch, D., Weber, E., and Voigt, K.-H.,** Met-enkephalin-like immunoreactivity in a cephalopod neurohemal organ, *Neuroscience Lett.,* 15, 253, 1979.
33. **Zipser, B.,** Identification of specific leech neurones immunoreactive to enkephalin, *Nature (London),* 283, 857, 1980.
34. **Alumets, J., Håkanson, R., Sundler, F., and Thorell, J.,** Neuronal localization of immunoreactive enkephalin and β-endorphin in the earthworm, *Nature (London),* 279, 805, 1979.
35. **Gros, C., Lafon-Cazal, M., and Dray, F.,** Présence de substances immunoréactivement apparantées aux enkephalines chez un Insecte, *Locusta migratoria, C. R. Acad. Sci. D,* 287, 647, 1978.
36. **Leung, M. and Stefano, G. B.,** Isolation of molluscan opioid peptides, *Life Sci.,* 33 (Suppl. 1), 77, 1983.
37. **Mancillas, J. R., McGinty, J. F., Selverston, A. I., Karten, H., and Bloom, F. E.,** Immunocytochemical localization of enkephalin and substance P in retina and eyestalk neurones of lobster, *Nature (London),* 293, 576, 1981.
38. **Jaros, P. P. and Keller, R.,** Localization of Leu-enkephalin-like material in the brain and neurohaemal organ of the brachyuran *Carcinus maenas* and the astacidean *Orconectes limosus,* presented at the 12th Congr. Eur. Comp. Endocrinol., Sheffield, July 31 to August 5, 1983.
39. **Jacobs, A. A. C. and Van Herp, F.,** Immunocytochemical localization of a substance in the eyestalk of the prawn, *Palaemon serratus,* reactive with an anti-FMRFamide rabbit serum, *Cell Tissue Res.,* 235, 601, 1984.

40. **Shaw, S. R. and Stowe, S.,** Photoreception, in *The Biology of Crustacea*, Vol. 3, Bliss, D. E. and Atwood, H. C., Eds., Academic Press, New York, 1982, 291.

41. **Hamori, J. and Horridge, G. A.,** The lobster optic lamina. IV. Glial cells, *J. Cell Sci.*, 1, 275, 1966.

42. **Hisano, S.,** Synaptic junctions in the sinus gland of the freshwater prawn, *Palaemon paucidens, Cell Tissue Res.*, 189, 435, 1978.

43. **May, B. A. and Golding, D. W.,** Aspects of secretory phenomena within the sinus gland of *Carcinus maenas* (L.). An ultrastructural study, *Cell Tissue Res.*, 228, 245, 1983.

44. **Glantz, R. M., Kirk, M. D., and Arechiga, H.,** Light input to crustacean neurosecretory cells, *Brain Res.*, 265, 307, 1983.

45. **Keller, R. and Andrew, E. M.,** The site of action of the crustacean hyperglycemic hormone, *Gen. Comp. Endocrinol.*, 20, 572, 1973.

46. **Price, D. A. and Greenberg, M. J.,** Purification and characterization of a cardioexcitatory neuropeptide from the central ganglia of a bivalve mollusc, *Prep. Biochem.*, 7, 261, 1977.

47. **Scharrer, B.,** Peptidergic neurones: facts and trends, *Gen. Comp. Endocrinol.*, 34, 50, 1978.

48. **Boer, H. H., Schot, L. P. C., Veenstra, J. A., and Reichelt, D.,** Immunoctyochemical identification of neural elements in the central nervous system of a snail, some insects, a fish, and a mammal with an antiserum to the molluscan cardioexcitatory tetrapeptide FMRF-amide, *Cell Tissue Res.*, 213, 21, 1980.

49. **Dockray, G. J., Vaillant, C., and Williams, R. G.,** New vertebrate brain-gut peptide related to a molluscan neuropeptide and an opioid peptide, *Nature (London)*, 293, 656, 1981.

50. **Jaros, P. P.,** Cellular rhythmicity of RNA and protein synthesis in the crustacean hyperglycemic hormone-producing neurosecretory perikarya of the X-organ studied by autoradiography, *Eur. J. Cell Biol.*, 22, 499, 1980.

51. **Hamann, A.,** Die neuroendokrine Steuerung tagesrhythmischer Blutzuckerschwankungen durch die Sinusdruse beim Flußkrebs, *J. Comp. Physiol.*, 89, 197, 1974.

52. **Jaros, P. P.,** in preparation.

LOCALIZATION AND CHARACTERIZATION OF OPIOID-LIKE PEPTIDES IN THE NERVOUS SYSTEM OF THE BLOWFLY *CALLIPHORA VOMITORIA*

Hanne Duve, Alan Thorpe, and Alan Scott

SUMMARY

Immunochemical studies on the blowfly, *Calliphora vomitoria*, have shown the presence of opioid-like peptides within the nervous tissues. In particular, an α-endorphin-like peptide, located by means of immunocytochemistry in certain of the median neurosecretory cells of the brain, has been isolated from acid extracts and partially purified by means of Sephadex gel filtration, ion-exchange chromatography, and high-performance liquid chromatography. Other peptides studied by means of immunocytochemistry and/or radioimmunoassay of extracts include both Met- and Leu-enkephalin, β-endorphin, and an ACTH-immunoreactive material.

INTRODUCTION

Knowledge of the structure and function of the endogenous opioid peptides and their receptors in vertebrates has greatly increased over the last decade. In contrast, there have been rather fewer studies of these peptides in invertebrates. Indeed, one phylogenetic study went so far as to suggest that below the level of the vertebrates there were neither morphine-like factors nor opioid receptors.[1] Subsequently, however, considerable evidence from several different lines of research has been presented to the contrary.

Immunocytochemical studies have pointed to the existence of a variety of invertebrate opioid-like substances in neurons of insects,[2-8] annelids,[9-12] crustaceans,[13,14] molluscs,[15-19] and protochordates.[20] In support of these cytological data, several studies have provided evidence of opioid-like peptidergic material in invertebrate tissue extracts as determined by radioimmunoassay.[21-23] Very recently, the structures of some of these peptides have been elucidated.[24-26] Opioid receptor-binding studies have been carried out in both insects and molluscs[27-31] and there are also studies that suggest a physiological role for opioid peptides in invertebrates.[32-34]

In the blowfly, *Calliphora vomitoria* we have previously demonstrated the presence of a variety of neuropeptides related to the gastroenteropancreatic peptides of vertebrates[35-38] and in the present account we report on those peptides of *Calliphora* that are related immunologically to the endorphins and enkephalins of vertebrates.

Earlier, by means of immunocytochemistry (ICC), we have described the localization of an α-endorphin-like peptide in adult specimens of *Calliphora* aged between 3 and 30 days.[7] In the present study, we have examined individuals from eclosion (day 0) through to the age of 6 to 7 weeks with the same α-endorphin antisera as used previously. We have made some preliminary observations with a range of Met- and Leu-enkephalin antisera and also with antisera directed against β-endorphin and ACTH.

We have carried out a partial purification of the α-endorphin-like peptide and report here on the presence of Met-enkephalin-, β-endorphin-, and ACTH-like peptides in the same extracts.

MATERIALS AND METHODS

Pupae of *Calliphora vomitoria* from a commercial supplier (Kent Bait Company) were maintained in the laboratory prior to eclosion. Flies for immunocytochemistry were isolated

Table 1
ANTISERUM DETAILS

Antiserum	Immunizing antigen	Use in present study		Donor/Ref.
		ICC	RIA	
αE1(7)	Synthetic α-endorphin	√	√	Dr. S. Jackson, Pituitary Hormone
αE2(7)	conjugated to	√	√	Laboratory, Dept. Chemical Pathol-
αE3(7)	thyroglobulin	√	√	ogy St. Bartholomew's Hospital,
				London (Reference 42)
LP 9	β-endorphin	√	—	As above
B 42	β-endorphin	√	—	As above
Reference letter or	β-endorphin	—	√	Dr. M. Fenger, Dept. Clinical Chem-
number not known				istry, Rigshospitalet, Copenhagen,
				Denmark
KA3	Met-enkephalin coupled	√	√	Prof. L.-I. Larsson, Rigshospitalet,
	to BSA			Copenhagen, Denmark
E₂II	Met-enkephalin	√	√	Dr. S. Jackson (address as above)
3LQ	Leu-enkephalin	√	—	Dr. S. Jackson (address as above)
Simon	ACTH	√	—	Prof. L.-I. Larsson (address as above)
Reference letter or	Synthetic human ACTH	—	√	Dr. M. Fenger (address as above)
number not known	(1-24) coupled to BSA			(Reference 43)

in small cages and fed sugar, water, and Ovaltine® until the required age (0 to 72 days). For the biochemical studies, large batches of flies were fed *ad libitum* on this diet for 1 week, after which they were killed and frozen by means of Dri-Ice®. They were held for periods of up to 1 month at $-20°C$ before extraction.

Immunocytochemistry

The procedure adopted for the immunocytochemical studies has been described in full elsewhere.[7,39,40] Briefly, animals were fixed in Bouin's fluid and various nervous tissues (brain, corpus cardiacum/corpus allatum complex, thoracic ganglion) dissected out, embedded in paraffin wax, and sectioned at 6 μm. Commonly, the peroxidase-antiperoxidase (PAP) technique was used, although on occasions it was found useful to use the indirect fluorescence method. Some details of the antisera used in the study are given in Table 1. Appropriate concentrations of antisera for ICC ranged from 1:500 to 1:2500. Controls carried out included the use of antigen-saturated antisera and replacement of the specific antiserum by normal or hyperimmune serum.

Purification Studies

The procedure for extraction and purification of peptides was designed initially for the α-endorphin-like material and was based on methods adopted in previously published vertebrate studies. During the course of the investigation, eluates from certain chromatographic steps were also assayed for other peptides. In this communication full details of the partial purification of the *Calliphora* α-endorphin-like peptide will be given, but reference will also be made to preliminary results of opioid-related peptides in order to give as complete an account of this group of peptides as possible and in order to give an indication of the future direction and scope of our studies on the blowfly.

For the initial extraction, either separated heads, bodies (thorax plus abdomen), or whole flies were used. Batches of up to 500 g of tissues were homogenized briefly using a Waring Blendor® (up to 1 min) in cold 0.2*M* HCl and extracted in the same fluid (1:10 w/v) overnight at 4°C. After filtration through muslin to remove crude debris (particles of cuticle, etc.) and through a Hyflo bed which served to remove some of the pigments and lipid materials, centrifugation at 28,000 × g produced an opalescent dark brown fluid that was then partially freeze-dried. By this means the initial extraction volume was reduced by about 80% and the charring of the residue, which occurred if taken to complete dryness, was avoided. The fluid remaining after this procedure was allowed to stand for several hours at 4°C. Further centrifugation at 70,000 × g removed the flocculent precipitate which formed on standing and the remaining clear brown supernatant was subjected to Sephadex G50-SF gel filtration. All chromatography was carried out at room temperature (22°C). Samples (200 μℓ) of the eluate (1% formic acid) were freeze-dried and subjected to RIA for α-endorphin and other opioid peptides.

The fractions containing α-endorphin immunoreactivity were pooled, freeze-dried, and subjected to CM 52 cellulose ion-exchange chromatography. In pilot experiments, both linear gradients and step-wise elution procedures were adopted, with NaCl in citrate buffer, pH 3.25, as the eluant. In the experiments described here, a step-wise gradient of 0 to 1.0 M NaCl in 0.05M increments was used. Most of the material eluted in the four stages between 0.15 and 0.35M NaCl. The α-endorphin immunoreactive material was desalted by means of Sephadex G10 chromatography either as a single pool of material if a linear gradient had been used or as separate peaks of material from the step-wise procedure. It was then subjected to HPLC. Both μ-Bondapak® C$_{18}$ and CN columns (Waters Associates) were used with linear gradients of either acetonitrile to water or methanol to water plus 0.0125% TFA. (Full details of all chromatographic steps are given in the figure legends.)

Radioimmunoassay

Peptides were identified and monitored throughout the purification procedure by means of RIA.

α-Endorphin

The tracer for the RIA of α-endorphin was prepared according to the method of Linde et al.[41] Synthetic human α-endorphin (10 μg; Cambridge Research Biochemicals) was subjected to lactoperoxidase (1.5 mCi ^{125}I; Amersham International) for 5 min (65% incorparation) after which the iodination mixture was applied to 10% polyacrylamide gels (20 × 0.5 cm) for electrophoretic separation. At the completion of electrophoresis (15 min at 2.5 mA per tube followed by 2 hr at 4 mA per tube), the gels were sliced (1.5 or 2.0 mm) and eluted at 4°C overnight with 500 μℓ 0.1 M ammonium bicarbonate buffer, pH 8.0, containing 0.5% human serum albumin. The procedure resulted in two peaks of labeled α-endorphin, both of which were suitable for RIA. The antiserum αEl(7) used in the assay is the same as that used by Duve and Thorpe in earlier ICC studies of *Calliphora*[7,8] and also in the present cytological studies. The characteristics of this antiserum in RIA have been described by its originators[42] who have also demonstrated its high specificity for α-endorphin. In the disequilibrium assay used here, 200 μℓ of column fractions were freeze-dried, redissolved in 200 μℓ of the assay buffer (0.5 M phosphate buffer + 5% BSA + 0.1% mercaptoethanol, pH 7.4), and incubated with 50 μℓ of a 1:2500 dilution of the antiserum for 24 hr. After a further 24 hr with the labeled α-endorphin (50 μℓ) containing approximately 10,000 cpm and giving a 60% maximum binding in the zero tubes, the bound and unbound fractions were separated by adding a mixture of 600 μℓ polyethylene glycol 6000 (20% w/v) and 200 μℓ of either horse or human serum (50% v/v in 0.9% NaCl) and centrifuging at 3000 × g for 1 hr at 4°C. The assay is sensitive over a range of 10 pg to 5 ng α-endorphin per milliliter and with variations in amounts of tracer and antibody it can be extended to almost 20 ng/mℓ.

Met-Enkephalin

The Met-enkephalin tracer was prepared by the same method as that described for α-endorphin. Incorporation of ^{125}I was approximately 40% and the maximum binding to a 1:1500 dilution of either antibody KA3 or E$_2$II (see Table 1) was about 20%. Despite this rather low % binding, it was possible to obtain a standard curve from which peaks of immunoreactivity in the elution profiles could be determined. There was negligible (less than 0.05%) binding of α-endorphin tracer to Met-enkephalin antibodies and vice versa for the Met-enkephalin tracer and α-endorphin antibody.

ACTH and β-Endorphin

An initial survey of the elution profile from Sephadex G50 was carried out by means of RIA for ACTH and β-endorphin in collaboration with Dr. M. Fenger, Department of Clinical Chemistry, Rigshospitalet, Copenhagen, Denmark. The ACTH tracer (1-39) was prepared by Dr. Fenger by means of the chloramine-T method and purified on a C$_{18}$ Sep-Pak® (Waters Associates) with a gradient of methanol and 0.1% TFA. The antiserum used is highly specific for ACTH, reading the 14-17 sequence.[43] ACTH standards (1-39) were in the range 1 to 90 pmol/mℓ. Separation of the free and bound forms was by means of activated charcoal.

RESULTS

Immunocytochemistry

A detailed study of the distribution of the α-endorphin-like peptide of adult *Calliphora* aged between 3 and 30 days has recently been published by Duve and Thorpe.[7] In such specimens, rather large amounts of immunoreactive material appear consistently to fill certain of the median neurosecretory cells (Figure 1A). Immunoreactivity also appears in the long neurosecretory axons within the median nerve bundle and in the cardiac-recurrent nerve which enters the corpus cardiacum. Although this neurohemal organ may be one of the sites from which the material is released into the hemolymph, we have shown that axons containing the peptide project even more posteriorly in the direction of both the heart and the gut. In the present study, we have observed that immediately following eclosion there is very little α-endorphin immunoreactivity in the median neurosecretory cells or their axons, and we have an indication that older flies too (6 to 8 weeks) have a greatly reduced content of this material. In the difficult search for the function of the α-endorphin-like material in *Calliphora*, age-related distribution patterns such as these are clearly of significance and warrant further study in the future.

The Met- and Leu-enkephalin-like immunoreactivities appear separately in different populations of cells, distinct from those that contain α-endorphin-like material. Their distribution within the brain of *Calliphora* is shown diagrammatically in Figure 1, and studies of individual cells and immunoreactive material in the neuropil are seen in Figure 1B to I. We have observed between 20 and 30 of each type of cell, symmetrically arranged in the manner of other vertebrate-type peptides studied in this species.[39,40] (Detailed information on the location of the cells is given in the legend to Figure 1.)

Immunocytochemical studies with antisera against β-endorphin and ACTH, although still at a preliminary stage, suggest that peptides related to these substances are present in *Calliphora*.

Purification of the α-Endorphin-Like Peptide

Results from Sephadex gel filtration (G50-SF) of extracts of whole flies are given in Figure 2. It can be seen that two major peaks of α-endorphin immunoreactivity (A and B) are produced by this chromatographic step. An apparent multiplicity of immunoreactive forms has been a consistent feature of the different α-endorphin purification programs we have followed. Sometimes as many as three distinct peaks have been observed. In some experiments, we have pooled the immunoreactive fractions from individual peaks and sub-

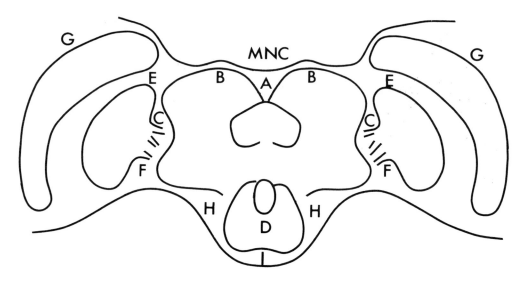

FIGURE 1.   Drawing of a transverse section of the brain of *Calliphora* to demonstrate regions that contain ce,
and axons immunoreactive to opioid peptide antisera. Micrographs (A) through (I) are studies of various region
containing opioid-like immunoreactivity; PAP immunocytochemical technique; Nomarski optics: Antibody (Ab.)
details given in Table 1. (A) Median neurosecretory cells (mnc: arrows) immunoreactive to α-endorphin antiserum
[Ab. αE1(7)]. (B) A lateral neurosecretory cell immunoreactive to Met-enkephalin antiserum (Ab. KA3) positioned
on edge of neuropil. (C) A Met-enkephalin immunoreactive cell (Ab. E₂II) positioned dorsally close to the anterior
optic tract. Large arrows indicate axon hillock and neurite running dorsally. Small arrows indicate beads of
immunoreactive material in neuropil. (D) Beads of neurosecretory material immunoreactive to Leu-enkephalin
antiserum (Ab. 3LQ) ventral to the esophagus. (E) Two cells immunoreactive to Leu-enkephalin antiserum (Ab.
3LQ) positioned dorsally close to the lobula. (F) Immunoreactive material to Met-enkephalin antiserum (Ab. KA3)
in a cell positioned ventrally to the posterior optic tract. Small arrows indicate the axon hillock. Large arrows
show immunoreactive material in axon. Beads of material can also be seen in the vicinity of the perikarya. (G)
Beads of Met-enkephalin immunoreactive material in the lamina indicated by arrows (Ab. KA3). Same section
plane as (F). (H) A Met-enkephalin immunoreactive cell (Ab. E₂II) in the lateral rind of the subesophageal ganglion.
(I) Leu-enkephalin immunoreactive cells (Ab. 3LQ) in the rind of the subesophageal ganglion. Arrows indicate
material in neurites. a opt, Anterior optic tract; mnc, median neurosecretory cells; n, neuropil; oe, esophagus; sog,
subesophageal ganglion. Scale bar: 18.5 μm.

jected them separately to high-performance liquid chromatogrpahy (HPLC). In the experiments described in this report, however, we deferred judgment on differences in the nature
of peaks A and B, assuming that identification could more easily be made at a later stage.
The two peaks of material were therefore pooled, freeze-dried, and subjected to ion-exchange
chromatography. Use of the ion exchanger DE-52 DEAE cellulose and a gradient of 20 to
400 mM ammonium acetate, pH 8.5, with or without 20% acetonitrile (previously used by
workers in the vertebrate field of the ACTH-related peptides), failed to separate the α-
endorphin-like immunoreactive material from the bulk of the contaminating peptides appearing in the void volume. The cation exchanger CM 52 cellulose was also not entirely
satisfactory in that a linear gradient of 0 to 0.5 M NaCl in 0.04 M sodium citrate buffer,
pH 3.25, resulted in a broad peak of α-endorphin immunoreactive material between 0.15
and 0.35 M NaCl. In subsequent experiments with stepwise elution (0.05 M steps) from 0
to 1.0 M NaCl, it was similarly found that the majority of the material eluted in four peaks
between 0.15 and 0.35 M NaCl (Figure 3A).

   The separate peaks A, B, C, and D were desalted by means of Sephadex G10, a step that
on each occasion produced two peaks of immunoreactive material (Figure 3C and D).
Immunoreactive material after the salt peak gave a deliquescent material on freeze-drying
that did not dilute in parallel with the synthetic α-endorphin standards. The immunoreactive
material before the salt peak did, however, dilute in parallel with two different α-endorphin

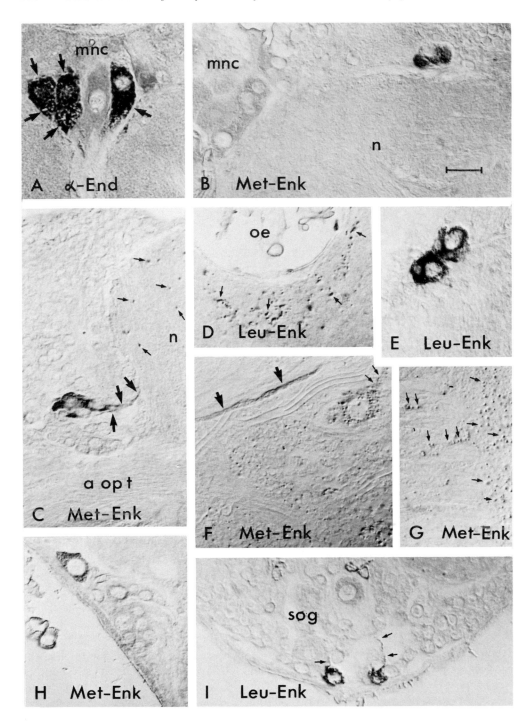

FIGURE 1 continued

antibodies (Figure 4) and this material was subjected to reverse-phase HPLC. The elution profile of material from G10 added to an analytical C_{18} column with a gradient of 16 to 50% acetonitrile with 0.0125% TFA and radioimmunoassayed for α-endorphin is seen in Figure 5A. The material figured is actually from peak C, but α-endorphin immunoreactivity from all of the four peaks behaved similarly, eluting between 29 and 33% acetonitrile.

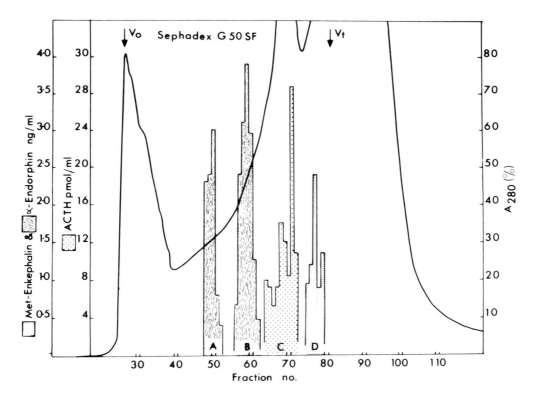

FIGURE 2. Sephadex G50-SF chromatography of 0.2 *M* HCl extract of 700 blowflies (35g), *Calliphora vomitoria*. (Column 1.5 × 54 cm; eluent 1% formic acid; flow rate 74 mℓ/hr; fractions 12.3 mℓ) Samples (200 μℓ) were freeze-dried and radioimmunoassayed for α-endorphin (peaks A and B), ACTH (peak C), and Met-enkephalin (peak D). Void volume V_o and salt peak V_t are arrowed. Absorbance profile A_{280} shown by continuous line.

Synthetic α-endorphin marker, in contrast, elutes at approximately 35% acetonitrile on an identical column (Figure 5A). On CN columns with the same gradient, the fly α-endorphin-like material is held up for a much longer time and elutes at approximately 60% acetonitrile in the second phase of the gradient which extends from 50 to 80% (Figure 5C). With a methanol gradient of 16 to 50% on a C_{18} column, the immunoreactive material elutes at 46 to 48% (Figure 5B).

With polyacrylamide gel electrophoresis, the R_f values of the fly α-endorphin-like material and the standard peptide are very similar, 0.40 compared with 0.41 (Figure 6).

Preliminary Results with Met-Enkephalin, ACTH and β-Endorphin RIAs

Preliminary radioimmunoassays for other opioid and opioid-related peptides have been carried out on the elution profile from the Sephadex G50 columns. The results for ACTH (peak C) and Met-enkephalin (peak D) are shown in Figure 2. Work on the purification of these peptides is now in progress.

We also have an indication that a β-endorphin-like peptide is present in the eluate from G50 fractionation of acid extracts of *Calliphora*. The elution position is between fractions 42 to 46. This initial result requires confirmation by further studies now being undertaken in our laboratory.

DISCUSSION

Evidence is presented here for the existence of several opioid-like peptides in the blowfly, *Calliphora*. In particular, an α-endorphin-like material, known from immunocytochemical

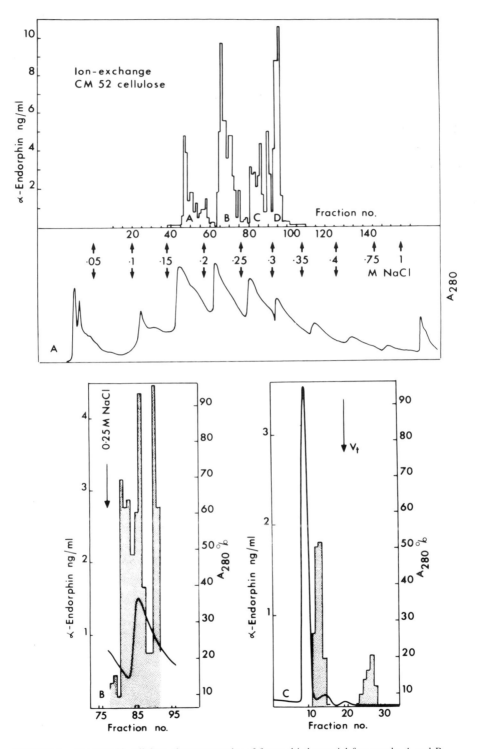

FIGURE 3.   (A) CM 52 cellulose chromatography of freeze-dried material from peaks A and B, Figure 2. (Column 1.1 × 45 cm; flow rate 44 mℓ/hr; fractions 1.35 mℓ.) Elution by a step-wise gradient of 0.05 to 0.4 M NaCl in 0.04 M sodium citrate buffer, pH 3.25, with increasing steps of 0.05 M NaCl. Immunoreactive α-endorphin-like material, top graph; absorbance profile A_{280} lower graph. (B) Detail of peak C obtained with the 0.25M NaCl step (from A). (C) Sephadex G10 chromatography of freeze-dried material in peak C. (Column 1.5 × 40 cm; eluent 1% formic acid; flow rate 70 mℓ/hr; fractions 5.2 mℓ.) Salt peak V_t arrowed. Absorbance profile A_{280} shown by continuous line.

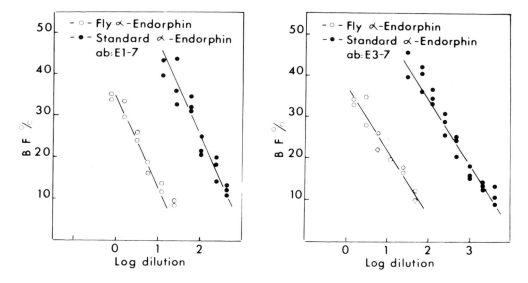

FIGURE 4.   Dilution curves for two α-endorphin antisera E1(7) and E3(7). Comparison of synthetic α-endorphin with corresponding neuropeptides from *Calliphora*. The parallelism indicates common antigenic determinants.

studies to be localized in specific neurosecretory cells in the brain, has been partially purified by a series of extraction and chromatographic procedures. α-Endorphin is known from studies on mammals to be one of the products derived from the processing of the precursor pro-opiocortin.[37,44,45] The three major products of this large common precursor are ACTH, β-LPH, and pro-α-MSH. The β-LPH fragment is further cleaved to give γ-LPH (1-58) and β-endorphin (61-91) which can be further processed to give α-endorphin (61-76).[42,44,46] In lower vertebrates too the sequence of α-endorphin has been demonstrated and shown to be derived from a precursor similar to that of mammals.[27]

Although the sequence of Met-enkephalin appears in β-endorphin (61-65), the Thr-Ser sequence by which it is attached is regarded as a poor cleavage site. It is possible to derive small amounts of the pentapeptide by digestion of β-LPH COOH-terminal fragments,[48] but from all available evidence, it seems more likely that Met-enkephalin has precursors separate from β-LPH or β-endorphin.[49-51] Likewise, Leu-enkephalin is thought to stem from a different precursor (α-neo-endorphin or dynorphin).

As in mammalian studies where the regional distribution of enkephalins is different from that of the endorphins,[52] we have observed in *Calliphora* that Met- and Leu-enkephalin immunoreactivities occur in cells different from those that contain the α-endorphin-like immunoreactivity. Of course, separate cellular localization, either in flies or mammals, does not necessarily indicate different precursor molecules since it could just as well be explained by different processing of a common precursor. Work on β-endorphin and ACTH-like immunoreactive materials both with ICC and RIA of extracts is still at a very early stage and restricted somewhat by availabilities of suitable antisera. Nevertheless, the indications are that substances immunologically related to these vertebrate molecules exist at these prevertebrate levels of evolution.

The question of chemical identity of the peptides and their relationship to mammalian molecules can be solved either by extraction, purification, and amino acid sequence studies or by the newer recombinant DNA techniques. For the moment, we are relying upon purification procedures with the use of mammalian RIAs for identifying and monitoring the peptides. Our studies on the opioid peptides reported here have established that an α-endorphin-like material exists in the tissues of *Calliphora*. By comparision with the performance of synthetic α-endorphin on HPLC and polyacrylamide gel electrophoresis, we have a strong indication that the material is rather similar to the vertebrate peptide. We have

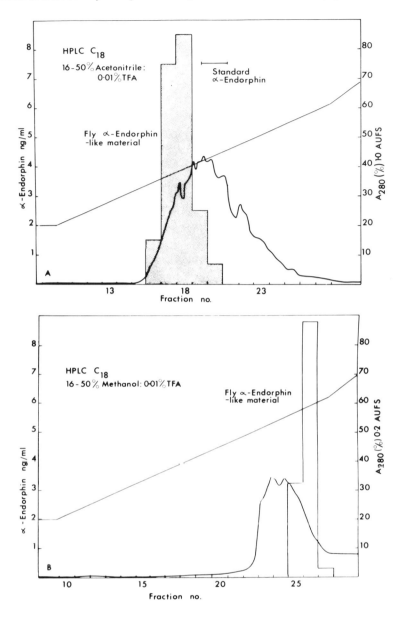

FIGURE 5.   Reverse-phase HPLC of samples of α-endorphin immunoreactivity resolved by cation-exchange chromotography. (A) Analytical HPLC of immunoreactive material from peak C (Figure 3). [Column; μ-Bondapak® C₁₈ Waters Associates; 0.39 × 30cm; flow rate 1.5 mℓ/min; fractions 1.5 mℓ; gradient 16 to 50% acetonitrile to water with 0.0125% TFA; A₂₈₀, 1.0 absorbance unit full scale (AUFS).] Elution position of synthetic α-endorphin indicated by ├──┤. (B) Analytical HPLC of part of freeze-dried material from fractions 17 and 18 from (A) above. All details as for (A), except that methanol replaces acetonitrile in the gradient and AUFS = 0.2. (C) Analytical HPLC of part of freeze-dried material from fractions 17 and 18 above. Details as for (A), except that the column is μ-Bondapak® CN and AUFS = 0.2. The second phase of the gradient is 50 to 80% acetonitrile. In (A), (B), and (C), α-endorphin immunoreactivity is shown as a histogram and absorbance profile A₂₈₀ as a continuous line.

not, however, been able to establish unequivocally that we have a single form of the substance. Indeed, the indications are to the contrary since more than one peak of immu-

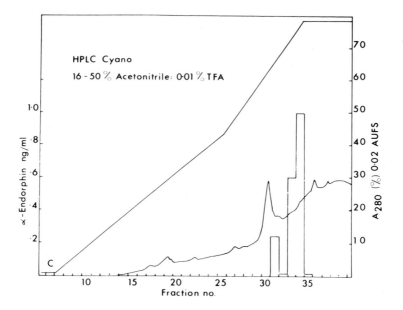

FIGURE 5C

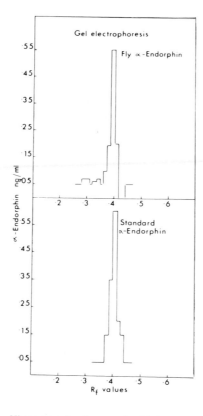

FIGURE 6. Histograms showing α-endorphin immunoreactivity after polyacrylamide gel electrophoresis. Top, fly material; bottom, synthetic α-endorphin. Gel slices (2 mm) eluted with 0.1 M NH$_4$HCO$_3$, pH 8.0, freeze-dried, and immunoassayed. Rf values: fly material, 0.40; synthetic α-endorphin, 0.41.

noreactive material is consistently seen with both Sephadex gel filtration and ion-exchange chromatography. This observation is by no means unusual and the diversity of the opioid peptides is the rule rather than the exception. As examples of this diversity, two forms of β-endorphin were reported by Jackson et al.[42] in a Sephadex gel filtration step in the purification of peptides from the rat pars intermedia, and Zakarian and Smyth[53] reported as many as six peptides having β-endorphin immunoreactivity from the rat pars intermedia and brain. These forms include acetylated derivatives of both β-endorphin$_{1-31}$ and β-endorphin$_{1-27}$.

Future studies with HPLC will enable us to resolve whether or not different forms of α-endorphin-like peptides are truly present in *Calliphora*. We already have some evidence (not given here) that the pattern of peptides seen in extracts of separated heads differs from that of the peptides seen in body extracts, suggesting perhaps that processing to give different forms takes place between the site of production (MNC in the brain) and the site of storage and release into the hemolymph (corpus cardiacum and maybe other neurohemal areas in the thorax and abdomen). This will certainly be an interesting problem for the future. For the moment, however, with the rather broad peaks of immunoreactivity that result from our current HPLC studies, we are not certain that we can differentiate between the individual peaks derived from ion-exchange chromatography.

For the other opioid-like peptides in *Calliphora*, although our results are preliminary, collectively they may be taken as evidence of a precursor molecule or molecules similar to the pro-opiomelanocortin precursor of vertebrates that is cleaved to give a similar range of peptides.

The function of opioid-like peptides in invertebrates is an intriguing problem to which some answers have been provided by studies of molluscs.[32-34] Immunocytochemical studies in *Calliphora* suggest that amounts of the α-endorphin-like peptide within the few specific cells that produce it are related to the age of the activities or developmental changes. A previous indication of this has been observed by Duve et al.[54] in studies of the CCK-like peptides of *Calliphora*. A continuation of work in this area is certainly warranted by these results and such a program may yield important clues as to the function of these peptides. For the other opioid-related peptides examined here, more detailed ICC studies are required. There appears to be a clear difference in the function of the two enkephalin-like peptides as compared with the α-endorphin-like peptide, the former being restricted to specific areas of the brain and the latter being transported by neurosecretory axons to distant regions of the body. It seems reasonable on these criteria to make the distinction between neurotransmitter/neuromodulator (the enkephalins) and neurohormone (endorphins).

Immunocytochemistry is a very useful technique for the initial localization of peptides within tissues. In order to relate the structure of the invertebrate peptides with their vertebrate counterparts, however, it is necessary to isolate and characterize them fully. One of the more obvious difficulties in working with small animals such as the blowfly is that amounts of active peptides are present in such small quantities. As an example, in the present study we have calculated that there are approximately 325 ng α-endorphin-like peptide in 33 g whole flies (approximately 600 animals) giving a value of 0.5 ng per fly. Nevertheless, it should certainly be possible by means of the highly developed analytical techniques now available to obtain composition and sequence data for some of the peptides in the not too distant future.

ACKNOWLEDGMENTS

We are pleased to acknowledge the support of the Science and Engineering Research Council of Great Britain for Grants GR/B88136 and GR/C20789 to A. Thorpe. We also wish to thank the Royal Society for the award of a Travel Grant to A. Thorpe to support participation in the Symposium and the organizing committee of the symposium for partial travel support to H. Duve.

REFERENCES

1. **Simantov, R., Goodman, R., Aposhian, D., and Snyder, S. H.,** Phylogenetic distribution of a morphine-like peptide "enkephalin", *Brain Res.,* 111, 204, 1979.
2. **Rémy, Ch., Girardie, J., and Dubois, M. P.,** Présence dans le ganglion sousoesophagien de la chenille processionaire du Pin (*Thaumetopoea pityocampa* Schiff) de cellules révélées en immunofluorescence par un anticorps anti-α-endorphine, *C.R. Acad. Sci. Ser. D,* 286, 651, 1978.
3. **Rémy, Ch., Girardie, J., and Dubois, M. P.,** Vertebrate neuropeptide-like substances in the suboeso-phageal ganglion of two insects: *Locusta migratoria* R. and F. (Orthoptera) and *Bombyx mori* L. (Lepidoptera): immunocytological investigation, *Gen. Comp. Endocrinol.,* 37, 93, 1979.
4. **Rémy, Ch. and Dubois, M. P.,** Immunohistological evidence of methionine enkephalin-like material in the brain of the migratory locust, *Cell Tissue Res.,* 218, 271, 1981.
5. **El-Salhy, M., Falkmer, S., Kramer, K. J., and Speirs, R. D.,** Immunohistochemical investigations of neuropeptides in the brain, corpora cardiaca, and corpora allata of an adult lepidopteran insect, *Manduca sexta* (L.), *Cell Tissue Res.,* 232, 295, 1983.
6. **Hansen, B. L., Hansen, G. N., and Scharrer, B.,** Immunoreactive material resembling vertebrate neuropeptides in the corpus cardiacum and corpus allatum of the insect *Leucophaea maderae, Cell Tissue Res.,* 225, 319, 1982.
7. **Duve, H. and Thorpe, A.,** Immunocytochemical identification of α-endorphin-like material in neurones of the brain and corpus cardiacum of the blowfly, *Calliphora vomitoria* (Diptera), *Cell Tissue Res.,* 233, 415, 1983.
8. **Duve, H. and Thorpe, A.,** Vertebrate-type brain/gut peptides in neurones of the blowfly, *Calliphora,* in *Proc. 9th Int. Symp. Comparative Endocrinol.,* Lofts, B. and Chan, D.K.O., Eds., Hong Kong University Press, 1985.
9. **Alumets, J., Håkanson, R., Sundler, F., and Thorell, J.,** Neuronal localisation of immunoreactive enkephalin and β-endorphin in the earthworm, *Nature (London),* 279, 805, 1979.
10. **Rémy, Ch. and Dubois, M. P.,** α-Endorphin-like cells in the infra-oesophageal ganglions of the earthworm *Dendrobaena subrubicunda* Eisen.: Immunocytological localization, *Experientia,* 35, 137, 1979.
11. **Zipser, B.,** Identification of specific leech neurones immunoreactive to enkephalin, *Nature (London),* 283, 857, 1980.
12. **Gesser, B. P. and Larsson, L.-I.,** Localisation of α-MSH-, endorphin- and gastrin/CCK-like immuno-reactivities in the nervous system of invertebrates, in *Integrative Neurohumoral Mechanisms,* Angelucci, L., de Wied, D., Endröczi, E., and Scapagnini, U., Eds., Elsevier Biomedical Press, Amsterdam, 1985.
13. **Mancillas, J. R., McGinty, J. F., Selverston, A. I., Karten, H., and Bloom, F. E.,** Immunocytochemical localization of enkephalin and substance P in retina and eyestalk neurones of lobster, *Nature (London),* 293, 576, 1981.
14. **Jaros, P. P. and Keller, R.,** Localization of Leu-enkephalin-like material in the brain and neurohaemal organs of the brachyuran *Carcinus maenas* and the astacidean *Oronectes limosus, Gen. Comp. Endocrinol.,* 53, 466, 1984.
15. **Stefano, G. B. and Martin, R.,** Enkephalin-like immunoreactivity in the pedal ganglion of *Mytilus edulis* (Bivalvia) and its proximity to dopamine-containing structures, *Cell Tissue Res.,* 230, 147, 1983.
16. **Martin, R., Frosch, D., Weber, E., and Voigt, K.-H.,** Met-enkephalin-like immunoreactivity in a cephalopod neurohemal organ, *Neurosci. Lett.,* 15, 253, 1979.
17. **Martin, R., Frosch, D., Kiehling, C., and Voigt, K.-H.,** Molluscan neuropeptide-like and enkephalin-like material coexists in octopus nerves, *Neuropeptides,* 2, 141, 1981.
18. **Schot, L. P. C., Boer, H. H., Swaab, D. F., and van Noorden, S.,** Immunocytochemical demonstration of peptidergic neurons in the central nervous system of the pond snail *Lymnaea stagnalis* with antisera raised to biologically active peptides of vertebrates, *Cell Tissue Res.,* 216, 273, 1981.
19. **Martin, R., Stefano, G. B., and Voigt, K. H.,** Neuronal distrubtion of immunoreactive enkephalin-like material in *Octopus, Mytilus,* and *Lymnaea* ganglia, *Gen. Comp. Endocrinol.,* 53, 468, 1984.
20. **Georges, D. and Dubois, M. P.,** Methionine-enkephalin-like immunoreactivity in the nervous ganglion and the ovary of a protochordate, *Ciona intestinalis, Cell Tissue Res.,* 236, 165, 1984.
21. **Gros, C., Lafon-Cazal, M., and Dray, F.,** Présence de substances immunoréactivement apparantées aux enképhalines chez un insecte *Locusta migratoria, C.R. Acad. Sci. Paris,* 287, 647, 1978.

22. **Williamson, M. E. and Emson, P. C.,** Peptides in the brain of the common snail, *Helix aspersa, Biochem. Soc. Trans.,* 10, 384, 1982.
23. **LeRoith, D., Liotta, A. S., Roth, J., Shiloach, J., Lewis, M. E., Pert, C. B., and Krieger, D. T.,** Corticotropin and β-endorphin-like materials are native to unicellular organisms, *Proc. Natl. Acad. Sci. U.S.A.,* 79, 2086, 1982.
24. **Leung, M. and Stefano, G. B.,** Purification and identification of enkephalins in molluscan ganglia, *Gen. Comp. Endocrinol.,* 53, 467, 1984.
25. **Stefano, G. B. and Leung, M.,** Isolation and identification of Met-enkephalin-Arg-Phe in molluscan ganglia, *Gen. Comp. Endrocrinol.,* 53, 467, 1984.
26. **Kiehling, C., Martin, R., Geis, R., Bickel, U., and Voigt, K. H.,** Cardioexcitatory and opioid activity in extracts from nervous tissue of *Octopus vulgaris, Gen. Comp. Endocrinol.,* 53, 467, 1984.
27. **Stefano, G. B., Kream, R. M., and Zukin, R. S.,** Demonstration of stereospecific opiate binding in the nervous tissue of the marine mollusc *Mytilus edulis, Brain Res.,* 181, 440, 1980.
28. **Kream, R. M., Zukin, R. S., and Stefano, G. B.,** Demonstration of two classes of opiate binding sites in the nervous tissue of the marine mollusc *Mytilus edulis, J. Biol. Chem.,* 255, 9218, 1980.
29. **Stefano, G. B., Scharrer, B., and Assanah, P.,** Demonstration, characterization and localization of opioid binding sites in the midgut of the insect *Leucophaea maderae* (Blattaria), *Brain Res.,* 253, 205, 1982.
30. **Stefano, G. B. and Scharrer, B.,** High affinity binding of an enkephalin analog in the cerebral ganglion of the insect *Leucophaea maderae* (Blattaria), *Brain Res.,* 225, 107, 1981.
31. **Scharrer, B. and Stefano, G. B.,** Opioid binding sites in the brain and midgut of the insect *Leucophaea maderae* (Blattaria), *Gen. Comp. Endocrinol.,* 53, 467, 1984.
32. **Stefano, G. B., S.-Rózsa, K., and Hiripi, L.,** Actions of methionine enkephalin and morphine on single neuronal activity in *Helix pomatia* L., *Comp. Biochem. Physiol.,* 66C, 193, 1980.
33. **Stefano, G. B. and Hiripi, L.,** Methionine enkephalin and morphine alter monoamine and cyclic nucleotide levels in the cerebral ganglia of the freshwater bivalve *Anodonta cygnea, Life Sci.,* 25, 291, 1979.
34. **Stefano, G. B. and Catapane, E. J.,** Enkephalins increase dopamine levels in the CNS of a marine mollusc, *Life Sci.,* 24, 1617, 1979.
35. **Duve, H. and Thorpe, A.,** Comparative aspects of insect-vertebrate neurohormones, in *Insect Neurochemistry and Neurophysiology,* Bořkovec, A. B. and Kelly, T. J., Eds., Plenum Press, New York, 1984, 171.
36. **Thorpe, A. and Duve, H.,** Immunochemical applications in the study of insect neuropeptides with special emphasis on the peptides of vertebrate type, in *Insect Neurochemistry and Neurophysiology,* Bořkovec, A. B. and Kelly, T. J., Eds., Plenum Press, New York, 1984, 197.
37. **Duve, H. and Thorpe, A.,** Immunochemical identification of vertebrate-type brain-gut peptides in insect nerve cells, in *Functional Neuroanatomy,* Strausfeld, N. J., Ed., Springer-Verlag, Heidelberg, 1983, chap. 13.
38. **Thorpe, A. and Duve, H.,** Neuropeptides of vertebrate type in the blowfly, *Calliphora,* in *Biosynthesis, Metabolism and Mode of Action of Invertebrate Hormones,* Hoffmann, J. A. and Porchet, M., Eds., Springer-Verlag, Heidelberg, 1984, 106.
39. **Duve, H. and Thorpe, A.,** Localisation of pancreatic polypeptide (PP)-like immunoreactive material in neurones of the brain of the blowfly, *Calliphora erythrocephala* (Diptera), *Cell Tissue Res.,* 210, 101, 1980.
40. **Duve, H. and Thorpe, A.,** Gastrin/cholecystokinin(CCK)-like immunoreactive neurones in the brain of the blowfly, *Calliphora erythrocephala* (Diptera), *Gen. Comp. Endocrinol.,* 43, 381, 1981.
41. **Linde, S., Hansen, B., and Lernmark, Å.,** Stable iodinated polypeptide hormones prepared by polyacrylamide gel electrophoresis *Anal. Biochem.,* 107, 165, 1980.
42. **Jackson, S., Hope, J., Estivariz, F., and Lowry, P. J.,** Nature and control of peptide release from the pars intermedia, in *Peptides of the Pars Intermedia: Ciba Foundation Symposium 81,* Pitman Medical, London, 1981, 141.
43. **Hummer, L.,** A radioimmunoassay of plasma corticotrophin, in *Radioimmunoassay and Related Procedures in Medicine 1977,* Vol. 2, International Atomic Energy Agency, Vienna, 1978, 391.
44. **Ling, N., Burgus, R., and Guillemin, R.,** Isolation, primary structure, and synthesis of α-endorphin and β-endorphin, two peptides of hypothalamic-hypophyseal origin with morphinmimetic activity, *Proc. Natl. Acad. Sci. U.S.A.,* 73, 3942, 1976.
45. **Jackson, S. and Lowry, P. J.,** Distribution of adrenocorticotrophic and lipotrophic peptides in the rat, *J. Endocrinol.,* 86, 205, 1980.
46. **Burbach, J. P. H., Loeber, J. G., Verhoef, J., Wiegant, V. M., de Kloet, E. R., and de Wied, D.,** Selective conversion of β-endorphin into peptides related to γ- and α-endorphin, *Nature (London),* 283, 96, 1980.
47. **Takahashi, A., Kawauchi, H., Mouri, T., and Sasaki, A.,** Chemical and immunological characterization of salmon endorphins, *Gen. Comp. Endocrinol.,* 53, 381, 1984.

48. **Cox, B. M., Goldstein, A., and Li, C. H.,** Opioid activity of a peptide, β-lipotropin-(61-91) derived from β-lipotropin, *Proc. Natl. Acad. Sci. U.S.A.,* 73, 1821, 1976.
49. **Huang, W-Y., Chang, R. C. C., Kastin, A. J., Coy, D. H., and Schally, A. V.,** Isolation and structure of pro-methionine-enkephalin: potential enkephalin precursor from porcine hypothalamus, *Proc. Natl. Acad. Sci. U.S.A.,* 76, 6177, 1979.
50. **Lewis, R. V., Stern, A. S., Kimura, S., Rossier, J., and Udenfriend, S.,** An about 50,000 dalton protein in adrenal medulla: a common precursor of (Met)- and (Leu)-enkephalin, *Science,* 208, 1459, 1980.
51. **Clement-Jones, V., Corder, R., and Lowry, P. J.,** Isolation of human Met-enkephalin and two groups of putative precursors (2K-Pro-Met-enkephalin) from an adrenal medullary tumour, *Biochem. Biophys. Res. Commun.,* 95, 665, 1980.
52. **Watson, S. J., Akil, H., Richard, C. W., and Barchas, J. D.,** Evidence for two separate opiate neuronal systems, *Nature (London),* 275, 226, 1978.
53. **Zakarian, S. and Smyth, D. G.,** β-endorphin is processed differently in specific regions of rat pituitary and brain, *Nature (London),* 296, 250, 1982.
54. **Duve, H., Thorpe, A., and Strausfeld, N. J.,** Cobalt-immunocytochemical identification of peptidergic neurons in *Calliphora* innervating central and peripheral targets, *J. Neurocytol.,* 12, 847, 1983.

IMMUNOCYTOCHEMICAL DEMONSTRATION OF A MATERIAL RESEMBLING VERTEBRATE ACTH AND MSH IN THE CORPUS CARDIACUM-CORPUS ALLATUM COMPLEX OF THE INSECT *LEUCOPHAEA MADERAE*

Bente Langvad Hansen, Georg Nørgaard Hansen, and Berta Scharrer

INTRODUCTION

This contribution contains the results of a follow-up study of our collaborative project concerned with the demonstration of "vertebrate neuropeptides" in invertebrates. As in a previous report,[1] the focus was on the corpus cardiacum of insects, part of a neuroendocrine apparatus that is comparable to the hypothalamic-hypophysial system of vertebrates. Like the posterior lobe of the pituitary gland, the corpus cardiacum stores and releases hormonal neuropeptides that are synthesized in neurosecretory cells located in the brain. In addition to playing the role of a neurohemal organ, this insect structure is the source of active neuropeptides originating in its intrinsic neuroglandular cells. The occurrence of a multiplicity of such compounds in this organ seems to be in keeping with its relative complexity.

Thus far we have been able to demonstrate the presence and differential distribution of reaction products of at least eight neuropeptides antigenically related to those of vertebrates.[1] The present report provides information on the occurrence in the corpus cardiacum-allatum of *Leucophaea* of a peptide containing an amino acid sequence characteristic of the human adrenocorticotropin (ACTH) molecule.

MATERIALS AND METHODS

Tissues Tested

Adult cockroaches (*Leucophaea maderae*) and rats (200 to 225 g, Wistar strain) of either sex were used. The corpus cardiacum-corpus allatum complex of *Leucophaea* was fixed *in situ* in Bouin's fluid.[1] The rat pituitaries were immersion fixed in Bouin's fluid and routinely embedded in paraffin.[2] Sections were cut at 5 to 6 μm and hydrated.

Antisera

ACTH (1-24) antiserum, and α-MSH antisera Albert VII and Charles were used. The ACTH (1-24) antiserum (kindly donated by Dr. L. Hummer, Glostrup Hospital, Denmark) was raised in guinea pigs against synthetic human ACTH (1-24) (Synacthen, short acting, Organon, The Netherlands) as described elsewhere.[3] The IgG fraction of this antiserum purified by the method of Harboe and Ingild[4] has been previously used at the immunocytochemical level.[5-7] The antiserum "Albert" (kindly donated by Dr. L.-I. Larsson, Institute of Pathology and Anatomy, University of Copenhagen, Denmark) and the antiserum Charles (originally obtained from Dr. C. Oliver, Southwestern Medical School. Dallas, Texas) were raised in rabbits against synthetic α-MSH as described elsewhere.[8,9] The antiserum Charles has been characterized previously at the radioimmunoassay (RIA) level.[9] The antiserum "Albert" has been studied in detail by L.-I. Larsson who kindly provided his data.[10]

Peptides

Peptides used in both model and tissue studies included synthetic ACTHs (1-17), (11-24), and (1-24); synthetic β-MSH; and synthetic α-MSH (all kindly donated by Drs. P. A. Desaulles and W. Rittel, Ciba-Geigy).

Antibody Characterization and Specificity Controls

In order to identify the amino acid sequences that can be presumed to react with antigens in the corpus cardiacum-allatum complex of *Leucophaea*, screening for the region specificity of the antisera used was performed in an immunocytochemical model system and in a tissue control system (rat pituitary).

Model System

Specificity tests were carried out according to Larsson[11] and a modification of his model system. A small amount (2 $\mu\ell$) of each of the peptides listed above (in various concentrations) was applied, at room temperature, to strips of filter paper. Type 1 (20H, Munktell, Sweden) paper strips were air-dried and fixed in formaldehyde vapor at 80°C for 1 hr before the immunocytochemical procedure. Type 2 (BA 85 cellulose nitrate filter paper, Schleicher and Schüll, West Germany) required no prior fixation. The results of immunocytochemical "staining" were the same in both instances.

Tissue System

Specificity tests at the tissue section level were performed as conventional liquid-phase absorption controls. All antisera studied were tested by absorption with the peptides mentioned above. The absorptions were conducted by allowing the optimally diluted antisera [anti-ACTH (1-24) 1:50, "Albert" and Charles, both 1:500] to react with different amounts of the peptides for 72 hr at 4°C prior to immunocytochemical staining. If the added peptides were without effect on the staining results, their concentrations were raised tenfold. All tests were run independently three times.

Immunocytochemistry

All steps of the immunocytochemical staining procedure were carried out at room temperature unless otherwise stated. After an initial rinse for 5 min in 0.05 M Tris-HCl, pH 7.2, 0.5 M, NaCl (TBS) strips and sections were incubated with normal swine serum diluted in TBS for 30 min. Subsequently, various dilutions of the primary antisera were applied. The primary antisera, diluted by TBS containing 0.25% crystalline human serum albumin (Behring Werke), were allowed to react for 72 hr at 4°C. All other antibody solutions were diluted in TBS only. The site of the antigen-antibody reaction was determined either by the peroxidase-antiperoxidase (PAP) technique[12] or by the labeled antibody-enzyme method.[12] The PAP technique was employed for α-MSH antisera. The second and third layers in this procedure consisted of swine anti-rabbit IgG (diluted 1:10) and PAP (diluted 1:80), both obtained from Dakopatts, Copenhagen, Denmark. Anti-ACTH (1-24) IgG was detected with horseradish peroxidase-conjugated rabbit anti-guinea pig IgG (Dakopatts), diluted 1:100. Peroxidase activity was visualized with 3,3'-diaminobenzidine tetrahydrochloride (DAB, Sigma). Controls included conventional staining controls as well as specific absorption controls. Staining controls included (1) sequential deletion of the various antibody layers, (2) omission of all antibody layers, (3) substitution of the primary antisera for normal rabbit or guinea pig serum, and (4) omission of DAB. Absorption controls are outlined in "Antibody Characterization and Specificity Controls".

RESULTS

Evaluation of Antisera

All antisera examined were initially tested in immunocytochemical model systems. The outcome of the various absorptions was tested on sections of rat pituitaries. The results obtained in both systems paralleled each other.

Filter Model System

All peptides used could be immobilized on filter strips as documented by positive immunocytochemical staining with the antisera tested. No nonspecific staining of peptides was observed, and all staining controls were negative. Moreover, antisera preabsorbed with homologous antigens failed to stain in the model system. In these experiments ACTH (1-24) antibodies produced a strong reaction with ACTHs (1-24), (11-24), and α-MSH [= N-acetylated, C-amidated ACTH (1-13)]. In addition, a weak reaction toward ACTH (1-17) was noted, but not toward β-MSH. Though the presence of multiple antibody subpopulations could not be ruled out, these results indicate binding of the antibodies to the (11-17) sequence of ACTH (1-24). Antiserum Charles reacted intensely with α-MSH and ACTH (1-24) and weakly with ACTHs (1-17) and (11-24), but did not react with β-MSH. These results indicate that this antiserum is heterogeneous and contains one subpopulation able to react with both α-MSH and ACTH (1-24), and another subpopulation entirely specific for α-MSH. The antiserum ''Albert'' produced a strong reaction with α-MSH, whereas the remaining peptides did not react, indicating the existence of at least one antibody population specifically recognizing α-MSH without reacting with ACTH.

Tissue Control System

In sections of rat pituitary, virtually all cells of the intermediate lobe reacted with ACTH (1-24) antibodies. The staining of scattered cells in the distal lobe was abolished by addition of the various ACTH congeners. Staining in the intermediate lobe was inhibited by ACTHs (1-24) and (1-17), and α-MSH. These results may indicate that this antiserum contains at least two antibody populations, one directed to the (11-17) region of ACTH and the other both to the NH_2 terminus and the (11-13) region of ACTH. Antiserum Charles stained all cells in the intermediate lobe and very few cells in the distal lobe. Absorption of this antiserum with α-MSH abolished staining in the intermediate lobe. Immunoreaction in the distal lobe was quenched by absorption with both α-MSH and ACTH (1-24). Thus this antiserum may contain two antibody populations, one which is specific for α-MSH and another which recognizes both (1-3) and (11-13) regions of ACTH and α-MSH. Antiserum ''Albert'' stained all cells in the intermediate lobe. Only absorption with α-MSH inactivated staining and confirmed the region specificity of this antiserum. In accordance, this antiserum has been shown[10] to contain one antibody population wholly specific for α-MSH (directed toward the N-acetylated and C-amidated terminal portion of α-MSH, and another recognizing an (as yet) unidentified sequence between amino acids 3 and 13 of α-MSH.

Immunocytochemical Results

Reaction products were obtained after exposure of corpus cardiacum-allatum sections to ACTH (1-24) antibodies (Figures 1 and 3) and to α-MSH antiserum Charles (Figures 2, 4, and 5). By contrast, no staining could be obtained with the α-MSH-specific antiserum ''Albert''.

Progressive dilutions of the positive antisera gradually decreased staining. At dilutions of ACTH (1-24) antibodies 1:100 and of Charles antiserum 1:2000, staining was no longer observable. Nonspecific staining of the tissue constituents studied did not occur. Also, classical staining controls were negative. The immunocytochemical response to antiserum Charles was more pronounced than that to ACTH (1-24) antibodies, but the deposits appeared in the same location (Figures 3 and 4). In addition, antiserum Charles was found, to some extent, to bind in areas that revealed no binding of ACTH (1-24) antibodies. As reported for the neuropeptides previously demonstrated in this composite organ,[1] the distribution of the binding sites for anti-ACTH (1-24) and anti-α-MSH Charles was found to be regional rather than diffuse. Immunocytochemistry revealed a substantial amount of reactive material in the central zone facing the dorsal blood vessel (Figure 2). In addition, a fan-like distinct

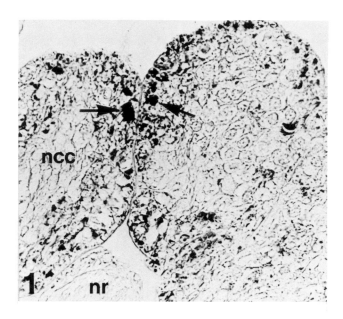

FIGURE 1.   Cross section of corpora cardiaca of *Leucophaea* showing reaction product of ACTH-like peptide (arrows). ncc, nervus corporis cardiaci; nr, nervus recurrens. (Magnification × 400.)

network of beaded immunoreactive fibers was detected in the path of the nervi corporis cardiaci (Figures 3 and 4). This localization bespeaks an extrinsic (cerebral) origin of this peptide material. However, the possibility that some of it may be derived from intrinsic neuroglandular cells cannot be excluded. By contrast, only a limited amount of immuno-reactive material was observed in the peripheral area of the corpus cardiacum (Figure 2) as well as in the corpus allatum (Figure 5).

The ACTH (1-24) antibodies showed optimal staining intensity at dilutions between 1:40 and 1:60. Staining with this antiserum was inactivated by the various ACTH congeners, but was not abolished by absorption with α-MSH and β-MSH. These results imply that the staining observed was caused by the antibody population specific for the (11-17) region of the ACTH (1-24) antigen. The antiserum Charles showed optimal staining intensity at dilutions between 1:400 and 1:600. Absorption with ACTHs (1-24), (1-17), and (11-24), and α-MSH was found to inhibit all staining, whereas addition of β-MSH was without effect. This, and the fact that the antiserum "Albert" did not show any immunoreaction in the tissues examined, indicate that the antiserum Charles possibly may have recognized the regions (1-3) and (11-13) of ACTH (1-24).

The presence of α-MSH-like material within the corpus allatum is in line with the generally accepted concept that this nonneural endocrine gland receives directives from peptidergic neurons whose axons enter it via the corpus cardiacum. Therefore, the site of production of this ACTH-like principle, like that of the peptides resembling substance P, β-endorphin, and LH-RF previously found to occur in the corpus allatum of *Leucophaea*,[1] can be presumed to be outside of this gland. Because of the localized character of their putative activity in the corpus allatum, these neuropeptides can be expected to be present in small amounts. Therefore, the chances of detecting a reaction product of anti-ACTH (1-24) in this organ are relatively small, since the response to this antiserum in the rest of the tissue examined turned out to be considerably weaker than that to anti-α-MSH Charles.

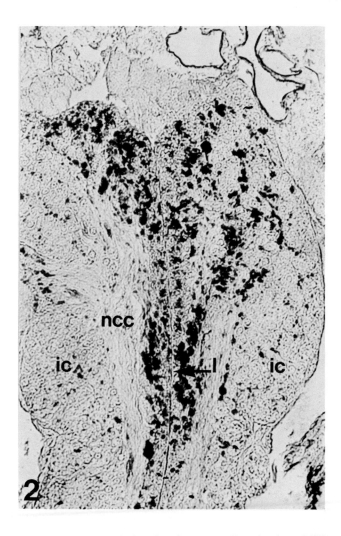

FIGURE 2. Longitudinal section of corpora cardiaca showing α-MSH-like material located primarily in area facing dorsal blood vessel (1). ncc, nervus corporis cardiaci; ic, intrinsic cell area. (Magnification × 216.)

DISCUSSION

The list of substances antigenically related to mammalian neuropeptides demonstrated thus far within the corpus cardiacum of *Leucophaea*[1] in varying amounts includes oxytocin, somatostatin, substance P, Met-enkephalin, bombesin, and neurotensin (abundant); vasopressin and β-endorphin (moderate); and LH-RF and calcitonin (marginal). The present study has added ACTH- and MSH-like reaction products to this list. The methods employed fulfill the criteria for a successful demonstration of these peptides. Since conclusions regarding immunocytochemical crossreactivity can be reached only from studies at the immunocytochemical level,[13] the region specificity of the antisera used was determined not only by conventional liquid-phase absorption tests, but also by use of immunocytochemical models. Rat pituitary sections were used for absorption controls because the intermediate lobe of the rat is known to contain large amounts of α-MSH and because rat ACTH closely resembles human ACTH.[14] The cytochemical model variant described, simulating a solid-phase ab-

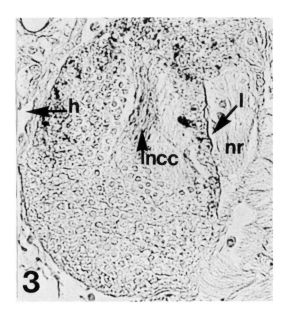

FIGURE 3.   Part of corpus cardiacum showing ACTH-like material in same location as that of α-MSH-like material. Compare with Figure 4. l, lumen of dorsal blood vessel; ncc, nervus corporis cardiaci; nr, nervus recurrens; h, hemocoel. (Magnification × 216.)

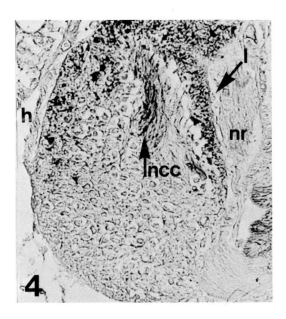

FIGURE 4.   Part of corpus cardiacum showing α–MSH-like material in nervus corporis cardiaci (ncc), facing lumen of dorsal blood vessel (l), and close to hemocoel (h). Compare with Figure 3. nr, nervus recurrens. (Magnification × 216.)

sorption system, offers not only an opportunity for evaluating effects of various fixatives on antigenicity, but also the possibility of screening antisera for specificity. The results obtained concerning the characterization of the antisera tested by means of the two model

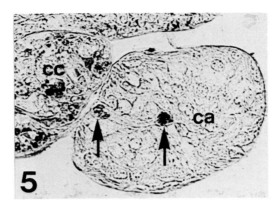

FIGURE 5.   Section through corpus allatum (ca) contiguous with corpus cardiacum (cc). Note presence of α-MSH-like material in corpus cardiacum and corpus allatum (arrows). (Magnification × 400.)

systems and the tissue control system (conventional liquid-phase absorption controls) parallel each other. This permits the conclusion that the antisera investigated are region-specific detection reagents, suitable for the identification of ACTH-like peptides in *Leucophaea* by use of immunocytochemistry. The results obtained show that ACTH (1-24) antibodies recognize the region (11-17) of the ACTH (1-24) molecule, while the α-MSH antiserum Charles recognizes the regions (1-3) and (11-13) common to both α-MSH and ACTH (1-24). Since the tissue under study did not show any immunoreaction with the antiserum "Albert", the results imply that the immunostained substances revealed in *Leucophaea* may have an antigenic determinant composed of all or part of the amino acid sequences (1-3) and (11-17) of human corticotropin. Furthermore, the vertebrate analog α-MSH [(= N-acetylated and C-amidated ACTH(1-13)], if at all present in the corpus cardiacum-corpus allatum complex, may be stored in exceedingly low concentration since the potent anti-α-MSH "Albert" failed to detect it in *Leucophaea*. It should, however, be noted that, although the data may suggest the presence of an ACTH-like molecule, they do not exclude the possibility that the sequences recognized are contained within a larger precursor or, alternatively, represent unrelated components containing the appropriate antigenic determinants.

The results presented here add a "new" peptide to the list of vertebrate-type peptides previously demonstrated in *Leucophaea*.[1] It seems reasonable to postulate that this peptide may originate from neurosecretory cells in the brain from which it is transported to the corpus cardiacum-allatum for storage and release into the hemolymph. Supporting evidence for this presumption comes from the distribution of the immunoreactive material revealing the pathway of the nervi corporis cardiaci and their terminals whose cells of origin lie in the brain.[15] By the same token, the presence of corticotropin-like material in a distribution highly reminiscent of that of β-endorphin and Met-enkephalin[1] may indicate an intimate topographic linkage between ACTH-opioid peptides in *Leucophaea* in parallel with data reported for vertebrate pituitaries.[16] One may wonder if the substances are carried by the same nerve fibers, and if a relationship exists between these substances during their biosynthesis. The fact that *Leucophaea* material crossreacts with antisera directed against the mammalian ACTH (1-24) molecule may indicate that this molecule has a long evolutionary history. At least the (1-17) region, or part of it, may have been firmly established early in evolution. These suggestions support the conclusion reached in vertebrates[7,17] that the (1-19) region of ACTH has remained relatively unaltered during evolution. Further data on the structure of this corticotropin-like peptide, its site or origin, and its distribution in *Leucophaea* will lead to greater certainty about the evolution of this peptide and its function as short-acting neurotransmitter or long-acting neurohormone. It remains to be seen whether the ACTH-like material demonstrated in *Leucophaea* indicates the existence of a pro-opiomelanocortin precursor in insects.

A comprehensive study in another insect species, *Manduca sexta*,[18] confirmed the presence of four of the peptides of *Leucophaea* (resembling substance P, somatostatin, and endogenous opioids) in the corpus cardiacum and added that of materials related to glucagon, insulin, gastrin, secretin, vasoactive intestinal peptide (VIP), and pancreatic peptide (PP). Tests for the existence of a material resembling ACTH in this insect were negative. However, the presence of such material in invertebrates other than insects has been reported for the mollusc *Lymnaea*,[19] the annelid *Lumbricus*,[20] the platyhelminth *Dugesia*,[21] and the ciliate *Tetrahymena*.[22]

A somatostatin-like product had been reported earlier also in the corpus cardiacum of *Locusta migratoria*[23] and an insulin-like substance in that of *Bombyx mori*.[24]

This array of substances is impressive even in an organ of such complexity as the corpus cardiacum, which combines the characteristics of a neurohemal structure accommodating neurosecretory products derived from various extrinsic cell groups and of a neuroglandular organ in its own right capable of producing additional neuropeptides. The different sites of biosynthesis of these active principles are reflected in their differential distribution within the organ. The classification of extrinsic neuropeptides is consistent with their preferential localization in the nervi corporis cardiaci and in areas known to store them close to sites of release into the dorsal blood vessel or the hemocoel. In addition, it is supported by the demonstration that the same substances occur in identified peptidergic cells outside of the corpus cardiacum. by contrast, intrinsic neuropeptides are localizable in the cell bodies and processes of the neuroglandular cells of the corpus cardiacum.

On the basis of these criteria, the majority of the principles demonstrated in *Leucophaea* (see Hansen et al.[1] and present data) and all of those reported for *Manduca*[18] seem to be of extrinsic origin. However, there appears to be a discrepancy with regard to the classification of somatostatin-like material, considered to be extrinsic in *Locusta*[23] and *Manduca*,[18] and intrinsic in *Leucophaea*.[1] One possible explanation for this is the existence of gene differences. However, there is also the possibility that one or another of these substances may arise in more than one location, and that priorities may vary among species. A comparable situation in vertebrates is the duality in the sites of biosynthesis of ACTH in the anterior lobe of the pituitary and the hypothalamus, respectively.[25]

The number of neuropeptides thus far known to enter the corpus allatum via neuronal pathways is smaller than that shown to be present in the corpus cardiacum. It includes materials resembling substance P, β-endorphin, LH-RF and α-MSH in *Leucophaea* (see Hansen et al.[1] and present data), and substances related to insulin, enkephalin, somatostatin, VIP, and substance P in *Manduca*.[18]

Regarding the functional significance of the vertebrate-type neuropeptides demonstrated in the corpus cardiacum-corpus allatum complex of insects, several general statements can be made based on previously obtained ultrastructural and physiological information.[26] The majority of these neuroregulators seem to be released into the circulation at the level of the corpus cardiacum to carry out hormonal functions a variety of which have been documented by endocrinological experimentation. Some of these specific roles can be assigned to extrinsic and others to intrinsic neurohormones.

All peptides demonstrable within the corpus allatum are of neural, i.e., extrinsic origin. They too may reach the circulation but, for the most part, their release sites are in close spatial relationship with the endocrine cells of the gland over which they are presumed to exert stimulatory and/or inhibitory local control. Comparable synaptoid junctions between extrinsic peptidergic fibers and intrinsic corpus cardiacum cells suggest a local control mechanism to operate also at the level of the corpus cardiacum. Within this general framework designed for multiple functions and different modes of operation, it is not yet possible to assign individual roles to the various neuropeptides whose presence has been demonstrated by immunocytochemical methods.

SUMMARY

Adrenocorticotropin (ACTH)-like and α-melanotropin (α-MSH)-like peptides have been demonstrated in the corpus cardiacum-corpus allatum complex of the insect *Leucophaea maderae* by region-specific immunocytochemistry. The distribution pattern of the immunoreactive deposits varies within the complex. A substantial amount of material that contains antigenic determinants recognizing at least the (1-3) and the (11-17) regions of the ACTH (1-24) molecule is primarily localized in the central release area of the neuroglandular corpus cardiacum and is therefore judged to be of extrinsic (cerebral) origin. A much smaller amount of this reactive material is found within peptidergic fibers entering the nonneural corpus allatum by way of the corpus cardiacum. These data suggest that a corticotropin-like peptide exists in an insect and supports the view that the mammalian ACTH molecule has a long evolutionary history. The differential localization of the ACTH-like material described is of interest since it is in line with that previously documented for other opioid peptides in the corpus cardiacum-corpus allatum complex of *Leucophaea*.

ACKNOWLEDGMENTS

This study was supported in part by NSF grant BMS 74-12456 (B.S.). We thank Drs. Lotte Hummer, Lars-Inge Larsson, and Jørgen Warberg for the generous gifts of anti-ACTH (1-24), antisera Albert VII, and Charles respectively.

REFERENCES

1. **Hansen, B. L., Hansen, G. N., and Scharrer, B.,** Immunoreactive material resembling vertebrate neuropeptides in the corpus cadiacum and corpus allatum of the insect *Leucophaea maderae, Cell Tissue Res.,* 225, 319, 1982.
2. **Hansen, B. L., Hansen, G. N., and Hagen, C.,** Immunoreactive material resembling prolactin in perikarya and nerve terminals of the rat hypothalamus, *Cell Tissue Res.,* 226, 121, 1982.
3. **Hummer, L.,** A radioimmunoassay of plasma corticotropin, in *Radioimmunoassay and Related Procedures in Medicine,* Vol. 2, International Atomic Energy Agency, Vienna, 1978, 391.
4. **Harboe, N. and Ingild, A.,** Immunization, isolation of immunoglobulins, estimation of antibody titre, *Scand. J. Immunol.,* 2(Suppl.), 161, 1973.
5. **Hansen, G. N., Hansen, B. L., and Hummer, L.,** The cell types in the adenohypophysis of the South-American lungfish, *Lepiodosiren paradoxa,* with special reference to immunocytochemical identification of the corticotropin-containing cells, *Cell Tissue Res.,* 209, 147, 1980.
6. **Hansen, B. L., Hansen, M., Hansen, G. N., and Hummer, L.,** Demonstration of autoantibodies against ACTH in serum from a patient with cancer of unknown origin, *Acta Pathol. Microbiol. Immunol. Scand. Sect. A.,* 91, 489, 1983.
7. **Hansen, G. N.,** Cell types in the adenohypophysis of the primitive actinopterygians, with special reference to immunocytochemical identification of pituitary hormone producing cells in the distal lobe, *Acta Zool. (Stockholm),* 1983(Suppl.), 87, 1983.
8. **Larsson, L-I.,** Adrenocorticotropin-like and α-melanotropin-like peptides in a subpopulation of human gastrin cell granules: bioassay, immunoassay, and immunocytochemical evidence, *Proc. Natl. Acad. Sci. U.S.A.,* 78, 2990, 1981.
9. **Usategui, R., Oliver, C., Vaudry, H., Lombardi, G., Rozenberg, I., and Mourre, A. M.,** Immunoreactive α-MSH and ACTH levels in rat plasma and pituitary, *Endocrinology,* 98, 189, 1976.
10. **Larsson, L-I.,** personal communication, 1983.

11. **Larsson, L.-I.,** A novel immunocytochemical model system for specificity and sensitivity screening of antisera against multiple antigens, *J. Histochem. Cytochem.*, 29, 408, 1981.
12. **Sternberger, L. A.,** *Immunocytochemistry*, 2nd ed., John Wiley & Sons, New York, 1979, 354.
13. **Larsson, L.-I. and Rehfeld, J. F.,** Characterization of antral gastrin cells with region-specific antisera, *J. Histochem. Cytochem.*, 25, 1317, 1977.
14. **Scott, A. P., Lowry, P. J., Ratcliffe, J. G., Rees, L. H., and Landon, J.,** Corticotrophin-like peptides in the rat pituitary, *J. Endocrinol.*, 61, 335, 1974.
15. **Scharrer, B.,** Neurosecretion. XIII. The ultrastructure of the corpus cardiacum of the insect *Leucophaea maderae*, *Z. Zellforsch.*, 60, 761, 1963.
16. **Watson, S. J. and Akil, H.,** Anatomy of β-endorphin-containing structure in pituitary and brain, in *Hormonal Proteins and Peptides*, Vol. 10, Li, C. H., Ed., Academic Press, New York, 1981, chap. 5.
17. **Lowry, P. J. and Scott, A. P.,** The evolution of vertebrate corticotropin and melanocyte stimulating hormone, *Gen. Comp. Endocrinol.*, 26, 126, 1975.
18. **El-Salhy, M., Falkmer, S., Kramer, K. J., and Speirs, R. D.,** Immunohistochemical investigations of neuropeptides in the brain, corpora cardiaca, and corpora allata of an adult lepidopteran insect, *Manduca sexta*, *Cell Tissue Res.*, 232, 295, 1983.
19. **Boer, H. H., Schot, L. P. C., Roubos, E. W., ter Maat, A., Lodder, J. C., Reichelt, D., and Swaab, D. F.,** ACTH-like immunoreactivity in two electronically coupled giant neurons in the pond snail *Lymnaea stagnalis*, *Cell Tissue Res.*, 202, 231, 1979.
20. **Aros, B., Wenger, T., Vigh, B., and Vigh-Teichmann, I.,** Immunohistochemical localization of substance P and ACTH-like activity in the central nervous system of the earthworm *Lumbricus terrestris* L., *Acta Histochem.*, 66, 262, 1980.
21. **Schilt, J., Richoux, J. P., and Dubois, M. P.,** Demonstration of peptides immunologically related to vertebrate neurohormones in *Dugesia lugubris* (Turbellaria, Tricladida), *Gen. Comp. Endocrinol.*, 43, 331, 1981.
22. **Le Roith, D., Liotta, A. S., Roth, J., Shiloach, J., Lewis, M. E., Pert, C. B., and Krieger, D. T.,** Corticotropin and β-endorphin-like materials are native to unicellular organisms, *Proc. Natl. Acad. Sci. U.S.A.*, 79, 2086, 1982.
23. **Doerr-Schott, J., Joly, L., and Dubois, M. P.,** Sur l'existence dans la pars intercerebralis d'un insecte (*Locusta migratoria* R. et F.) de cellules neurosécrétrices fixant un antisérum antisomatostatine, *C.R. Acad. Sci. Paris*, Ser. D, 286, 93, 1978.
24. **Yui, R., Fujita, T., and Ito, S.,** Insulin-, gastrin-, pancreatic polypeptide-like immunoreactive neurons in the brain of the silkworm, *Bombyx mori*, *Biomed. Res.*, 1, 42, 1980.
25. **Krieger, D. T.,** Brain peptides: what, where and why?, *Science*, 222, 975, 1983.
26. **Scharrer, B.,** Peptidergic neurons: facts and trends, *Gen. Comp. Endocrinol.*, 34, 50, 1978.

IMMUNOCYTOCHEMICAL AND PHYSIOLOGICAL STUDIES ON (NEUROENDOCRINE) NEURONS OF THE POND SNAIL *LYMNAEA STAGNALIS*, WITH PARTICULAR REFERENCE TO COEXISTENCE OF BIOLOGICALLY ACTIVE PEPTIDES (BAP) AND BIOGENIC AMINES, TO CARDIOACTIVE PEPTIDES AND TO CORELEASE OF BAP FROM THE OVULATION HORMONE-PRODUCING CAUDODORSAL CELLS

H. H. Boer, W. P. M. Geraerts, L. P. C. Schot, and R. H. M. Ebberink

INTRODUCTION

The Peptidergic Neuron

During the past decades, the distinction between neurosecretory and nonneurosecretory (conventional) neurons seems to have faded.[1] Not only neurosecretory neurons, but also conventional neurons can use biologically active peptides (BAP) as chemical messengers, as became clear from behavioral,[2] immunocytochemical,[3,4] and physiological[5-7] studies. The number of putative neurotransmitters has increased considerably and comprises, in addition to the classical transmitters, also neuropeptides and gut peptides.[8]

Immunocytochemical and physiological studies have indicated that conventional neurons may produce and release more than one BAP or a BAP and a classical neurotransmitter.[9-11] Biochemical and physiological studies on the ovulation hormone-producing neuroendocrine cells of the opistobranch *Aplysia californica*[12] and the pulmonate *Lymnaea stagnalis*[13] lend support to the hypothesis that multiple release of BAP is a common phenomenon. Some of the released peptides obviously act at neurotransmitters and are invovled in the regulation of neuronal networks that control egg-laying behavior.[14,15]

Research with recombinant-DNA techniques suggest that the BAP released from a particular neuron are derived from a large precursor molecule. Apparently several cell types produce the same precursor from which each type makes its own end product(s).[16,17] Work on *Aplysia* furthermore shows that in a species a (small) number of genes may exist that code for related precursors, each of which contains one common BAP — in the case of *Aplysia* the egg-laying hormone (ELH) — and other (variable) peptides. Thus a number of peptides could be released in different combinations by the same cell at different stages of physiological activity.[12,18]

The release of BAP and of classical transmitters does not seem to be restricted to synapses and neurohemal endings.[14] Apparently, release can also take place from unspecialized sites along the axons (nonsynaptic release). This phenomenon has recently been shown at the ultrastructural level.[19]

Not only classical transmitters, but also BAP seem to have a wide distribution in the animal kingdom,[1,20-25] which suggests that BAP and the precursors from which they are cleaved, are phylogenetically ancient molecules.[1] In immunocytochemical and biochemical studies on invertebrates the presence of opioid and related peptides has been indicated.[23,25-27]

From all this evidence "the neuron" emerges as a multifunctional cell. The fact that the new concepts are based on results obtained with a variety of techniques illustrates that a multidisciplinary approach is needed in the study of neurobiological problems. The observation that the "peptidergic" neuron has a general occurrence in the animal kingdom renders it conceivable that the study of invertebrate model systems (*Aplysia, Lymnaea*) will be of great value for the further elucidation of fundamental aspects of the functioning of the brain.

Immunocytochemistry

Immunocytochemistry has become an important technique in the study of peptidergic cells and recently also in that of aminergic cells. However, several authors have argued that it is rather difficult, if not impossible, to deduce from immunocytochemical results the molecular structure of the immunoreactive substance.[28-30] Polyclonal antisera contain usually numerous populations of IgG molecules with different specificities. Purification of the antiserum and characterization of its specificity improve the quality of the results, as does the use of monoclonal antibodies. However, although with these specified antisera the presence of a particular or closely related (crossreaction) antigenic determinant can be shown, it cannot be concluded from a positive reaction whether the antigenic determinant forms part of the molecule to which the antiserum was raised or whether it is present as part of another, unknown, molecule.[10] Thus, immunocytochemical results should be interpreted cautiously. This holds in particular for invertebrates, as little is known about the chemical structure of invertebrate BAP.

Lymnaea stagnalis

The central nervous system of the pond snail *Lymnaea stagnalis* has extensively been studied in our laboratory with endocrinological, biochemical, electrophysiological, and morphological methods.[31,32] With immunocytochemistry neurons have been identified with antisera to a large number of vertebrate BAP.[23] In addition, results obtained with an antiserum to FMRFamide have been reported.[10,21,34,35] Coexistence of immunoreactivity to antisera raised to chemically unrelated peptides as well as to antisera to a peptide and to a biogenic amine has been observed.[10,35]

In the present paper, a survey of the immunocytochemical observations on *L. stagnalis* will be presented. In addition results will be reported of experiments on the isolation and characterization of cardioactive peptides and on the biosynthesis of peptides by the ovulation hormone-producing caudodorsal cells (CDC).[13,32]

In recent years a number of small cardioactive neuropeptides have been isolated from a variety of molluscs.[35,36-38] Only two of these have been fully purified and sequenced, i.e., the tetrapeptide Phe-Met-Arg-Phe-NH$_2$ (FMRFamide), isolated from the bivalve *Macrocallista nimbosa*,[23] and a nonapeptide, isolated from the opisthobranch *Aplysia brasiliana*.[38] It is interesting to note that FMRFamide shows structural similarity with Met-enkephalin (Tyr-Gly-Gly-Phe-Met) and with the mammalian heptapeptide Met-enkephalin-Arg-Phe.[17,25] Various studies suggest that the small cardioactive peptides form a group of related peptides, which perhaps also occur in other phyla.[5,21,22,24,38]

The approximately 100 peptidergic CDC are located in two symmetrical clusters in the cerebral ganglia.[32] The ovulation hormone (CDCH), a basic peptide with a molecular weight (MW) of ~4700 daltons, is released from the contents of elementary granules in the CDC axon terminals in the periphery of the intercerebral commissure, during a pacemaker driven discharge, i.e., a synchronous series of action potentials in all CDC, lasting 30 to 70 min.[13,32,39-41]

Egg laying in *L. stagnalis* lasts several hours and involves a series of complicated internal events (ovulation, egg formation, egg mass formation, and oviposition), which closely correspond to stereotyped stages of the overt egg-laying behavior (resting, turning, oviposition, and inspection phases[42,43]). The intricacy and fixedness of the pattern of behaviors raises the issue whether the CDC possibly synthesize and release other peptides, besides CDCH, with a role in egg laying. Here we wish to report on experiments concerning the release of the ovulation hormone and various other peptides after stimulation by cAMP.[32,44,45] Furthermore, we present circumstantial evidence for the production of a high molecular weight precursor from which CDCH and other peptides are derived.

MATERIALS AND METHODS

For the experiments mature specimens (shell height >25 mm) of laboratory-bred *Lymnaea stagnalis* were used.

Immunocytochemistry

For light microscopy fixation of the central nervous system (CNS) was carried out in one of the following fixatives: (1) paraformaldehyde vapor (PV), $1^1/_2$ hr at 70°C;[10,46] (2) glutaraldehyde (25%, 1 vol)-picric acid (sat. aqueous sol, 3 vol)-acetic acid (to give 1%) (GPA), 2 to 16 hr, room temperature; (3) *p*-benzoquinone vapor (BV), 3 hr at 60°C;[46] and (4) glutaraldehyde (G), 0.5% buffered with 0.05 *M* Na-cacodylate, pH 7.4, overnight at 4°C. The material was embedded in paraffin and 6-μm sections were cut. For electron microscopy fixative (G) was used; embedding in araldite followed. Staining was carried out with the unlabeled antibody-enzyme method (peroxidase-antiperoxidase (PAP)[29]). For further details, cf. Reference 10.

The following antisera were used (abbreviations in brackets[23]): (1) antivasopressin (aVP); (2) antivasotocin (aVT); (3) antioxytocin; (4) anti-α-melanocyte stimulating hormone (aMSH); (5) anti-β-endorphin; (6) antidynorphin 1-8 (aDyn); (7) antidynorphin 1-17; (8) anti-α-neoendorphin (aneoEnd); (9) anti-Met-enkephalin (aME$_1$, C-terminal); (10) anti-Met-enkephalin (aME$_2$); (11) anti-Leu-enkephalin (aLE); (12) antiglucagon (C-terminal); (13) antiinsulin; (14) antiglucose-dependent insulinotropic peptide (aGIP, midportion on the molecule); (15) antivasoactive intestinal polypeptide (aVIP, C-terminal); (16) antigastrin (aGas, C-terminal); (17) anticholecystokinin (aCCK, mid portion of the molecule); (18) antisecretin; (19) antibovine pancreatic polypeptide (aPP); (20) antisomatostatin; (21) anticalcitonin; (22) antisubstance P (C-terminal); (23) antineurophysin I; (24) antineurophysin II; (25) anticorticotropin-releasing factor (aCRF); (26) anti-ACTH (aACTH; 1-39; 1-24); (27) anti-FMRFamide (645, 646, 647, 648; aFM); (28) antiserotonin (a5HT); (29) antidopamine (aDA); (30) antioctopamine; (31) antihistamine. Antisera 5, 9, 12-24 were obtained from Dr. J. Polak (London); antisera 1-8, 10, 11 from Dr. F. van Leeuwen (Netherlands Institute of Brain Research, Amsterdam); antiserum 25 from Dr. F. Vandesande (Louvain, Belgium); and antisera 28-31 from Dr. H. W. M. Steinbusch (Vrije Universiteit, Amsterdam; cf. Reference 47). Antisera 27 were raised by one of us.[10] The incubation time with the first antiserum varied from 1 to 48 hr depending on the dilution (1:500 to 1:16,000). The second antiserum used was swine-anti-rabbit γ-globulin (Nordic, 1:60).

The controls of specificity included the use of nonimmune serum and adsorption of the antiserum with the homologous antigen (liquid-phase adsorption, overnight, by incubating the diluted antiserum with 10 nmol peptide per milliliter, or solid-phase adsorption.[48]).

Coexistence of immunoreactivity was studied by comparing consecutive sections mounted on different slides (grids) and stained with different antisera. The following combinations were studied (fixation: PV, except combination 1: GPA, and 6: BV): (1) aVP-aVT; (2) aVP-aVT-aMSH-aFM; (3) aVT-aFM; (4) aVT-aGas; (5) aVT-aFM-aGas; (6) aVT-aGas-a5HT; (7) aVT-a5HT-aDA; (8) aVT-aMSH-aFMRF-aGas-a5HT; (9) aMSH-aME; (10) aMSH-aME-aLE; (11) aMSH-aME-aDA; (12) aMSH-aneoEnd-aLE-aDyn-aCRF; (13) aACTH-aCRF; (14) aME-aLE; (15) aFM-a5HT-aDA; (16) aGas-aPP; (17) a5HT-aDA. Combinations 3, 4, 5, and 17 were also studied after adsorbing (liquid phase, solid phase) the antisera with the heterologous antigens (FMRFamide, vasotocin, C-terminal part of gastrin: t-gastrin). Furthermore, aFM (646) was adsorbed (solid phase) with the following fragments of FMRFamide: Phe-NH$_2$, Arg, Arg-Phe-NH$_2$, Phe-Met, Met-Arg-Phe-COOH.

Physiology
Cardioactive Peptides

In vitro muscle preparations — The muscles (auricle, ventricle, penis retractor muscle, esophagus) were dissected out, attached to a displacement transducer, suspended in a 750-$\mu\ell$ gassed (1.72% CO_2 in O_2) organ bath, and superfused with snail saline.[49]

Purification and characterization of *Lymnaea* FMRFamide-like peptide(s) — CNS (1200) with adhering nerves were dissected out, extracted, and subjected to gel permeation, ion-exchange, and thin-layer chromatography.[42] In a second series of experiments, 1000 CNS were extracted with 80% acetone. The supernatant was dried, dissolved in distilled water, and subjected to reverse-phase HPLC on Radial-PAK C18. The solvent system consisted of buffer A: 2mM TEA/HAC (pH 4.6) in 10% methanol, and of buffer B: 20 mM TEA/HAC (pH 6.7) in 80% methanol. The program consisted of running isocratically at 0% B for 10 min and then employing a linear gradient from 0% B to 100% B in 45 min. The flow was 1.5 mℓ/min. In the second reverse-phase system, with similar running conditions, buffer A consisted of 10 mM TEA/HAC (pH 4.5), and buffer B of 10 mM TEA/HAC (pH 5.9) in 60% acetonitrile. After acid hydrolysis (6 N HCl, 24 hr, 108°C) amino acid composition was performed with reverse-phase HPLC after a precolumn derivatization of the amino acids with *o*-pthalaldehyde. (This method does not detect proline.)

CDC System

Incubation and stimulation procedure — CG/COM preparations (Figure 1) were dissected out and all nerves of the CG were cut at the origin. In experiments on release from the isolated DB, the medio-DB (MDB) were carefully removed from the CG. The preparations were collected in sterilized physiological *Lymnaea* saline.[13] The CG/COM were incubated, at 20°C, in 250 $\mu\ell$ of Millipore® filtered (0.22 μm) fresh saline. In the experiments of labeled release the saline contained 50 μCi of radioactive amino acids. These included L-[4,5-^{3}H]-leucine (130 Ci/mmol), L-[U-^{14}C]-arginine (344 mCi/mmol), or a mixture of 15 tritiated amino acids (Amersham). After 20 hr of incubation the preparations were rinsed collectively for 6 hr in 6 successive 2-mℓ rinses of fresh saline. Next, the CG/COMS were placed in 0.4 mℓ of fresh saline to which the protease inhibitor aprotinin (Sigma) was added in a final concentration of 1 TIU/mℓ (control solution). After 1 hr this solution was replaced with 0.4 mℓ of fresh saline, containing the protease inhibitor and 2 mM 8-cpt-cAMP [8-(4 chlorophenylthio)adenosine-3′-5′ monophosphate; Boehringer, Mannheim] (stimulation solution). Control and stimulation solutions were stored frozen until gel chromatography. For pulse-chase studies, exposure to the label (100 μCi of the mixture of tritiated amino acids) was terminated after 10 min and the CG/COM transferred to fresh saline containing 1 mM unlabeled amino acids. All incubations were terminated by dissection of the CDC somata and the central third parts (between the MDB, Figure 1) of COM. The tisuses were separately homogenized in 4 M guanidine HCl containing 1 TIU/mℓ of aprotinin.

Surgical procedures — Anaesthetization and recovery of the snails and cauterization of the CDC and sham operations were carried out as described before.[50] The operations of animals that had not produced egg masses during a 5-week experimental period were considered successful[50] and only CG/COM of these animals were used in the incubation experiments.

Chromatographical procedures — Bio-Gel® P-6 chromatography was carried out as described previously.[13] High-performance gel permeation chromatography of labeled peptides and of internal standard peptides was performed as described by Montelaro et al.[51] with the following protein columns in tandem: Protein-PAK 300 sw, Protein-PAK 125 (2x), and Protein-PAK 60 (Waters). The eluant was 4 M guanidine hydrochloride. For the determination of radioactivity, each column fraction was added to 10 mℓ of Dynagel® (Baker) and counted for 10 min on a Mark III. All data are corrected for background and quenching.

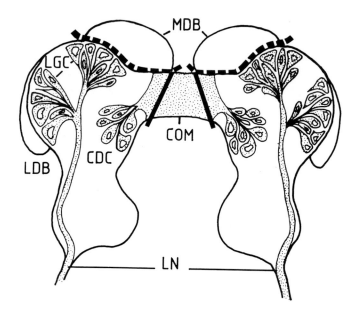

FIGURE 1.   Diagram of the cerebral ganglia of *Lymnaea stagnalis* showing: (1) the caudo-dorsal cells (CDC) and their neurohemal area at the
periphery of the intercerebral commissure (COM); (2) the light green cells
(LGC) and their neurohemal area at the periphery of the median lip nerves
(LN); and (3) the medio- and laterodorsal bodies (MDB, LDB). The preparations and parts dissected are indicated: (1) the cerebral ganglia connected
by the COM, without the nerves (GC/COM preparations); (2) paired MDB,
and (3) central third COM part.

RESULTS

Immunocytochemical Observations

Table 1 shows that in the central nervous system (CNS) of *L. stagnalis* neurons can be
distinguished with antisera to 19 vertebrate BAP, to 5 biogenic amines, and to FMRFamide.
Maps of localization of the majority of the cell types have been published previously.[23,34,35]
With many of the antisera, neurons are observed in most of the 11 central ganglia; with
others only a few neurons stain in one or two of the ganglia. With some antisera (anti-CCK,
antineurophysin I and II, anti-β-endorphin) negative results were obtained. However, the
conclusion that substances related to the peptides to which these latter antisera were raised
are absent from the brain of *Lymnaea* seems premature, as it appeared in the course of the
investigations that the type of fixation may affect immunoreactivity, and, as shown in Table
1, the antisera giving negative results have as yet only been used after fixation in one fixative.
It seems quite possible that with other fixatives, positive results will be obtained.

Studying consecutive sections of the CNS stained with different antisera showed that many
cells are immunoreactive to two or even to three antisera (Table 1). In addition to these
cells, there are also neurons that stain only with one of the antisera involved in multiple
staining ("separate cells", Table 1). Table 1 shows that 13 different combinations of multiple
staining have been observed. However, at this moment many of the possible combinations
of antisera (and fixatives) have not (yet) been studied.

FMRFamide

The four different aFMs used gave fully comparable results. On the basis of differences
in reaction to fixation and to various antisera, 12 types of aFM-positive cells can be distin-

Table 1
CNS OF *LYMNAEA STAGNALIS*

Immunoreactivity		No reaction
Oxytocin 3	Glucagon 3	CCK3
CRF 1	Insulin 3	β-endorphin 3
Somatostatin 3-(LGC)	Secretin 3	Neurophysin I 3
Dynorphin 1	VIP 3	Neurophysin II 3
Octopamin 1	GIP 3	α-Neoendorphin 1
Histamin 1	Calcitonin 3	

Separate cells   **Coexistence of immunoreactivity**

Separate cells:

VP 1,2,4
VT 1,2,4
ACTH 1,2,4
MSH 1,2
MetE 1.3,4
LeuE 1
FMRF 1,2,4
Gas 1,3
PP 3
SubP 3
5HT

Coexistence of immunoreactivity:

I — $\boxed{\text{VP+VT 1,2,4}}$

I — VT + VP 1,2,4 *II* — $\boxed{\text{VT + FMRF 1,2,4}}$ *III* — $\boxed{\text{VT + FMRF + Gas 1}}$ *IV* — $\boxed{\text{VT + 5HT + DA 1}}$

V — $\boxed{\text{ACTH + MetE 1,4}}$ (CDC)

VI — $\boxed{\text{MSH + MetE + LeuE 1}}$ *VII* — $\boxed{\text{MSH + 5HT + DA 1}}$

V — MetE + ACTH 1,4 (CDC) *VIII* — MetE + LeuE I *VI* — MetE + LeuE + MSH 1

VIII — LeuE + MetE I *VI* — LeuE + $\boxed{\text{MetE + MSH 1}}$

II — FMRF + VT 1,2,4 *IX* — $\boxed{\text{FMRF + Gas 1,2,4}}$ *III* — FMRF + VT + Gas 1

IX — Gas + FMRF 1,2,4 *III* — Gas + VT + FMRF I *X* — Gas + PP 3 *XI* — Gas + 5HT + DA 1

X — PP + Gas 3 *XII* — $\boxed{\text{Gas + SubsP 3}}$

XII — SubsP + Gas 3

XIII — $\boxed{\text{5HT + DA 1}}$ *IV* — 5HT + DA + VT I *VII* — 5HT + DA + MSH 1 *XI* — 5HT + DA + Gas 1

Note: Immunocytochemical results obtained with antisera to BAP and biogenic amines (for abbreviations, see text). Arabic numbers indicate fixatives: (1) paraformaldehyde vapor; (2) GPA (glutaraldehyde-picric acid-acetic acid); (3) *p*-benzoquinone; (4) glutaraldehyde. Roman numbers: types of coexistence of immunoreactivity. CDC, caudodorsal cells (ovulation hormone); LGC, light green cells (growth hormone).

FIGURE 2. Localization and staining characteristics of aFM-reactive elements in the CNS of the pond snail *Lymnaea stagnalis*. A-Z and Aa, cell groups, single cells, and fiber tracts; closed stars, cells reactive to aFM after PV, GPA, or G fixation; open stars, cells reactive to aFM and aVT after fixation in PV or G; open circle, cell (J) reactive to aFM and aGas after fixation in PV, GPA, or G; closed circles, cells reactive to aFM and aVT after fixation in PV, GPA, or G; open squares, cells reactive to aFM, aVT, and aGas after fixation in PV; star in closed circles, cells reactive to aFM after fixation in PV, BU, buccal ganglion; CE, cerebral ganglion; DB, dorsal body; LL, lateral lobe; PE, pedal ganglion; PL, pleural ganglion; LP, left parietal ganglion; RP, right parietal ganglion; VI, visceral ganglion.

guished (Figure 2; Table 2). As far as fixation is concerned, the table shows that types 6, 7, 10, 11, and 12 are only aFM immunoreactive after PV fixation, whereas type 2 can be stained after PV or GPA, types 3 and 8 after PV or G, and types 1, 4, 5, and 8 after fixation in either PV or G or GPA. Obviously PV fixation gives the most extensive results. Further distinction of the cell types is based on their reactivity to aVT and aGas, on that to these antisera adsorbed with the heterologous antigens, and on that to aFM adsorbed with fragments of FMRFamide.

The results can be interpreted as follows. Since all cells were negative after staining with aFM adsorbed with FMRFamide (not in the Table), it can be concluded that all IgG molecules in aFM that are invovled in staining are bound by FMRFamide. (This holds, *mutatis mutandis*, for aVT adsorbed with vasotocin and for aGas adsorbed with t-gastrin.) Apparently fragments of FMRFamide and t-gastrin can remove from aFM those IgG molecules that are responsible for staining a number of cells (types 5, 6, and 10 to 12). Similarly, vasotocin can remove the IgG molecules from aFM that are responsible for staining cell types 6 and 10. It thus

Table 2
CLASSIFICATION OF ANTI-FMRFAMIDE (aFM)-REACTIVE ELEMENTS ON THE BASIS OF FIXATION EFFECTS AND STAINABILITY WITH ANTIVASOTOCIN (aVT) AND ANTIGASTRIN (aGAS)

Cell type	Fixation	Antisera aFM	Antisera aVT	Antisera aGas	aFM+ fragments	aFM+ VT	aFM+ tGas	aVT+ FM	aVT+ tGas	aGas+ FM	aGas+ VT
1 (B)	IV	+	+	−	+	+	+	+	+	−	−
2 (C)	II	+	+	−	+	+	+	−	−	−	−
3 (E)	III	+	+	−	+	+	+	−	−	−	−
4 (H,I,L,M)	IV	+	+	−	+	+	+	+	+	−	−
5 (J)	IV	+	−	+	− (1,3,4)	+	−	−	−	−	+
6 (U)	I	+	+	+	− (2—5)	−	−	−	+	+	−
7 (N,Q,V,W)	I	+	−	−	+	+	+	−	−	−	−
8 (Y)	III	+	−	−	+	+	+	−	−	−	−
9 (A,F,G,K,S,Z,Aa)	IV	+	−	−	− (1—3,5)	+	+	−	−	−	−
10 (D,R,T)	I	+	−	−	− (1,3)	−	−	−	−	−	−
11 (O,P)	I	+	−	−	− (1,3)	+	−	−	−	−	−
12 (X)	I	+	−	−	− (2,3)	+	−	−	−	−	−

Note: (1) Paraformaldehyde vapor (PV); (II) PV or glutaraldehyde-picric acid-acetic acid (GPA); (III) PV or glutaraldehyde (G); (IV) PV or GPA or G. Solid-phase adsorptions were carried out with FMRFamide (FM), with vasotocin (VT), with the C-terminal part of gastrin (tGas), or with fragments of FMRFamide: Phe-NH$_2$ (1), Arg (2), Arg-Phe-NH$_2$ (3), Phe-Met (4), Met-Arg-Phe-COOH (5). Letters in parentheses indicate localization of cells (cf. Figure 1).

seems that cell types 5, 6, and 10 to 12 are stained by less selective IgG molecules present aFM.

On the other hand, cell types 1 to 4 and 7 to 9 are apparently reactive to IgG molecules in aFM that are not bound by either vasotocin or by t-gastrin, or by fragments of FMRFamide, as adsorptions of aFM with these substances do not abolish immunoreactivity. These more selective IgG molecules probably recognize the entire sequence of FMRFamide or at least a large part of it — they are removed from aFM by FMRFamide, not by its fragments, as mentioned — which means that the cell types concerned contain substances with FMRFamide-like determinants.

Following the same line of reasoning, it can be concluded that cell types 2 to 4 and 6 are stained by less selective IgG molecules in aVT: staining with aVT is abolished by adsorbing this antiserum with FMRFamide. (In the case of cell types 2 and 3, even t-gastrin can remove from aVT the IgG molecules involved in staining.) Thus is seems unlikely that these cell types contain substances with a vasotocin-like antigenic determinant. On the other hand, cell type 1 probably does, as the IgG molecules in aVT involved in staining this cell type cannot be removed by FMRFamide (neither can t-gastrin).

Cell type 5 is aFM/aGas and cell type 7 aFM/aVT/aGas positive. As mentioned, these cell types probably do not contain an FMRFamide-like determinant. The staining results indicate that cell type 5 neither contains a gastrin-like determinant: staining with aGas can be abolished by adsorbing the antiserum with FMRFamide (or with vasotocin). Similarly cell type 6 does not contain a vasotocin-like sequence (staining with aVT is abolished by adsorptions with FMRFamide). On the other hand, this cell type may well contain a gastrin-like determinant (staining with aGas is neither abolished by adsorption with FMRFamide nor with vasotocin).

Summarizing these results, it can be concluded that of the 12 cell types that are immunoreactive to aFM, 7 probably contain substances possessing an FMRFamide-like antigenic determinant (types 1 to 4 and 7 to 9). One of these types (1) may furthermore contain a vasotocin-like sequence. The other aFM-immunoreactive cell types are stained by less selective IgG molecules in aFM. Finally, cell type 6 probably contains a gastrin-like sequence.

Enkephalins, α-MSH, ACTH

Table 1 shows that in addition to neurons stained by only one of the antisera to these peptides, there are also cells that are immunoreactive to two or three (Figure 3A to F). Although these neurons have as yet less extensively been investigated than the aFM-positive cells, the results sustain the conclusion that fixation affects immunoreactivity. Thus, whereas after GPA fixation only two aACTH-positive cells were observed, one in the right parietal and one in the visceral ganglion, after PV fixation additional neurons (1 to 2, diameter approximately 30 μm) were found in the dorsal part of the cerebral ganglia; moreover after PV the axon endings of the CDC appeared to be immunoreactive (Figure 4).

In several ganglia, neurons reactive to both aME and aLE were observed. It seems plausible that staining is due to crossreaction. In addition to these cells, other neurons were found to be reactive to either aME (e.g., CDC axon terminals) or aLE. Probably it is less selective IgG molecules that are responsible for the immunoreactivity of these cells: (1) they stain rather weakly (this might, however, also be due to the fact that only little immunoreactive substance is present in the cells); (2) certain elements, e.g, the axon endings of the CDC, were only stained by one of the aMEs used (staining was relatively weak, although the axon endings are usually crowded with secretory granules[32]); and (3) with RIA no Met-enkephalin activity could be measured in snail Ringer after *in vitro* stimulation of CDC to release their products (unpublished). With immunoelectron microscopy (G fixation) all secretory granules in the axon endings appeared to be immunoreactive to aME (Figure 5).

In addition to cells immunoreactive to aMSH alone (Figure 3E), cells were found to be

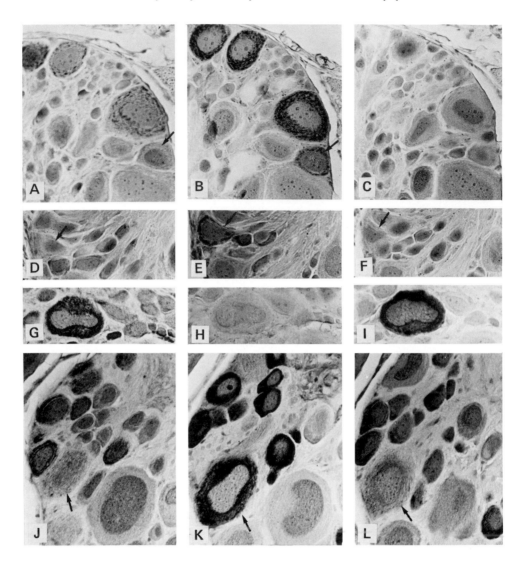

FIGURE 3.   Immunocytochemically stained consecutive sections of pedal ganglion (A: aMSH, (B: aME, (C: a5HT), pedal ganglion (D: aME, (E: aMSH, (F: a5HT), cerebral ganglion (G: aMSH, H: aME, I:aDA), and parietal ganglion (J: aDA, K: a5HT, L: aDA) showing coexistence of immunoreactivity to aMSH and aME (A, B; D, E), to aMSH and aDA (G, I), and to aDA and a5HT (J to L). Figures A and B; D to F; and J to L: at arrows cells only stained with aME, aMSH, and a5HT, respectively. (Figures A to F and J to L: magnification × 220; Figures G to I: magnification × 280.)

reactive to aMSH, aME and aLE (Figure 3A to F). Until heterologous adsorptions are carried out, it cannot be concluded whether these cells are immunoreactive to more selective or to less selective IgG molecules present in the antisera. The neurons always were intensely stained.

Immunoreactivity to aACTH was found to only coexist with that to aME, viz., in the axon endings of the CDC. Since, as mentioned, these axon endings were only aACTH positive after PV fixation, it may be assumed that it was less selective IgG molecules in aACTH that caused staining.

Biogenic Amines and Peptides

In various ganglia, neurons were observed to stain with antisera to 5HT, DA, histamine,

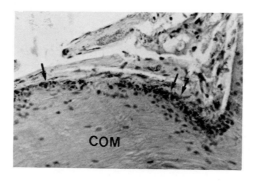

FIGURE 4. Cerebral commissure (COM), aACTH; at arrows axon endings of CDC. (Magnification × 220.)

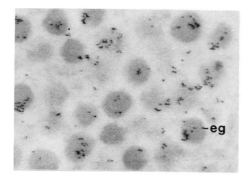

FIGURE 5. Electron micrograph of CDC axon ending immunocytochemically stained with aME, 2nd antibody; goat-anti-rabbit-gold. All elementary granules (eg) appear immunoreactive. (Magnification × 40,000.)

and octopamine. Studying consecutive sections stained with a5HT, aDA, aVT, aMSH, and aGas suggested that coexistence of two biogenic amines and a peptide occurs (Figure 3G to L). That coexistence of immunoreactivity to a5HT and aDA is not due to crossreaction can be concluded from (1) the observation that in addition to a5HT/aDA-positive cells, also neurons occur that only stained with a5HT (Figure 3J to L; cells only aDA positive were not found) and (2) the results obtained with the antisera after homo- and heterologous adsorptions (Table 3). After homologous adsorptions the reactions become negative, which indicates that 5HT and DA remove all IgG molecules involved in staining from a5HT and aDA, respectively. Apparently DA cannot bind to IgG molecules in a5HT (staining intensity was not diminished with a5HT + aDA). On the other hand, aDA probably contains crossreacting molecules (with aDA + 5HT staining intensity was reduced) in addition to more selective IgG molecules (with the adsorbed serum staining was not entirely abolished).

Physiology
Cardioactive Peptides

The isolated auricle of *L. stagnalis* was used as a routine assay system during the purification and characterization of *Lymnaea* peptides. A cardioactive peptide coeluted with commerical FMRFamide on Sephadex G-15 columns. Both substances were strongly retarded by Sephadex G-15 and eluted just after the salts. The *Lymnaea* peptide was not bound by the ion exchanger CM-Sephadex C-25. FMRFamide, on the other hand, was strongly bound,

Table 3
COEXISTENCE OF IMMUNOREACTIVITY TO ANTISERA AGAINST PEPTIDES (aVT, aMSH) AND BIOGENIC AMINES (a5HT, aDA)

Cell type	Antisera					Antisera and adsorbentia			
	aVT	aMSH	aGas	a5HT	aDA	a5HT + 5HT	a5HT + DA	aDA + DA	aDA + 5HT
CGN	+ +	–	–	+ + +	+	–	+ + +	–	+
CGC	–	+ + +	–	+ + +	+ + +	–	+ + +	–	+ +

Note: Results after solid-phase adsorption of antisera to the biogenic amines with the homologous and heterologous antigens. CGN[52]: cerebral giant neuron; CGC: cerebral ganglion cells.

but could be eluted by a linear gradient of ammonium acetate at pH 8. The *Lymnaea* peptide has a lower Rf value (0.60) on thin-layer chromatography on cellulose plates than FMRFamide (0.75).[42]

Low threshold (e.g., 10^{-14} to 10^{-6} *M* FMRFamide) dose-response relationships could be established with all muscle preparations studied. Both the *Lymnaea* peptide (prepurified on Sephadex G-15 and CM-Sephadex C-25) and FMRFamide increased: (1) the beating rate and amplitude of the auricle, (2) the amplitude of the ventricle, (3) the contraction of the penis retractor muscle, and (4) the contractile activity of the esophagus.

A Sep-Pak C_{18} prepurified acetone extract of the CNS was analyzed using the first solvent system. Eight cardioactive regions were detected. Amino acid composition analysis indicated that only one region contained peptide material in sufficient amounts to allow further analysis. This material appeared to be homogenous and to elute well after FMRFamide on the second HPLC system. Amino acid analysis suggested the following composition: Phe (2), Arg (1), Gly (1), and Ile (1).

The observations indicate that this *Lymnaea* peptide is nonidentical to FMRFamide. Both substances coelute on Sephadex G-15 columns, but behave differently during ion-exchange and thin-layer chromatography. On the other hand, the great similarity between the excitatory effects on the *Lymnaea* peptide and FMRFamide on four different muscle preparations indicates that these compounds must be closely related. Since it has been shown that the requirements for biological activity are very stringent at the C-terminal of the molecule,[53] it may be assumed that the C-terminal of the *Lymnaea* peptide is (nearly) identical to that of FMRFamide. The preliminary results on the amino acid composition support this supposition: in both peptides Phe and Arg are present in the ratio 2:1.

Multiple Peptide Release by the CDC

In control experiments with intact CG/COM, bioassays of fractions obtained by gel chromatography indicated that during CDC discharges evoked by 8-cpt-cAMP CDCH bioactivity were released into the saline, whereas virtually no CDCH bioactivity was released during the preceding control period (Figure 6A). Released CDCH bioactivity corresponded with a molecular weight (MW) in the range of 3500 to 5200. Figure 6A also shows the radioactivity pattern of released peptides by intact CG/COM. During the control period, there only was background leakage from the preparations (i.e., radioactive material in peaks I and V), whereas during CDC discharges peaks I to IV appeared, corresponding with MWs of ≥6000, ~5500, ~4500, and ~2500, respectively. CDCH bioactivity coeluted with peak III material (MW range 3700 to 4800), which suggests that a peptide of this peak is identical to CDCH. After stimulation of prelabeled CG/COM preparations from which the CDC were previously

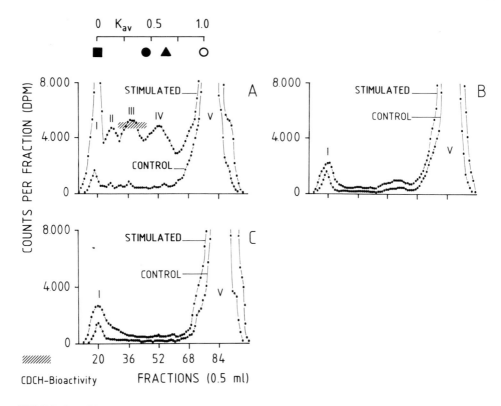

FIGURE 6.   Bio-Gel® P-6 chromatography of radioactive peptides released by various preparations prelabeled during 20 hr with ¹⁴C Arg. Markers, not run concurrently, are (■) blue dextran 2000 (MW 2.10^6 dalton); (●) glucagon (MW 3500); (▲) insulin A (MW 2530); (○) [⁵H] Leu (MW 141). (A) Radioactivity profiles of control and stimulation solutions are shown (20 CG/OM preparations). Release was stimulated with 2 m*M* 8-cpt-cAMP. No CDCH bioactivity could be detected in the control solution. (B) Radioactivity profiles of CG/COM preparations (20 per experiment) from which the CDC were previously removed. No CDCH bioactivity could be detected in control and stimulation solutions. (C) Radioactive profiles of isolated MDB (40 per experiment).

removed, no CDCH bioactivity could be detected, and there was virtually no release of radioactivity in the regions of peaks I, II, III, and IV (Figure 6B). (The residual small peak I and peak V probably represent background leakage.) Similarly, after stimulation of the isolated prelabeled endocrine MDB, no radioactivity appeared in the regions of peaks I to IV, but there was background leakage (Figure 6C).

The fact that CG/COM without CDC are not capable of releasing CDCH and other peptides clearly shows that it is the CDC that are the source of these substances. This conclusion is sustained by the fact that they are not released by isolated MDB, although these organs contributed considerably to the nonstimulus-dependent background leakage from CG/COM preparations. Also the neuroendocrine light green cells can be excluded, because (1) their neurohemal area is located in the median lip nerves[37] (that had been cut off), and (2) the stimulus used is not capable of activating the light green cells.[62] The fact that the Bio-Gel® P-6 elution patterns obtained after stimulation with 8-cpt-cAMP are similar to those after electrical stimulation (see Reference 13) very likely indicates that also in the first case peptides of the CDC system are preferentially released.

The results indicate that the peaks represent a limited number of molecular weight classes, and not necessarily distinct peptides. One on the peptides of peak III is identical to CDCH.

Biosynthesis

HPLC separation of extracts of CDC and COM exposed to labeled amino acids for long periods (20 hr) reveals eight major peaks of labeled peptide (data not shown). These correspond to apparent MW of ~35,000 (peak 1), ~20,000 (2), ~10,000 (3), ~7000 (4), ~6000 (5), ~4500 (6), ~3500 (7), and ~2000 (8). Pulse-chase experiments reveal a clear precursor-product relationship between these peaks. After a 10-min pulse of labeled precursor, only peaks 1 and 2 are present in the CDC somata and none in the COM. After 1 hr of chase, peak 1 has decreased and peaks 2 to 4 have appeared. With longer chase periods more peaks appear, and after 4 hr, peaks 5, 6, and 8 also are present in the COM.

The results suggest that a protein precursor (MW 35,000) is synthesized in the CDC somata, which gives rise to intermediate products (MW ~20,000, ~10,000, and ~7000), and a number of apparent end products with MWs of ~6000, ~4500 (probably CDCH), and ~2000. Recent unpublished results have shown that along with the end products, a peptide is released that is capable of exciting the CDC (autotransmitter,[15] ~1500).

DISCUSSION

The results confirm the view that on the basis of immunocytochemical observations, unequivocal conclusions about the chemical structure of an immunoreactive substance are hardly possible. Obviously, fixation influences immunoreactivity. Furthermore, the results obtained with aFM adsorbed with heterologous antigens or with fragments of FMRFamide showed that a (polyclonal) antiserum may contain rather unselective IgG molecules. Thus, the IgG molecules involved in staining of a few cell types could be removed from aFM by adsorbing the antiserum with a single amino acid (Arg). Immunoadsorptions of (polyclonal and monoclonal) antisera with fragments of the molecule to which the sera were raised, or, if unexpected crossreactions occur (e.g., aFM/aVT/aGas), with heterologous antigens, clearly allow a closer specification of the results. Nonetheless, also with these "characterized" antisera various substances will be stained, if they share the antigenic determinant. The fact that in the CNS of *Lymnaea* seven histochemically different cell types are immunoreactive to "selective" IgG molecules in aFM probably indicates that in this snail several substances occur with an antigenic determinant identical or closely related to the Phe-Met-Arg-Phe-NH$_2$ sequence.

Although the results suggest that unselective IgG molecules may play a part in various multiple stainings, the fact that in most cases in addition to double- or triple-stained cells single-immunoreactive neurons also occur indicates that the neurons represent different cell types. Whether the multiple-stained cells contain one peptide with two of three antigenic determinants, or more than one peptide, each possessing one of the determinants, cannot be concluded. Another possibility would be that several cell types produce the same precursor and that multiple stainings are rather due to immunoreactivity of the precursor than to that of the end product(s). This seems, however, not very likely, as not only the cell bodies of the neurons were stained, i.e., the production sites of the precursor, but also the axons and the axon terminals, suggesting that the multiple-stained materials are transported and released from the neurons.

It was argued that one of the aFM-positive cells (B) contains an FMRFamide-like and a vasotocin-like antigenic determinant. Since FMRFamide and vasotocin seem unrelated, it could also be speculated that in this cell type two precursors are synthesized, each yielding a particular end product. A similar conclusion might be drawn from the observations on the rat neurohypophysis, showing that vasopressin and oxytocin terminals may contain in addition Met- and Leu-enkephalin, respectively.[54] Evidently, in the case of coexistence of immunoreactivity to a peptide and a classical transmitter, different substances are involved.

The presence in and the release from neurons of multiple peptides and/or classical transmitters poses intriguing cell biological questions. Are these substance packaged in the same secretory granules, are they released simultaneously or separately, and can a neuron make different types of synapse? It has often been reported[55] that different types of granule occur in the same cell. This might indicate separate packaging, transport, and release of substances. On the other hand, the presence of aME immunoreactivity in all secretory granules of the CDC of *Lymnaea* indicates cotransport and corelease of BAP (cf. Reference 54). Differential packaging, transport, and release of biologically active substances would allow a neuron to adapt its reaction under varying physiological conditions. Another mode of achieving this is probably realized in the bag cells of *Aplysia,* which possess, as mentioned, multiple genes, encoding for different combinations of BAP.[12,18]

It would seem impossible to conclude on the basis of the present results whether in *L. stagnalis* one or more precursors of opioid peptides[17] occur. A suggestion for the possible production of proopiomelanocortin on the basis of coexistence of immunoreactivity to aMSH, aME, and aLE and of that to aACTH and aME seems premature, as there are also contraindications, e.g., with anti-β-endorphin negative results were obtained, and the reactions to aME and aACTH in the CDC axon endings are most probably due to binding of less selective IgG molecules. Similarly, no strong indications for the presence of Pro-enkephalin A and/ or B are available: neither was coexistence of immunoreactivity to aFM — FMRFamide is structurally related to Met-enkephalin and heptapeptide — and aME, observed, nor to aDyn and aME/aLE, whereas reactions with a-neoEnd were negative. It should be mentioned in this regard that our results with aFM differ to some extent from those of Martin et al.[56] who observed in *Lymnaea* coexistence of immunoreactivity to aFM and aEnkephalin and to aFM and aMSH. The discrepancy might be due to differences in the specificity of the antisera used, or to differences in fixation procedures. It should furthermore be mentioned that in contrast to *Lymnaea* in two other molluscs (*Octopus, Aplysia*), coexistence of immunoreactivity to aMSH, aEnkephalin, and aFM has been observed.[56,57]

Since it has been shown that the biological activity of FMRFamide-like peptides is largely dependent on the structure of the C-terminal part of the molecule,[53] it may be assumed that all seven FMRF-like substances of *Lymnaea* will possess biological activity. It seems likely that they are identical to the cardioactive peptides separated by HPLC. The fact that the numbers of substances identified with immunocytochemistry (7) and HPLC (8) are in good agreement, supports this view.

Observations on different species have indicated that aFM reacts with PP, neuropeptide Y (NPY), and LPLRFamide.[58-61] This latter peptide was isolated from chicken brain. Reaction is apparently due to the similarity of the C-terminal parts of the four peptides. Further investigations (the study of alternating sections stained with antisera to these peptides has not yet been carried out) may show whether or not PP, NPY, and LPLRFamide are members of the family of FMRFamide-like substances present in *Lymnaea*. The same might be true for Met[5]-enkephalin-Arg-Phe-NH$_2$. This cardioactive peptide has been isolated from *Octopus*.[25,56] From this species another three cardioactive peptides were extracted.[56] Since these probably possess as C-terminal the sequence Arg-Phe-NH$_2$, they might well be immunoreactive to aFM.

Supposedly some of the FMRFamide-like BAP serve as neurohormones and others as neurotransmitters. This can be concluded from light as well as electron microscope observations showing that aFM-positive axon terminals can be present in neurohemal regions,[10] which suggests the release of the aFM-positive substance into the blood (neurohormone), whereas in other cases aFM-positive axons end synaptically on muscles (heart, penis retractor muscle[4]) or on other neurons[4] (neurotransmitter). The neurotransmitter function of FMRFamide-like substances is supported by physiological results[5] (see also ter Maat, this volume).

The results indicate that the CDC produce a precursor from which several BAP are cleaved. With immunocytochemistry positive reactions were observed in the axon endings of the CDC with aME and aACTH. It has already been argued that it seems unlikely that the immunoreactive substances are structurally identical to the vertebrate peptides concerned.

The Met-enkephalin-like substance might structurally be related to the CDC autotransmitter (ter Maat, this volume). ACTH (39 amino acids) has about the same size as CDCH (MW ~4700). It seems, however, highly unlikely that there is a structural relationship between these two peptides, as can be concluded from investigations on the amino acid sequence of CDCH.[63] On the other hand, CDCH shows similarities with *Aplysia* ELH.

SUMMARY

With antisera to a great number (19) of vertebrate biologically active peptides (BAP) and to biogenic amines (5) immunoreactive neurons were identified in the central nervous system (CNS) of the pond snail *Lymnaea stagnalis*. Coexistence was observed of immunoreactivity to antisera raised against related as well as to unrelated BAP, to antisera against 2 different biogenic amines, and to antisera against 1 and/or 2 biogenic amines and a BAP (13 different combinations). Fixation appeared to affect immunoreactivity.

With an antiserum to the molluscan cardioactive tetrapeptide FMRFamide (Phe-Met-Arg-Phe-NH$_2$), 12 histochemically different cell types were distinguished. On the basis of results obtained with anti-FMRFamide adsorbed with the homologous and with heterologous antigens and with fragments of FMRFamide, it is concluded that seven cell types contain substance(s) with an antigenic determinant, which is identical or closely related to the sequence Phe-Met-Arg-Phe-NH$_2$.

With gel permeation, ion-exchange, thin-layer chromatography, and HPLC techniques eight cardioactive substances were isolated from the CNS of *L. stagnalis*. The results indicate that at least one of these substances is structurally closely related, although not identical, to FMRFamide. It is suggested that the eight cardioactive peptides are produced by the seven different anti-FMRFamide immunoreactive cell types.

Release of peptides by the ovulation hormone-producing caudodorsal cells (CDC) was studied in experiments in which preparations were incubated with radioactive-labeled amino acids. Release of peptides from the CDC axon endings was induced by 8-cpt-cAMP. Analysis of results obtained with high-performance gel permeation showed that in addition to the ovulation hormone (MW ~4700), 3 additional (groups of) peptide(s) are released (MW ≥6000, ~5500, and ~2500). Studies on the biosynthesis (pulse-chase experiments) of these peptides indicated that they are cleaved from a precursor protein with a MW of ~35,000. The axon endings of the CDC are immunoreactive to anti-Met-enkephalin and to anti-ACTH. It is argued that it is unlikely that any of the products released by the CDC is structurally identical to either Met-enkephalin or ACTH.

ACKNOWLEDGMENTS

The authors would like to thank Ms. C. Montagne, Ms. D. Reichelt, Ms. T. Hogenes, Mr. H. van Loenhout, Mr. T. Otter, and Mr. A. van der Leest for skillful technical assistance and Ms. T. Laan for typing the manuscript.

REFERENCES

1. **Scharrer, B.,** Peptidergic neurons: facts and trends, *Gen. Comp. Endocrinol.,* 34, 50, 1978.
2. **de Wied, D. and Gispen, W. H.,** Behavioral effects of peptides, in *Peptides in Neurobiology,* Gainer, H., Ed., Plenum Press, New York, 1977, 397.
3. **Buijis, R. H. and Swaab, D. F.,** Immunoelectronmicroscopical demonstration of vasopressin and oxytocin synapses in the rat limbic system, *Cell Tissue Res.,* 204, 355, 1979.

4. **Boer, H. H., Schot, L. P. C., Reichelt, D., Brand, H., and ter Maat, A.,** Ultrastructural immunocytochemical evidence for peptidergic neurotransmission in the pond snail *Lymnaea stagnalis, Cell Tissue Res.,* 238, 197, 1984.
5. **Cottrell, G. A., Schot, L. P. C., and Dockray, G. J.,** Identification and probable role of a single neurone containing the neuropeptide *Helix* FMRFamide, *Nature (London),* 304, 638, 1983.
6. **Otsuka, M. and Konishi, S.,** Substance P — the first peptide neurotransmitter?, *TINS,* 6, 317, 1983.
7. **Swanson, L. W.,** Neuropeptides — new vistas on synaptic transmission, *TINS,* 6, 294, 1983.
8. **Iversen, L. L.,** Neuropeptides — what next?, *TINS,* 6, 293, 1983.
9. **Hökfelt, T., Johansson, O., Ljungdahl, Å., Lundberg, J. M., and Schultzberg, M.,** Peptidergic neurones, *Nature (London),* 284, 515, 1980.
10. **Schot, L. P. C.,** Peptidergic Neurons in the Pond Snail *Lymnaea stagnalis,* Ph.D. thesis, Free University, Amsterdam, 1984, 1.
11. **Everitt, B. J., Hökfelt, T., Terenius, L., Tatemoto, K., Mutt, V., and Goldstein, M.,** Differential coexistence of neuropeptide Y (NPY)-like immunoreactivity with catecholamines in the central nervous system of the rat, *Neuroscience,* 11, 443, 1984.
12. **Arch, S.,** Posttranslational routing of neuropeptides in the bags cells of *Aplysia,* in *Neuro-Endocrinology,* Lever, J. and Boer, H. H., Eds., North-Holland, Amsterdam, 1983, 44.
13. **Geraerts, W. P. M., Tensen, C. P., and Hogenes, Th. M.,** Multiple release of peptides by electrically active neurosecretory caudo dorsal cells of *Lymnaea stagnalis, Neurosci. Lett.,* 41, 151, 1983.
14. **Mayeri, E. and Rothman, B. S.,** Nonsynaptic peptidergic neurotransmission in the abdominal ganglion of *Aplysia,* in *Neurosecretion: Molecules, Cells, Systems,* Farner, D. S. and Lederis, K., Eds., Plenum Press, New York, 1981, 305.
15. **Jansen, R. F.,** Neuronal and Hormonal Control of the Egg-Laying Behaviour in the Pond Snail *Lymnaea stagnalis,* Ph.D. thesis, Free University, Amsterdam, 1984, 1.
16. **Nakanishi, S., Inoue, A., Kita, T., Nakamura, M., Chang, A. C. Y., Cohen, S. N., and Numa, S.,** Nucleotide sequence of cloned cDNA for bovine corticotropin-β-lipotropin precursor, *Nature (London),* 278, 423, 1979.
17. **Höllt, V.,** Multiple endogenous opioid peptides, *TINS,* 6, 24, 1983.
18. **Scheller, R. H., Jackson, J. F., McAllister, L. B., Rothman, B. S., Mayeri, E., and Axel, R.,** A single gene encodes multiple neuropeptides mediating a stereotyped behavior, *Cell,* 32, 7, 1983.
19. **Buma, P. and Roubos, E. W.,** Ultrastructural demonstration of non-synaptic release sites in the brain of the snail *Lymnaea stagnalis,* the insect *Periplaneta americana,* and the rat, *Neuroscience,* in press.
20. **Schaller, H. C., Flick, U., and Darai, G.,** A neurohormone from *Hydra* is present in brain and intestine of rat embryos, *J. Neurochem.,* 29, 393, 1977.
21. **Boer, H. H., Schot, L. P. C., Veenstra, J. A., and Reichelt, D.,** Immunocytochemical identification of neural elements in the central nervous system of a snail, some insects, a fish and a mammal with an antiserum to the molluscan cardio-excitatory tetrapeptide FMRFamide, *Cell Tissue Res.,* 213, 21, 1980.
22. **Dockray, G. J., Valliant, C., and Williams, R. C.,** New vertebrate brain-gut peptide related to a molluscan neuropeptide and an opioid peptide, *Nature (London),* 292, 656, 1981.
23. **Schot, L. P. C., Boer, H. H., Swaab, D. F., and van Noorden, S.,** Immunocytochemical demonstration of peptidergic neurons in the central nervous system of the pond snail *Lymnaea stagnalis* with antisera raised to biologically active peptides of vertebrates, *Cell Tissue Res.,* 216, 273, 1981.
24. **Weber, E., Evans, C. J., Samuelson, S. J., and Barchas, J. D.,** Novel peptide neuronal system in rat brain and pituitary, *Science,* 214, 1248, 1981.
25. **Voigt, K. H., Kiehling, C., Frösch, D., Bickel, U., Geis, R., and Martin, R.,** Identity and function of neuropeptides in the vena cava neuropil of *Octopus,* in *Molluscan Neuro-Endocrinology,* Lever, J. and Boer, H. H., Eds., North-Holland, Amsterdam, 1983, 228.
26. **Leung, M. and Stefano, G. B.,** Purification and identification of enkephalins in molluscan ganglia, 12th Conf. Eur. Comp. Endocrinol., Abstr. 70, Sheffield, 1983.
27. **Scharrer, B. and Stefano, G. B.,** Opioid binding sites in the brain and midgut of the insect *Leucophaea maderae* (Blatteria), 12th Conf. Eur. Comp. Endocrinol., Abstr. 69, Sheffield, 1983.
28. **Swaab, D. F., Pool, C. W., and van Leeuwen, F. W.,** Can specificity ever be proved in immunocytochemical staining?, *J. Histochem. Cytochem.,* 25, 388, 1977.
29. **Sternberger, L. A.,** *Immunocytochemistry,* 2nd ed., John Wiley & Sons, New York, 1979.
30. **van Leeuwen, F. H.,** An introduction to the immunocytochemical localization of neuropeptides and neurotransmitters, *Acta Histochim. Suppl.,* 24, 49, 1981.
31. **Joosse, J., de Vlieger, T. A., and Roubos, E. W.,** Nervous systems of lower animals as models, with particular reference to peptidergic neurons in gastropods, in *Chemical Transmission in the Brain, Progress in Brain Research,* 55, Elsevier Biomedical Press, Amsterdam, 1982, 379.
32. **Roubos, E. W.,** Cytobiology of the ovulation-neurohormone producing neuroendocrine caudo-dorsal cells of *Lymnaea stagnalis, Int. J. Cytol.,* 89, 295, 1984.

33. **Greenberg, M. J. and Price, D. A.,** Cardioregulatory peptides in molluscs, in *Peptides: Integrators of Cell and Tissue Function,* Bloom, F. E., Ed., Raven Press, New York, 1980, 107.

34. **Schot, L. P. C. and Boer, H. H.,** Immunocytochemical demonstration of peptidergic cells in the pond snail *Lymnaea stagnalis* with an antiserum to the molluscan cardioactive tetrapeptide FMRFamide, *Cell Tissue Res.,* 225, 347, 1982.

35. **Boer, H. H. and Schot, L. P. C.,** Phylogenetic aspects of peptidergic systems, in *Molluscan Neuro-Endocrinology,* Lever, J. and Boer, H. H., Eds., North-Holland, Amsterdam, 1983, 9.

36. **Lloyd, P. E.,** Cardioactive neuropeptides in gastropods, *Fed. Proc. Fed. Am. Soc. Exp. Biol.,* 41, 2948, 1982.

37. **Joosse, J. and Geraerts, W. P. M.,** Endocrinology, in *The Mollusca, Vol. 4, Physiology, Part I,* Wilbur, K. M. and Saleuddin, A. S. M., Eds., Academic Press, New York, 1983, 317.

38. **Morris, H. R., Panico, M., Karplus, A., Lloyd, P. E., and Riniker, B.,** Elucidation by FAB-MS of the structure of a new cardioactive peptide from *Aplysia, Nature (London),* 300, 634, 1982.

39. **de Vlieger, T. A., Kits, K. S., ter Maat, A., and Lodder, J. C.,** Morphology and electrophysiology of the ovulation hormone producing neuroendocrine cells of the freshwater snail *Lymnaea stagnalis* (L.), *J. Exp. Biol.,* 84, 259, 1980.

40. **Geraerts, W. P. M. and Joosse, J.,** Freshwater snails (Basommatophora), in *The Mollusca, Vol. 7, reproduction,* Wilbur, K. M. and Saleuddin, A. S. M., Eds., Academic Press, New York, 1984, 141.

41. **Kits, K. S. and Bos, N. P. A.,** Pacemaking mechanism of the afterdischarge of the ovulation hormone producing caudo-dorsal cells in the gastropod mollusc *Lymnaea stagnalis, J. Neurobiol.,* 12, 425, 1981.

42. **Geraerts, W. P. M., ter Maat, A., and Hogenes, Th. M.,** Studies on the release activities of the neurosecretory caudo-dorsal cells of *Lymnaea stagnalis,* in *Biosynthesis, Metabolism and Mode of Action of Invertebrate Hormones,* Hoffmann, J. A. and Porchet, M., Eds., Springer-Verlag, Heidelberg, 1984, 44.

43. **Goldschmeding, J. T., Wilbrink, M., and ter Maat, A.,** The role of the ovulation hormone in the control of egg-laying behaviour in *Lymnaea stagnalis,* in *Molluscan Neuro-Endocrinology,* Lever, J. and Boer, H. H., Eds., North-Holland, Amsterdam, 1983, 251.

44. **Buma, P., Roubos, E. W., and Pieters, F. A. L.,** Significance of calcium and cAMP for the control of neurohormone release by the neuroendocrine caudo-dorsal cells of the freshwater snail *Lymnaea stangalis,* in *Molluscan Neuro-Endocrinology,* Lever, J. and Boer, H. H., Eds., North-Holland, Amsterdam, 1983, 74.

45. **ter Maat, A. and Jansen, R. F.,** The egg-laying behaviour of the pond snail: electrophysiological aspects, in *Biosynthesis, Metabolism and Mode of Action of Invertebrate Hormones,* Hoffmann, J. A. and Porchet, M., Eds., Springer-Verlag, Heidelberg, 1984, 57.

46. **Pearse, A. G. E. and Polak, J. M.,** Bifunctional reagents as vapour-phase and liquid phase fixatives for immunocytochemistry, *Histochem. J.,* 7, 179, 1975.

47. **Steinbusch, H. W. M. and Tilders, F. J. H.,** Localization of dopamine, noradrenalin, adrenalin, serotonin and histamin in the central nervous system. A light-microscopal immunohistochemical study, in *IBRO-Handbook Series: Methods in Neuroscience, Vol. 6, Histochemical and Ultrastructural Identification of Monoamine Neurons,* Furness, J. and Costa, M., Eds., John Wiley & Sons, Chichester, in press.

48. **Swaab, D. F. and Pool, C. W.,** Specificity of oxytocin and vasopressin immunofluorescence, *J. Endocrinol.,* 263, 1975.

49. **Geraerts, W. P. M., van Leeuwen, J. P. Th. M., Nuyt, K., and de With, N. D.,** Cardioactive peptides of the CNS of the pulmonate snail *Lymnaea stagnalis, Experientia,* 37, 1168, 1981.

50. **Gerearts, W. P. M. and Bohlken, S.,** The control of ovulation in the hermaphrodite freshwater snail *Lymnaea stagnalis* by the neurohormone of the caudo-dorsal cells, *Gen. Comp. Endocrinol.,* 28, 350, 1976.

51. **Montelaro, R. C., West, M., and Issel, C. J.,** High performance gel permeation chromatography of proteins in denaturing solvents and its application to the analysis of enveloped virus polypeptides, *Anal. Biotheor.,* 114, 398, 1981.

52. **Goldschmeding, J. T., van Duivenboden, Y. A., and Lodder, J. C.,** Axonal branching pattern and coupling mechanisms of the cerebral giant neurones in the snail *Lymnaea stagnalis, J. Neurobiol.,* 12, 405, 1981.

53. **Price, D. A. and Greenberg, M. J.,** Pharmacology of the molluscan cardioexcitatory neuropeptide FMRFamide, *Gen. Pharmacol.,* 11, 237, 1980.

54. **Martin, R., Geis, R., Holl, R., Schäfer, M., and Voigt, K.-H.,** Co-existence of unrelated peptides in oxytocin and vasopressin terminals of rat neurohypoyses: immunoreactive methionine-enkephalin-, leucine-enkephalin-, and cholecystokinin-like substances, *Neuroscience,* 8, 213, 1983.

55. **Golding, D. W. and May, B. A.,** Duality of secretory inclusions in neurons. Ultrastructure of the corresponding sites of release in invertebrate nervous systems, *Acta Zool.,* 63, 229, 1982.

56. **Martin, R., Stefano, G. B., and Voigt, K.-H.,** Neuronal distribution of immunoreactive enkephalin-like material in *Octopus, Mytilus* and *Lymnaea* ganglia, 12th Conf. Eur. Endocrinol., Abstr. 72, Sheffield, 1983.

57. **Haas, C., Voigt, K.-H., and Martin, R.,** Peptidergic neurons in *Aplysia* ganglia, 12 Conf. Eur. Comp. Endocrinol., Abstr. 74, Sheffield, 1983.
58. **Dockray, G. J. and Williams, R. G.,** FMRFamide-like immunoreactivity in rat brain: development of a radioimmunoassay and its application in studies of distribution and chromatographic properties, *Brain Res.,* 266, 295, 1983.
59. **Veenstra, J. and Schooneveld, H.,** Immunocytochemical localization of neurons in the nervous system of the Colorado potato beetle with antisera against FMRFamide and bovine pancreatic polypeptide, *Cell Tissue Res.,* 235, 303, 1984.
60. **Moore, R. Y., Gustafson, E. L., and Card, J. P.,** Identical immunoreactivity of afferents to the rat suprachiasmatic nucleus with antisera against avian pancreatic polypeptide, molluscan cardioexcitatory peptide and neuropeptide Y, *Cell Tissue Res.,* 236, 41, 1984.
61. **Dockray, G. J., Reeve, J. R., Jr., Shively, J., Gayton, R. J., and Barnard, C. S.,** A novel active pentapeptide from chicken brain identified by antibodies to FMRFamide, *Nature (London),* 305, 328, 1983.
62. **de Vlieger, Th. and Roubos, E. W.,** personal communication.
63. **Ebberink, R. H. M., van Loenhout, H., Geraerts, W. P. M., and Joosse, J.,** Purification and amino acid sequence of the ovulation neurohormone of *Lymnaea stagnalis, Proc. Natl. Acad. Sci. U.S.A.,* in press.

Peptide-Monoamine Interaction

PEPTIDE-MONOAMINE INTERACTIONS IN PLANARIA AND HYDRA

Giorgio Venturini, Antonio Carolei, Guido Palladini, and Vito Margotta

SUMMARY

Planaria and hydra nervous systems have been studied with neuropharmacological and neurochemical methods. Dopamine, norepinephrine, and serotonin are present in planaria neurons. Dopaminergic hyperstimulation rises cAMP and induces typical hyperkinesias, whereas dopaminergic blocking agents decrease cAMP and motility. Dopamine also stimulates adenylate cyclase in vitro. Naloxone induces in planaria a rise in cAMP levels and dopaminergic-like hyperkinesias. Both actions are antagonized by dopaminergic blocking agents. Morphine, on the contrary, decreases cAMP and motility. Met-enkephalin-like immunoreactivity is present in planaria neurons. A model for interactions between dopaminergic and enkephalinergic neurotransmission and/or neuromodulation has been proposed via an inhibitory modulation of opiates on dopamine release.

The same amines found in planaria are present in hydra, without any evidence of dopaminergic effects on behavior. The amines also are ineffective on hydra adenylate cyclase. No enkephalin-like immunoreactivity has been detected with radioimmunoassay and immunocytochemistry in hydra. Nevertheless, the evidence of a dopaminergic interaction with GSH-induced feeding response is reported, and the interaction of some opiates with the GSH receptor is documented.

INTRODUCTION

Invertebrates are suitable research models in neurobiology because of their relatively simple hierarchical organization. Functional neuronal interactions can be detected in invertebrates and theories can be validated in these simple systems easier than in more evolved animals. Among invertebrates, planaria look particularly interesting, since they are considered the first occurring example in the phylogenetic scale which exhibit centralization and cephalization of the nervous system. The planarian nervous system consists of a cerebrum and longitudinal cords with ladder-like interconnections and thus represents the first organized stage in the phylogenetic evolution of cerebral and spinal neurons present in vertebrates.

The motor system of *Dugesia gonocephala* (cerebral ganglia and nervous cords) possesses nervous pathways and centers, responsible for translation of the nervous impulse into active muscular movement, which use mainly dopamine as a synaptic neurotransmitter.[1] This flatworm may be considered a sensitive animal model in which dopaminergic agonists induce hyperkinesias (abnormal screw-like movements), whereas dopaminergic blocking drugs antagonize such motor performance. The utilization of this model is useful in eventually discriminating interferences of other pathways or levels of action and other neurotransmitters or neuromodulators. The evaluation of planaria behavior is easy when compared to the difficulties met in the more complicated animal models commonly employed. Moreover it seems worth mentioning that some peptides (melanocyte-stimulating hormone release-inhibiting factor: MIF, oxytocin, Met- and Leu-enkephalin) exhibit in planaria behavioral effects, namely depending on a peptidergic neurotransmission or neuromodulation and/or endocrine-like permissive actions.[2-5]

The physiological and pharmacological characteristics of this flatworm and the capacity to respond to polypeptide substances supported the existence, in planaria, of opiate receptors and related endogenous opiates.[5] Recent data supporting the hypothesis that enkephalins and related receptors are present in the nervous system of many invertebrates[6-8] and are strictly related to dopaminergic neurons[9] have been confirmed in planaria.[10] These encouraging results induced us to study, with a similar approach, the nervous system of hydra. Coelenterates have an uncertain evolutionary position, possibly not leading to Platyhelminthes, and includes both very primitive forms, like hydra, and more evolved ones also. In hydra the first example of nervous system is found, organized as a continuous network, whose cells are polymorph and continuously differentiating from totipotent interstitial elements. Nerve cells seem to be more abundant at levels of the head and foot. Recent histochemical data support the presence, in hydra, of peptidergic putative neuromodulators, neurotransmitters, or neurohormones.[11-15] Due to lacking suitable behavioral parameters in hydra, the glutathione-induced feeding response has been preferred to spontaneous motor performances as a potentially scrutinizable pattern. In fact the reduced glutathione, naturally coming from a prey stricken by nematocysts, punctually stimulates tentacles bending and mouth opening through an interaction with an ectodermal surface receptor.[16,17] In addition to pharmacological treatments possibly affecting behavior, histochemical and biochemical amine assays, together with detection of cAMP levels and adenylate cyclase characterization, have been performed both in planaria and in hydra. Radioimmunoassay and immunocytochemistry of endogenous opiates have also been performed in both species.

MATERIALS AND METHODS

Animals

Planaria (*Dugesia gonocephala*) were kept in glass containers, filled with tap water, and located in a dimly lighted thermostatic chamber at 18°C. All specimens were fed raw calf liver. Hydra (*Chlorohydra viridissima*) were cultured in M solution after Lenhoff and Brown[18] at 18°C and fed, twice a week, with nauplia of *Artemia salina*. Before every experimental session planaria and hydra specimens were kept unfed for 5 days.

Pharmacological Treatments

Behavioral observations were performed comparing normal and treated specimens by a blind randomized procedure, where treatment and dose given were unknown. Drugs were dissolved in the same medium used to grow the experimental animals. Each treatment was performed in petri dishes under microscopical control.[1]

Histochemistry

For histochemical observations planaria and hydra specimens were frozen in isopenthane, cooled in liquid nitrogen, and then freeze-dried in an Edward-Pearse Tissue Dryer (2 hr at $-20°C$, 10^{-2} torr). After paraffin embedding 10-μm sections were used for the fluorescence method of Falck et al.[19] for catecholamines and for immunofluorescence with anti-Met-enkephalin serum.[10]

Biochemical Determinations

In hydra all biochemical determinations were carried out after elimination of symbiotic algae. Specimens were gently homogenized by hand in saline solution. Homogenates were deprived of algae by centrifugation (150 g for 5 min). A microscopical test of supernatant was used as a control of algae absence. All biochemical assays were peformed both on the animal supernatant and on algae pellet. Protein content was determined by the method of Lowry et al.[20] using bovine serum albumin (BDH) as standard. In order to measure catecholamines and serotonin on the same samples, the extraction procedures and the liquid chromatographic analysis with electrochemical detection (LCED) described by Keller et al.[21] and by Ponzio and Jonnson,[22] modified after Algeri et al.,[23] were used.

For cAMP determinations, planaria specimens were homogenized in ice-cold trichloroacetic acid (5% w/v). The supernatant after 10 min 10,000 rpm centrifugation was washed 5 times with ethyl ether, after addition of HCl to a final concentration of 0.1 N, and then lyophilized. Dried samples were used for cAMP assay according to Gilman.[24] cAMP-binding protein was prepared from bovine brain according to Miyamoto et al.[25] Adenylate cyclase was determined according to Krishna et al.[26] on planaria specimens and on hydra homogenates deprived of symbiotic algae, sonified in 2 mM Tris-maleate buffer, pH 7.4, containing 2 mM EGTA. Incubation buffer contained 80 mM Tris-maleate buffer, 25 mM theophylline, and 0.6 μCi of ^{14}C-ATP in a final volume of 500 μℓ. AG 50 W-X4 (Bio-Rad®) columns and Zn-Ba precipitation were used for ^{14}C-cAMP purification.

Radioimmunoassay and Immunocytochemistry of Met-Enkephalin

The Met-enkephalin antiserum was raised in rabbits against synthetic Met-enkephalin coupled to *Helix pomatia* hemocyanin according to Childers et al.[27] For enkephalin extraction, planaria and hydra specimens were heated at 100°C in 50 mM phosphate buffer, pH 7.4, for 10 min, then homogenized and centrifuged (40,000 g for 120 min at 4°C). Radioimmunoassay was peformed as previously described.[10] The same immune serum was used for immunocytochemical observations using goat, anti-rabbit serum, coupled to FITC (Miles) for the indirect immunofluorescence technique.

GSH-Induced Feeding Behavior

For each observation 10 hydra specimens were kept in a Petri dish (3.5 cm in diameter) containing 3 mℓ M solution.[18] Feeding behavior was induced by adding GSH to a final concentration of 10^{-6} M. The effect of added drugs was studied by recording the number of specimens showing the typical feeding behavior (tentacles curling and mouth opening) for each minute, during 15 min, under a Wild binocular microscope. In each experiment one treated and one untreated group were observed. Statistical analysis was performed according to Armitage.[28]

RESULTS

Behavioral Observations

Treatment with dopaminergic agonists (L-dopa, apomorphine, and amphetamine) induced the typical screw-like hyperkinesias, whereas after treatment with reserpine and haloperidol, all the planaria specimens became motionless.[1] The hyperkinetic pattern induced by dopaminergic agonists was prevented in the planaria specimens pretreated with reserpine or haloperidol. In hydra, after similar pharmacological treatments, no gross behavioral changes were observed.

Histochemistry

In sections of untreated planaria, the Falck et al.[19] fluorescence method for catecholamines visualized, at the cephalic level, many green-yellow fluorescent cells, the greater part of which lies between the eyes and the proximal cephalic intestinal branch. After L-dopa treatment a marked increase of fluorescence intensity was observed due to increased dopamine content, whereas reserpine treatment reduced fluorescence according to the depleting action on monoamine content operated by this drug[1] (Table 1). In hydra, the same method revealed a diffuse fluorescence within all the cellular layers, more evident in the nematocysts. Neither network-like patterns, nor fluorescence of filaments or single cells were observed.

LCED Amine Assay

In whole planaria extracts, LCED analysis revealed the presence of dopamine (5 to 10 pmol per planaria), norepinephrine (0.3 to 1.2 pmol per planaria), and serotonin (10 to 30

Table 1
CATECHOLAMINE HISTOCHEMISTRY OF TREATED
PLANARIA

Treatment	Dose (mg/mℓ)	Time (hr)	Fluorescence intensity
L-Dopa	200	48	Increased
Apomorphine	0.6	3	Unchanged
Haloperidol	1	24	Unchanged
Reserpine	8	24	Reduced
Naloxone	4	24	Reduced

Note: Fluorescence intensity is referred to neuronal fluorescence due to catecholamines.

Table 2
CATECHOLAMINE LEVELS IN NORMAL AND
TREATED PLANARIA

Treatment	Dose (mg/mℓ)	Time (hr)	Dopamine (ng/planaria)	Norepinephrine (ng/planaria)
None	—	—	2.53 ± 0.03	0.22 ± 0.02
Reserpine	0.008	12	0.30 ± 0.02 (+)	0.10 ± 0.03 (+)
Haloperidol	0.001	12	1.75 ± 0.02 (+)	0.16 ± 0.01 (+)
Apomorphine	0.0006	12	2.60 ± 0.04	0.21 ± 0.01
L-Dopa	0.200	48	5.90 ± 0.4 (+)	0.35 ± 0.06 (+)

Note: (+) = $p < 0.01$ from controls by Student's t-test.

Table 3
MONOAMINE LEVELS IN NORMAL AND TREATED HYDRA

Treatment	Dose (mg/mℓ)	Time (hr)	Dopamine (ng/10 hydra)	Norepinephrine (ng/10 hydra)	Serotonin (ng/10 hydra)
None	—	—	0.40 ± 0.08	0.25 ± 0.05	1.30 ± 0.06
Reserpine	0.004	24	0.06 ± 0.04	0.10 ± 0.03	0.50 ± 0.05 (+)

Note: (+) = $p < 0.01$ from controls by Student's t-test (n = 10).

pmol per planaria). In hydra extracts, dopamine ranged from 0.2 to 0.3 pmol per hydra, norepinephrine from 0.1 to 0.3 pmol per hydra, and serotonin from 0.6 to 1 pmol per hydra. LCED analysis performed on hydra homogenates deprived of symbiotic algae gave similar results. Reserpine treatment of planaria and hydra induced a significant decrease of amine levels in both animals (see Tables 2 and 3). L-dopa treatment of planaria specimens increased monoamine concentrations, while haloperidol reduced them. (Table 2).

cAMP Levels

In planaria, apomorphine and amphetamine significantly enhanced cAMP levels, whereas haloperidol significantly reduced the nucleotide levels (see Table 4). Chronic morphine treatment significantly decreased cAMP levels in planaria, whereas naloxone significantly enhanced them. The increase of cAMP levels after naloxone treatment was dose dependent (Table 4). Naloxone-treated specimens showed a decrease of neuronal fluorescence when examined with the Falck et al.[19] method. Morphine treatment of planaria decreased motility, whereas naloxone treatment induced screw-like hyperkinesias similar to those observed upon treatment with dopaminergic agonists. Both naloxone hyperkinesias and rise in cAMP levels were undetectable in haloperidol-pretreated specimens. Spontaneous motility of hydra was unaffected by naloxone or morphine treatment.

Adenylate Cyclase Assay

A basal cyclase activity, in the presence of 5 mM Mg^{2+}, of 18 ± 3 pmol/mg protein/15 min at 37°C was present in planaria homogenates and of 27 ± 6 pmol/mg protein/15 min at 37°C in hydra homogenates. With Mg-free incubation media, adenylate cyclase activity was barely detectable. Both Mn and F ions stimulated enzymatic activity and the guanosine nucleotide analog, guanosine 5′-(beta-gamma-imino)triphosphate [Gpp(NH)p], greatly enhanced activity in both planaria and hydra. Planaria cyclase activity was stimulated by serotonin and by dopamine, and this effect was more evident in the presence of Gpp(NH)p (Table 5). Dopamine, serotonin, glutamate, or GSH did not affect hydra adenylate cyclase, both without and with Gpp(NH)p. (See Table 5.)

Met-Enkephalin-Like Immunoreactivity

Radioimmunoassay of Met-enkephalin performed on planaria extracts revealed the presence of peptide concentrations between 100 and 400 pmol/g of planaria. Immunohistochemical observations demonstrated Met-enkephalin-like immunoreactivity in neuronal perikarya and in neuropil of planaria specimens. ^{3}H-Met-enkephalin binding to the antiserum was not significantly inhibited by hydra extracts and no immunoreactivity was histochemically observed in hydra sections.

GSH-Induced Feeding Behavior

Hydra specimens treated with 10^{-6} M GSH showed, within 3 to 5 min, their typical feeding behavior, consisting of tentacles curling (so called tentacles concert), mouth opening, and inhibition of body contractile responses. Pretreatment with dopaminergic agonists significantly reduced the response to GSH administration, whereas dopaminergic blocking agents significantly increased the same response. (See Table 6.) GSH sensitivity was reduced or absent after naloxone treatment. Nalorphine (*N*-allylnormorphine) enhanced the response to GSH and, in higher doses (0.08 mg/mℓ), induced per se a typical feeding behavior, also without GSH pretreatment. Nalorphine and GSH-induced feeding responses were prevented by 10^{-4} M glutamate administration. Naloxone inhibition of feeding behavior was reverted by high GSH doses (10^{-5} M). Theophylline reduced GSH sensitivity and, after 24 hr pretreatment, the feeding response was completely inhibited. This inhibition was not reverted by high GSH doses, and living preys, given to theophylline-pretreated hydra, were captured but not swallowed. Living preys on the contrary, when given to naloxone-pretreated hydra, were swallowed, even if more slowly than in control specimens.

DISCUSSION

In planaria, behavioral observations, histochemical studies, LCED assays, cAMP levels, and adenylate cyclase activity all substantiated the presence and the relevance of dopamine as an active neurotransmitter. Ultrastructural studies, focusing on the morphological characterization of the synapsis, showed the presence of dense-cored vesicles, strictly related to the dopamine content.[29] These techniques confirmed that the neurons of planaria, situated at the first levels of the phylogenetic scale, were basically the same as those of vertebrates

Table 4
MOTILITY AND cAMP LEVELS IN NORMAL AND TREATED PLANARIA

Treatment	Dose (mg/mℓ)	Time	Motility	cAMP ± S.E.M. (pmol/planaria)
None	—	—	Normal	5.25 ± 0.15
Amphetamine	0.01	30 min	Hyperkinesia	6.75 ± 0.12 (+)
Apomorphine	0.007	24 hr	Hyperkinesia	9.05 ± 0.43 (+)
Haloperidol	0.001	24 hr	Motionless	2.65 ± 0.19 (+)
Reserpine	0.008	24 hr	Motionless	2.90 ± 0.22 (+)
Morphine	0.2	20 days	Decreased	3.19 ± 0.16 (+)
Naloxone	0.004	3 hr	Hyperkinesia	7.40 ± 0.40 (+)
	0.016	3 hr	Hyperkinesia	10.18 ± 0.35 (+)
Naloxone + haloperidol	0.008 + 0.001	3 hr + 24 hr	Motionless	2.85 ± 0.21 (+)

Note: (+) = $p < 0.001$ from controls by Student's t-test (n = 10).

Table 5
EFFECTS OF ADDED DRUGS ON ADENYLATE CYCLASE ACTIVITY IN HYDRA AND IN PLANARIA

Addition	Dose (μM)	Hydra		Planaria	
		Basal activity	$10^{-4}\ M$ Gpp(NH)p	Basal activity	$10^{-4}\ M$ Gpp(NH)p
None	—	27 ± 6	130 ± 20	18 ± 3	170 ± 12
Dopamine	10	24 ± 5	120 ± 17	22 ± 3	230 ± 15
Serotonin	10	22 ± 6	135 ± 25	28 ± 5	300 ± 18
Glutamate	10	28 ± 5	122 ± 12	n.t.	n.t.
GSH	100	30 ± 6	115 ± 11	n.t.	n.t.

Note: Values in pmoles cAMP/mg protein/15 min at 30°C.
n.t., Not tested.

Table 6
EFFECTS OF ADDED DRUGS ON GSH-INDUCED FEEDING BEHAVIOR

Pretreatment	Dose (mg/mℓ)	Feeding response
Apomorphine	0.001 — 0.005	Decreased-absent
Haloperidol	0.0007 — 0.001	Increased
Morphine	0.03—0.15	Normal
Leu-enkephalin	0.5	Normal
Met-enkephalin	0.5	Normal
Naloxone	0.003 — 0.008	Decreased
Nalorphine	0.04	Increased
Theophylline	0.02—0.05	Decreased-absent

Note: Feeding response was induced with $10^{-6}\ M$ GSH. Treatment effects were evaluated with the sequential trials method (Armitage[28]).

from the ultrastructural and molecular standpoint. Moreover, neuropharmacological investigations of the motor system of *Dugesia gonocephala* showed a striking similarity with the extrapyramidal system of mammals with the evidence of correlations between dopaminergic activity and motor performances.[1,5] As suggested for mammals by Kebabian and Calne,[30] dopamine probably acts in planaria through an interaction with a specific D1-like receptor linked to adenylate cyclase. The increase of cAMP levels after treatment with dopaminergic drugs and the fall of the same levels after treatment with dopaminergic blocking agents confirm the relationships between dopaminergic receptors and adenylate cyclase. A dopamine-stimulated adenylate cyclase has been described also in molluscs,[31] and our data support the observations of Franquinet et al.[32] on *Polycelis tenuis*.

The strong screw-like hyperkinesias, characteristic of dopaminergic hyperstimulation, and the increase of cAMP levels are the most striking features displayed by naloxone, at the same doses used by other authors in both vertebrates and invertebrates.[9] In mammals, naloxone displays an antagonistic opiate activity due to its greater affinity for the same receptor at physiological concentrations of extracellular Na.[33] In planaria, both the sequential and simultaneous administration of morphine and naloxone produce behavioral effects similar to those obtained with naloxone alone, thereby demonstrating the presence and the prevalence of the antagonistic action also in this flatworm. This peculiar effect of naloxone, evident in our observations, is possible through the serial involvement of both dopaminergic presynaptic structures and endogenous opiates. The removal by naloxone of the natural ligands from the specific receptors, by interrupting the enkephalinergic inhibitory modulation, induces the dopamine release responsible, in our model, for the hyperkinesias and the increase of cAMP levels, according to the model proposed by Costa et al.[34] in mammals. Morphine, on the contrary, induces a decrease in cAMP levels and a reduction of motor activity, due to an inhibition of dopamine release, as found also by Stefano and Catapane[9] in *Mytilus edulis*. The observed decrease of dopamine neuronal content in naloxone-treated specimens supports the hypothesis of a presynaptic activity. This possible mechanism of action is further confirmed by the absence of naloxone-induced hyperkinesias in haloperidol- or reserpine-pretreated specimens.

Immunohistochemical observations demonstrate the presence of Met-enkephalin in neuronal perikarya and in the neuropil. The high Met-enkephalin content revealed also by radioimmunoassay is in agreement with our neuropharmacological observations.[10] A similar interaction between dopaminergic and enkephalinergic neurons, through an inhibitory modulation of dopamine release operated by opiates, is also well described in molluscs.[9,35,36]

Gross hydra behavior did not respond to treatments with dopaminergic agonists and antagonists. Adenylate cyclase in this coelenterate, while present, did not respond to dopamine, serotonin, and glutamate, which in many invertebrates stimulate this enzyme activity.[37] Opiate agonists and antagonists, even in high concentrations, did not affect hydra spontaneous motility. Moreover, no enkephalin-like immunoreactivity has been detected, thus suggesting the absence of an opiate neuromodulatory mechanism in this coelenterate.

GSH-induced feeding behavior, on the contrary, can be assumed as a suitable research model in detecting pharmacological effects of both dopaminergic drugs and opiates. GSH sensitivity is reduced by naloxone. This reduction is dose dependent and can be reverted by high doses of glutathione. On the contrary, nalorphine (*N*-allylnormorphine) enhances hydra GSH sensitivity and induces per se the typical feeding response, also without GSH addition. These drugs seem to interact aspecifically with the GSH receptors, thus excluding an interaction with specific opiate receptors. Glutamate, a competitive inhibitor of feeding response,[17] also inhibits nalorphine-induced response, supporting further our hypothesis. Moreover, neither morphine nor Met- and Leu-enkephalin, even in high doses, is active on hydra behavior or on GSH sensitivity. The low specificity of the glutathione receptor is also demonstrated by the possibility to elicit a feeding response with several GSH analogs[17] and with single amino acids (arginine, isoleucine, leucine, ornithine, tryptophan, and valine).[38] Haloperidol and apomorphine, respectively, acting as dopaminergic blocking and stimulating drugs, suggest an inhibitory dopaminergic modulation on the control of feeding response. In addition, cyclic nucleotides seem to be implied in this inhibitory mechanism. Chronic theophylline administration, in fact, completely inhibits feeding response, while its action is not reverted by very high GSH doses. Besides, our data support the observations of Cobb et al.[39] on changes of cyclic nucleotides levels in hydra upon treatment with GSH. Cobb et al.[39] suggested that neither cAMP nor cGMP directly mediates GSH action, due to the time course of these changes. The lack of GSH activity on adenylate cyclase in our model supports further this interpretation.

As a general conclusion, we can assert that while in planaria neurotransmission and neurosecretion[4,40] as distinct features of the functional specialization of the nervous system have been well identified, in hydra both morphological and neurochemical data are at present not conclusive. The contrasting data on chemical synaptic structures[41] together with the strong evidence for electrical and dye coupling[42] of hydra cells are in favor of a very primitive neuronal organization. The presence and the neuronal localization of several peptides[11-15] may be indicative of a neurosecretory activity, with endocrine-like or morphogenic functions, as described for the head activator peptide.[43] Our observations on feeding response are in favor of a possible implication of a dopamine-linked control mechanism. Moreover, the interactions between GSH receptor and opiates confirm a possible evolutionary connection of neurosecretion-, neuromodulation-, or neurotransmission-linked receptors and primitive structures implied in the reception of chemical environmental signals as proposed by Lenhoff.[17] The functional and evolutive interpretations of the GSH-induced feeding response need more investigations and work is in progress to clarify these problems. At present, it is quite difficult to clarify the mechanisms operating in the feeding response of hydra. In fact, such a response could be mediated by a true chemical synaptic transmission or by the higher conduction speed of nerve cells, as compared with epithelial conduction.

REFERENCES

1. **Carolei, A., Margotta, V., and Palladini, G.,** Proposal of a new model with dopaminergic-cholinergic interactions for neuropharmacological investigations, *Neuropsychobiology*, 1, 355, 1975.
2. **Carolei, A., Margotta, V., and Palladini, G.,** Melanocyte-stimulating hormone release-inhibiting factor (MIF): lack of dopaminergic and anticholinergic activity, *Neuroendocrinology*, 23, 129, 1977.
3. **Polleri, A., Carolei, A., and Fazio, C.,** A simple invertebrate model as a tool for critical evaluation of behavioural MIF effects in neuropsychiatry, in *Proc. 2nd World Congr. Biol. Psychiat.*, Barcelona, 1978, 135.
4. **Palladini, G., Medolago-Albani, L., Margotta, V., Conforti, A., and Carolei, A.,** The pigmentary system of planaria. II. Physiology and functional morphology, *Cell Tissue Res.*, 199, 203, 1979.
5. **Venturini, G., Carolei, A., Palladini, G., Margotta, V., and Cerbo, R.,** Naloxone enhances cAMP levels in planaria, *Comp. Biochem. Physiol.*, 69C, 105, 1981.
6. **Stefano, G. B., Kream, R. M., and Zukin, S. R.,** Demonstration of stereospecific opiate binding in the nervous system of the marine mollusc *Mytilus edulis, Brain Res.*, 181, 440, 1980.
7. **Rémy, C. and Dubois, M. P.,** Localisation par immunofluorescence de peptides analogues a la alpha-endorphine dans les ganglions intra-aesophagiens du lombricide *Dendrobaena subrubiconda* Eisen, *Experientia*, 35, 137, 1979.
8. **Alumets, J., Håkanson, R., Sundler, F., and Thorell, J.,** Neuronal localisation of immunoreactive enkephalin and beta-endorphin in the earthworm, *Nature (London)*, 279, 805, 1979.
9. **Stefano, G. B. and Catapane, E. J.,** Enkephalins increase dopamine levels in the CNS of a marine mollusc, *Life Sci.*, 24, 1617, 1979.
10. **Venturini, G., Carolei, A., Palladini, G., Margotta, V., and Lauro, M. G.,** Radioimmunological and immunocytochemical demonstration of Met-enkephalin in planaria, *Comp. Biochem. Physiol.*, 74C, 23, 1983.

11. **Grimmelikhuijzen, C. J. P., Sundler, F., and Rehfeld, J. F.,** Gastrin/CCK-like immunoreactivity in the nervous system of coelenterates, *Histochemistry,* 69, 61, 1980.
12. **Grimmelikhuijzen, C. J. P., Balfe, A., Emson, P. C., Powell, D., and Sundler, F.,** Substance P-like immunoreactivity in the nervous system of hydra, *Histochemistry,* 71, 325, 1981.
13. **Grimmelikhuijzen, C. J. P., Carraway, R. E., Rökaeus, A., and Sundler, F.,** Neurotensin-like immunoreactivity in the nervous system of hydra, *Histochemistry,* 72, 199, 1981.
14. **Grimmelikhuijzen, C. J. P., Dockray, G. J., and Yanaihara, N.,** Bombesin-like immunoreactivity in the nervous system of hydra, *Histochemistry,* 73, 171, 1981.
15. **Grimmelikhuijzen, C. J. P., Dockray, G. J., and Schot, L. P. C.,** FMRFamide-like immunoreactivity in the nervous system of hydra, *Histochemistry,* 73, 499, 1982.
16. **Loomis, W. F.,** Glutathione control of the specific feeding reactions of hydra, *Ann. N.Y. Acad. Sci.,* 62, 209, 1955.
17. **Lenhoff, H. M.,** Biology and physical chemistry of feeding response of hydra, in *Biochemistry of Taste and Olfaction,* Cagan, R. H. and Kare, M. R., Eds., Academic Press, New York, 1981, 475.
18. **Lenhoff, H. M. and Brown, R. D.,** Mass culture of hydra: an improved method and its application to other aquatic invertebrates, *Lab. Anim.,* 4, 139, 1970.
19. **Falck, B., Hillarp, N. A., Thieme, G., and Torp, A.,** Fluorescence of catecholamines and related compounds condensed with formaldehyde, *J. Histochem. Cytochem.,* 10, 348, 1962.
20. **Lowry, O. H., Rosebrough, N. J., Farr, A. L., and Randall, R. J.,** Protein measurement with the Folin phenol reagent, *J. Biol. Chem.,* 193, 265, 1951.
21. **Keller, R., Oke, A., Mefford, I., and Adams, R. N.,** Liquid chromatographic analysis of catecholamines. Routine assay for regional brain mapping, *Life Sci.,* 19, 995, 1976.
22. **Ponzio, F. and Jonnson, G.,** A rapid and simple method for the determination of picogram levels of serotonin in brain tissue using liquid chromatography with electrochemical detection, *J. Neurochem.,* 32, 129, 1979.
23. **Algeri, S., Carolei, A., Ferretti, P., Gallone, C., Palladini, G., and Venturini, G.,** Effects of dopaminergic agents on monoamine levels and motor behaviour in planaria, *Comp. Biochem. Physiol.,* 74C, 27, 1983.
24. **Gilman, A. G.,** A protein binding assay for adenosine $3',5'$-cyclic monophosphate, *Proc. Natl. Acad. Sci. U.S.A.,* 67, 305, 1970.
25. **Miyamoto, E., Kuo, J. F., and Greengard, P.,** Cyclic nucleotide-dependent protein kinases. III. Purification and properties of adenosine $3',5'$-monophosphate-dependent protein kinase from bovine brain, *J. Biol. Chem.,* 244, 6395, 1969.
26. **Krishna, G., Weiss, B., and Brodie, B. B.,** A simple, sensitive method for the assay of adenyl cyclase, *J. Pharmacol. Exp. Ther.,* 163, 379, 1968.
27. **Childers, S. R., Schwartz, R., Coyle, J. T., and Snyder, S. H.,** Radioimmunoassay of enkephalins: levels of methionine and leucine-enkephalin in morphine-dependent and kainic acid lesioned rat brains, in *Advances in Biochemical Psychopharmacology,* Vol. 18, Costa, E. and Trabucchi, M., Eds., Raven Press, New York, 1978, 161.
28. **Armitage, P.,** *Sequential Medical Trials,* Blackwell Scientific, Oxford, 1960.
29. **Palladini, G., Margotta, V., Carolei, A., Chiarini, F., Del Piano, M., Lauro, G. M., Medolago-Albani, L., and Venturini, G.,** The cerebrum of *Dugesia gonocephala* s.l. Platyhelminthes, Turbellaria, Tricladida. Morphological and functional observations, *J. Hirnforsch.,* 23, 165, 1983.
30. **Kebabian, J. W. and Calne, D. B.,** Multiple receptors for dopamine, *Nature (London),* 277, 93, 1979.
31. **Stefano, G. B., Catapane, E. J., and Kream, R. M.,** Characterization of the dopamine stimulated adenylate cyclase in the pedal ganglia of *Mytilus edulis*: interactions with etorphine, beta-endorphin, DALA, and methionine enkephalin, *Cell. Mol. Neurobiology,* 1, 57, 1981.
32. **Franquinet, R., Le Moigne, A., and Hanoune, J.,** The adenylate cyclase system of planaria *Polycelis tenuis.* Activation by serotonin and guanine nucleotides, *Biochim. Biophys. Acta,* 539, 88, 1978.
33. **Perth, C. B. and Snyder, S. H.,** Opiate receptor binding of agonists and antagonists affected differentially by sodium, *Mol. Pharmacol.,* 10, 868, 1974.
34. **Costa, E., Fratta, W., Hong, J. S., Moroni, F., and Yang, Y. T.,** Interactions between enkephalinergic and other neuronal systems, in *Advances in Biochemical Psychopharmacology,* Vol. 18, Costa, E. and Trabucchi, M., Eds., Raven Press, New York, 1978, 217.
35. **Stefano, G. B., Hall, B., Makman, M. H., and Dvorkin, B.,** Opioid inhibition of dopamine release from nervous tissue of *Mytilus edulis* and *Octopus bimaculatus, Science,* 213, 928, 1981.
36. **Stefano, G. B. and Hiripi, L.,** Methionine enkephalin and morphine alter monoamine and cyclic nucleotide levels in the cerebral ganglia of the freshwater bivalve *Anodonta cygnea, Life Sci.,* 25, 291, 1979.
37. **Robinson, N. L., Cox, P. M., and Greengard, P.,** Glutamate regulates adenylate cyclase and guanylate cyclase activities in an isolated membrane preparation from insect muscle, *Nature (London),* 296, 354, 1982.

38. **Hanai, K.,** A new quantitative analysis of the feeding response in, *Hydra japonica:* stimulatory effects of amino acids in addition to reduced glutathione, *J. Comp. Physiol.,* 144, 503, 1981.

39. **Cobb, M. H., Heagy, W., Danner, J., Lenhoff, H. M., and Marshall, G. R.,** Effect of glutathione on cyclic nucleotide levels in *Hydra attenuata, Comp. Biochem. Physiol.,* 65C, 111, 1980.

40. **Lender, T.,** Endocrinologie des planaires, *Bull. Soc. Zool. Fr.,* 105, 173, 1980.

41. **Martin, S. M. and Spencer, A. N.,** Neurotransmitters in coelenterates, *Comp. Biochem. Physiol.,* 74C, 1, 1983.

42. **Fraser, S. E. and Bode, H. R.,** Epithelial cells of *Hydra* are dye coupled, *Nature (London),* 294, 356, 1981.

43. **Bodenmuller, H. and Schaller, H. C.,** Conserved amino acid sequence of a neuropeptide, the head activator, from coelenterates to humans, *Nature (London),* 293, 579, 1981.

EFFECT OF MORPHINE ON CYCLIC AMP LEVELS IN INVERTEBRATES

László Hiripi

SUMMARY

The in vivo effects of morphine and Met-enkephalin on cyclic AMP level in the ganglia of *Mytilus edulis, Anodonta cygnea,* and *Helix pomatia* were studied. The treatment of the animal with morphine and Met-enkephalin resulted in a dose-dependent decrease in the cyclic AMP level. The inhibitory effects of Met-enkephalin and morphine were antagonized by pretreating the animals with naloxone. The results suggest the decrease in the cyclic AMP concentration caused by opiates may be the result of inhibition of adenylate cyclase.

INTRODUCTION

Results of biochemical, electrophysiological, and immunocytochemical studies from various laboratories indicate the existence of an opiate mechanism in invertebrate brain similar to the one reported in mammalian neural tissues. The molluscan brain has high-affinity binding sites for opioid peptides[1] and these binding characteristics are altered during aging.[2,3] Available evidence demonstrates an interrelationship between the opioid and dopaminergic system not only in the mammalian brain, but in the molluscan ganglia too.[4-8] The general importance of cyclic nucleotides in biological system is well documented. It is well known that cyclic AMP (cAMP) is involved in receptor and agonist interaction. The hypothesis that opiate agonists regulate the adenylate cyclase in the brain has been demonstrated.[9-13] Our present results suggest that the cAMP system also is involved in the in vivo opiate effect in the molluscan ganglia.

MATERIALS AND METHODS

The effect of the opiate-like peptides on the cAMP system was demonstrated in the central nervous system of *Mytilus edulis, Anodonta cygnea,* and *Helix pomatia*. Specimens of *Mytilus edulis* were collected from Long Island Sound at Northport, N.Y. The in vivo treatments of whole animals with pharmacological agents were performed via intracardiac injection through a hole drilled in the shell above the heart. The injecting needle was supported by putty and held at a sharp angle, which facilitated penetration of the ventricle while avoiding the rectum which runs through the heart. The 100-$\mu\ell$ syringe barrel was positioned by a micromanipulator. Pharmacological agents were dissolved in artificial seawater (Instant Ocean, ASW) containing 0.1% ascorbic acid and administered during a 10-min infusion period in a volume of 100 $\mu\ell$. Vehicle-injected animals served as controls. At the appropriate time after treatments, the pedal ganglia (Pg) of eight animals were excised on ice, pooled, deproteinized in acidic ethanol (1 mℓ of 1 N HCL per 100 mℓ ethanol), and centrifuged. The supernatant was washed 5 times with 4 volumes of water-saturated ether, and the residues then processed according to the detailed procedure of the Amersham/Searle cAMP kit RIA TrK 432 for cAMP estimations.

The fresh-water mussel *Anodonta cygnea* were collected locally from Lake Balaton in Hungary. Animals used for these experiments were of the same shell lengths (10 cm). The animals were maintained in filtered, aerated, fresh running water from Lake Balaton. Several doses of either Met-enkephalin, morphine, or naloxone were administered to the right cerebral ganglion of various groups of *Anodonta cygnea* by topical applications in a volume of 100 $\mu\ell$. This method of drug administration localized the drug to the cerebral ganglia better than other routes of administration. The total 100 $\mu\ell$ volume was administered during a 20-min time period at 5 $\mu\ell$/min. The animals were then left undisturbed for various time periods.

The Amersham/Searle cAMP RIA kit Trk 432 were employed to determine cyclic nucleotide levels. Eight cerebral ganglia which weigh about 16 mg total were pooled, and deproteinized in acidic ethanol (1 mℓ of 1 N HCl per 100 mℓ ethanol). After deproteination and centrifugation, the supernatants were washed with 4 volumes of water-saturated ether 5 times and the residues then processed according to the detailed procedure outlined in the kit. Met-enkephalin was obtained from Calbiochem. Naloxone was a gift from Endo Labs, Inc. Morphine was acquired from the Hungarian Academy of Sciences.

Helix pomatia were collected locally from the land surrounding the Biological Research Institute and maintained under standard conditions. Morphine was dissolved in distilled water and administered to the animal in a volume of 50 $\mu\ell$ using subcutaneously injection just superior to the internal ganglionic structures. For the estimation of cAMP level, the subesophageal ganglia were used. The ganglia were homogenized in acified ethanol and the isolation as well as the estimation of cAMP were carried out by the same method as in the case of *Mytilus* and *Anodonta* according to Gilman.[14]

RESULTS

Mytilus edulis

The intracardiac injection of Met-enkephalin (10 μmol), D'Ala^2-Met^5-enkephalinamide (DAMA, 10 μmol), and etorphine (1 μmol) produced time-dependent decreases in the basal level of cAMP in the pedal ganglia of *Mytilus* (Table 1). Met-enkephalin and DAMA caused a 50% decrease in cAMP levels, while etorphine decreased levels by 61% at 20 min. Etorphine appeared to have a longer duration of action than Met-enkephalin and DAMA. The reduction of cAMP levels by Met-enkephalin, DAMA, β-endorphin, and etorphine was dose-dependent (Table 2). Half-maximal depression of cAMP levels occurred at approximately 0.5 μmol for etorphine and β-endorphin, and 5 μmol for DAMA and Met-enkephalin. The inhibitory effects of the previously mentioned agents on cAMP levels were antagonized by treating the animals 5 min prior to agonist administration with naloxone (Table 3). Etorphine and β-endorphin were more resistant to naloxone's action than DAMA and Met-enkephalin. The 5-μmol dose of naloxone completely blocked the activity of the pentapeptides, but only decreased the activity of etorphine and β-endorphin by 50%.

Anodonta cygnea

Met-enkephalin and morphine depressed the cAMP concentration in the cerebral ganglia by 49 and 43%, respectively (Table 4). Prior treatment of the ganglia with naloxone for 5 min before applying either Met-enkephalin or morphine prevented the changes in the cyclic nucleotide levels (Table 4).

Helix pomatia

Chronic treatment of animals with morphine decreased the cyclic AMP level in the ganglia by 30% at 18 hr and 42 hr after the last injection. At 66 hr treatment, the cAMP level increased by 50% over the control value (Table 5). This latter effect may be a manifestation of a compensatory mechanism.

DISCUSSION

The general importance of cyclic nucleotides in biological systems is well documented. Cyclic nucleotide system has been strongly implicated as a regulatory factor concerned with the synthesis, release, and receptor interaction of neurotransmitters. In the vertebrates it is

Table 1
THE IN VIVO EFFECT OF MET-ENKEPHALIN
(10 μMOL), DAMA (10 μMOL), AND ETORPHINE
(1 μMOL) ON *MYTILUS EDULIS* PEDAL GANGLIA
CYCLIC AMP LEVELS WITH RESPECT TO TIME

	Time (min)						
	0	**10**	**20**	**30**	**40**	**50**	**60**
Met-enkephalin	15.3	11.5	8.0	10.2	14.0	14.8	14.2
DAMA	15.0	10.2	7.2	11.2	14.6	14.2	15.2
Etorphine	15.0	9.8	5.8	8.0	11.8	13.6	13.8

Note: Values are given in pmoles cyclic AMP/mg protein.

Table 2
THE IN VIVO EFFECT OF MET-ENKEPHALIN,
DAMA, β-ENDORPHIN, ETORPHINE ON
CYCLIC AMP LEVELS IN THE PEDAL GANGLIA

	Injected amount of drugs (μmol)				
	0.01	**0.1**	**1.0**	**10.0**	**100.0**
Met-enkephalin	100	100	99.9	56.8	55.8
DAMA	99.5	91.7	67.5	43.4	43.4
β-Endorphin	84.8	73.8	46.2	36.5	—
Etorphine	88.9	67.7	42.3	38.6	—

Note: The animals were treated by intracardiac injection of drugs. The control level of cyclic AMP was 14.5 pmol/mg protein. The estimation of cyclic AMP was carried out 20 min after drug injection and the values are given in percent of control.

Table 3
THE EFFECTS OF NALOXONE ON THE
INHIBITORY ACTIONS OF 10.0 μMOL MET-
ENKEPHALIN, DAMA, β-ENDORPHIN, AND
ETORPHINE

	Amount of naloxone (μmol)					
	0	**0.1**	**1.0**	**5.0**	**10.0**	**100.0**
Met-enkephalin	56.8	61.5	77.6	93.0	89.5	88.1
DAMA	43.4	46.8	50.3	89.5	89.5	87.4
β-Endorphin	36.5	37.7	37.7	66.4	81.1	84.6
Etorphine	38.6	42.6	41.3	62.2	79.7	83.2

Note: Animals were treated with naloxone 5 min before the drugs were injected. Values are expressed as percent of control. The control value was 14.2 pmol and cyclic AMP/mg protein.

demonstrated that opiates regulate the intracellular cAMP level by inhibiting the adenylate cyclase enzyme.[9-13] In *Mytilus edulis, Anodonta cygnea,* and *Helix pomatia,* Met-enkephalin and morphine depressed the cAMP level in vivo. The lowering of cAMP was antagonized

Table 4
**THE EFFECT OF MET-ENKEPHALIN AND
MORPHINE ON CYCLIC AMP LEVELS IN THE
CEREBRAL GANGLIA OF *ANODONTA CYGNEA*,
AND THEIR ANTAGONISM BY NALOXONE 20
MIN AFTER DRUG APPLICATION**

	Dose (μmol)	Cyclic AMP (pmol/mg wet wt)
Control		2.6
Met-enkephalin	0.8	2.4
	1.6	2.1
	6.2	1.5
	12.4	1.3
Morphine	1.7	2.2
	3.5	2.1
	14.0	1.6
	28.0	1.5
Naloxone + Met-enkephalin	1.6 + 6.2	1.6
	3.2 + 6.2	2.2
	3.2 + 12.4	2.1
Naloxone + morphine	1.6 + 14.0	1.6
	3.2 + 14.0	2.3
	3.2 + 28.0	2.1

Note: Naloxone when used in combination with other agents was administered 5 min prior to the other agent.

Table 5
**CYCLIC AMP LEVELS IN SNAIL GANGLIA AFTER
PROLONGED TREATMENT WITH MORPHINE**

Time after last treat-ment (hr)	Control value (pmol cAMP/ganglia)	Morphine treated in % of control
3	4.0	118.4
18	4.2	69.2
42	4.3	69.0
66	4.0	152.5

Note: The snail was treated with 0.15 μmol of morphine every 6 hr for 4 days.

by naloxone. The decreasing of the cAMP level in the ganglia may be caused by inhibition
of adenylate cyclase enzyme as demonstrated in vertebrates[9-13] and in invertebrates.[8]

In addition to the effect of opiate on adenylate cyclase in *Mytilus* ganglia, opiates antagonize dopamaine-stimulated cyclic AMP increases. Dopamine is implicated as a neurotransmitter in vertebrate nervous systems.[15-20] A dopamine-sensitive adenylate cyclase usually is
associated with dopamine metabolism in various mammalian brain regions[15,16] and in invertebrate neural tissues.[21-23] Previous reports describe in part the pharmacology of a dopamine-stimulated adenylate cyclase in vivo and in vitro in nervous tissues of *M. edulis.*
Dopamine, epinine, and to lesser extent apomorphine increased cyclic AMP levels both in
vivo and in vitro. The doses required to elicit a similar response amplitude were higher in
vivo than in vitro. The agonist-induced elevations of the cAMP content were antagonized
by haloperidol, fluphenazine, chlorpromazine, and to a lesser extent BOL, thus indirectly

demonstrating the existence of a dopamine receptor mechanism.[15,16] The cAMP levels measured in these studies are attributed to adenylate cyclase activity, because phosphodiesterase activity always was inhibited by theophylline.[24] In slices of the caudate nucleus, dopamine and apomorphine increase cAMP content half-maximally at 60 and 150 μM, respectively,[25] as compared to 35 and 100 μM in *M. edulis* (in vitro). In *M. edulis*, epinine stimulation was equal to that of dopamine, as is also true in zona reticulate homogenates of the substantia nigra.[16] The similarities between dopamine-stimulated adenylate cyclase of mammals and of other invertebrates[23] thus appear also to exist for *M. edulis*. The dopamine receptor described in *M. edulis* appears to fit the criteria for the D-1 receptor described by Kebabian and Caine.[26] This receptor is linked to adenylate cyclase and is stimulated by micromolar concentrations of various dopamine agonists as well as dopamine. Independently, Malanga et al.[27] found that fluphenazine and haloperidol antagonize peripheral dopamine-stimulated adenylate cyclase activity in the gill of *M. edulis*, while phenoxybenzamine is ineffective in altering it. They also characterized the peripheral dopamine receptors in the gill of *M. edulis* as D-1.

Stereospecific opiate-binding sites are found in the crude membrane fraction of homogenates of *M. edulis*.[28] This binding is implicated with changes in dopamine levels and changes in cyclic nucleotide levels.[29] Met-enkephalin, DAMA, and FK 33 824, when applied topically to the Pg of *M. edulis*, raise dopamine levels, and are antagonised by naloxone and levallorphan. Etorphine and β-endorphin were highly potent and acted similarly to DAMA and Met-enkephalin in their actions on adenylate cyclase activity.

Data from various laboratories strongly indicate the involvement of dopamine in the actions of opiates[4,30-32] However, studies dealing with the activities of opiates on dopamine-stimulated cyclase often present conflicting results.[33-37] Wilkening et al.[38] reported that in the primate amygdala, morphine inhibits dopamine-stimulated adenylate cyclase and this action of morphine was reversible or blocked by naloxone. Walczak et al.[24] extended this to include etorphine, Met-enkephalin, and DAMA in vitro antagonists of the dopamine-stimulated adenylate cyclase. In *Mytilus* opiates inhibit dopamine-stimulated adenylate cyclase.[8] The etorphine- and β-endorphin-induced inhibition of cyclase activity increased with higher doses; however, the inhibition induced by Met-enkephalin and DAMA did not.[8] At concentrations of 10^{-9} to 10^{-7} M, the dopamine-stimulated adenylate cyclase activity decreased; however, at 10^{-7} to 10^{-5} M, the degree of inhibition decreased. At 10^{-7} M, approximately 50% of the dopamine-stimulated adenylate cyclase activity was inhibited, while at 10^{-5} M, the activity was inhibited by only 30% ($p < 0.01$; comparing the 10^{-7} M value with the 10^{-5} M of DAMA).[8] Walczak et al.[24] pointed out that various investigators[39,40] have found two classes of opiate receptors, or one receptor with different receptor conformations. Simantov and Snyder[40] reported both high and low affinity binding sites for Met-enkephalin. Based upon this, therefore, the pentapeptides may be reacting either with two receptors or with different conformations of the receptors, at different concentrations. Two varying affinity binding sites for opiates have been detected in *M. edulis*. (^{3}H)-etorphine, ^{125}I-FK-33-824, and ^{125}I-levallorphan bind stereospecifically, with high affinity, and reversibly to Pg membranes. The ligands exhibited noncooperative binding to a class of high affinity sites (Kd = 1 to 3 nM) and positive homotropic cooperative binding to a class of lower affinity sites (Kd = 6 to 11 nM).[1] The inhibition of dopamine-stimulated cyclase by pentapeptides exhibited a biphasic response, which might be accounted for by two opiate receptor-binding sites.

In *Drosophila* head membranes, an opiate receptor mechanism that is GTP-resistant has been characterized as a type 2 receptor.[41] In *M. edulis*, the order of opiate agonist potency described (etorphine > β-endorphin > DAMA > Met-enkephalin) is what is expected of this type 2 receptor. The present study demonstrated that relatively high doses of naloxone were required to block opiate actions, as is also true of the type 2 opiate receptor. Based on the pharmacological data, *M. edulis* appears to have the type 2 receptor, however, this conclusion is somewhat premature and will be the subject of future investigations.

M. edulis and other invertebrates may represent a new area for the investigation of opiate mechanisms. These animals may prove to be an invaluable aid in gaining insights into the actions and mechanisms of opiates as well as other neuroactive substances.

REFERENCES

1. **Kream, R. M., Zukin, R. S., and Stefano, G. B.,** Demonstration of two classes of opiate binding sites in the nervous tissue of the marine mollusc *Mytilus edulis, J. Biol. Chem.,* 265, 9218, 1980.
2. **Stefano, G. B.,** Decrease in the number of high affinity opiate binding sites during the aging process in *Mytilus edulis* (Bivalvia), *Cell. Mol. Neurobiol.,* 4, 343, 1984.
3. **Chapman, A., Gonzales, G., Burrowes, W. R., Assanah, P., Iannoe, B., Leung, M., and Stefano, G. B.,** Alteration in high affinity binding characteristics and levels of opioids in invertebrate ganglia during aging: evidence for a opioid compensation mechanism, 4, 143, 1984.
4. **Eidelberg, E. and Erspamer, J.,** Dopaminergic mechanisms of opiate actions in the brain, *J. Pharmacol. Exp. Ther.,* 192, 50, 1975.
5. **Biggio, G., Casn, M., Corda, M. G., DiBello, C., and Getsa, G. L.,** Stimulation of dopamine synthesis in caudate nucleus by intrastriatal enkephalins and antagonism by naloxone, *Science,* 200, 552, 1978.
6. **Deyo, S., Swift, R., and Miller, R. F.,** Morphine and endorphins modulate dopamine turnover in rat median eminence, *Proc. Natl. Acad. Sci. U.S.A.,* 77, 4341, 1979.
7. **Venturini, G., Carolei, A., Palladini, G., Margotta, V., and Cerbo, R.,** Naloxone enhances cAMP levels in planaria, *Comp. Biochem. Physiol.,* 69C, 105, 1981.
8. **Stefano, G. B.,** Comparative aspects of opioid dopamine interaction, *Cell. Mol. Neurobiol.,* 2, 167, 1982.
9. **Sharma, S. K., Klee, W. A., and Nirenberg, M.,** Opiate-dependent modulation of adenylate cyclase, *Proc. Natl. Acad. Sci. U.S.A.,* 74, 3365, 1977.
10. **Blume, A. F.,** Interaction of ligands with the opiate receptors of brain membranes: regulation by ions and nucleotides, *Proc. Natl. Acad. Sci. U.S.A.,* 75, 1713, 1978.
11. **Blume, A. F., Lichtshtein, D., and Boone, G.,** Coupling of opiate receptors to adenylate cyclase: requirement for Na^+ and GTP, *Proc. Natl. Acad. Sci. U.S.A.,* 76, 5626, 1979.
12. **Koski, G. and Klee, W. A.,** Opiates inhibit adenylate cyclase by stimulating GTP hydrolysis, *Proc. Natl. Acad. Sci. U.S.A.,* 78, 4185, 1981.
13. **Law, P. Y., Wu, J., Koehler, J. E., and Loh, H. H.,** Demonstration and characterization of opiate interaction of the striatal adenylate cyclase, *J. Neurochem.,* 36, 1834, 1981.
14. **Gilman, A. G.,** A protein binding assay for adenosine 3'5'-cyclic monophosphate, *Proc. Natl. Acad. Sci. U.S.A.,* 67, 305, 1970.
15. **Kebabian, J. W. and Greengard, P.,** Dopamine sensitive adenyl cyclase: possible role in synaptic transmission, *Science,* 174, 1346, 1979.
16. **Kebabian, J. W. and Saavedra, J. M.,** Dopamine sensitive adenylate cyclase occurs in a region of substantia nigra containing dopinergic dendrites, *Science,* 193, 683, 1976.
17. **Berry, M. S. and Cottrell, G. A.,** Dopamine: excitatory and inhibitory transmission from a giant dopamine neurone, *Nature New Biol.,* 242, 250, 1973.
18. **Osborne, N. N., Hiripi, L., and Neuhoff, V.,** The in vitro uptake of Biogenic amines by snail (*Helix pomatiai*) nervous tissue, *Biochem. Pharmacol.,* 24, 2141, 1975.
19. **Catapane, E. J., Stefano, G. B., and Aiello, E.,** Pharmacological study of the reciprocal dual innervation of the lateral ciliated gill epithelium by the CNS of *Mytilus edulis* (Bivalvia), *J. Exp. Biol.,* 74, 101, 1978.
20. **Catapane, E. J., Stefano, G. B., and Aiello, E.,** Neurophysiological correlates of the dopaminergic cilio-inhibitory mechanism of *Mytilus edulis, J. Exp. Biol.,* 83, 315, 1979.
21. **Cedar, H., Kandel, E. R., and Schwartz, J. H.,** Cyclic adenosine monophosphate in the nervous system of *Aplysia californica.* Increased synthesis in response to synaptic stimulation, *J. Gen. Physiol.,* 60, 558, 1972.
22. **Treistan, S. N. and Levitan, I. B.,** Alteration of electrical activity in molluscan neurons by cyclic nucleotide and peptide factors, *Nature (London),* 261, 62, 1976.
23. **Osborne, N. N.,** Adenosine 3', 5'-monophosphate in snail (*Helix pomatia*) nervous system: analysis of dopamine receptors, *Experientia,* 33, 917, 1977.
24. **Walczak, S. A., Wilkening, D., and Makman, M. H.,** Interaction of morphine, etorphine and enkephalins with dopamine-stimulated adenylate cyclase of monkey amygdala, *Brain Res.,* 160, 105, 1979.
25. **Forn, J., Krueger, B. K., and Greengard, P.,** Adenosine 3',5' monophosphate content in rat caudate nucleus: demonstration of dopaminergic and adrenergic receptors, *Science,* 186, 1118, 1974.

26. **Kebabian, J. W. and Caine, D. B.,** Multiple receptors for dopamine, *Nature (London),* 277, 93, 1979.
27. **Malanga, C. J., Poll, K. A., and O'Donnell, J. P.,** Agonist and antagonist effects of dopaminergic stimulation of C′AMP in the ciliated cell epithelium of the marine mussel *Mytilus edulis, Fed. Proc. Fed. Am. Soc. Exp. Biol.,* 39(3) (Abstr.), 3179, 1980.
28. **Stefano, G. B., Kream, R. M., and Zukin, R. S.,** Demonstration of stereospecific opiate binding in the nervous tissue of the marine mollusc *Mytilus edulis, Brain Res.,* 181, 440, 1980.
29. **Stefano, G. B. and Catapane, E. J.,** Enkephalins increase dopamine levels in the CNS of a marine mollusc, *Life Sci.,* 24, 1617, 1979.
30. **Saamivaara, L.,** Analgesic activity of some sympathetic drugs and their effect on morphine analgesia in rabbits, *Ann. Med. Exp. Biol. Fenn.,* 47, 180, 1976.
31. **Puri, S. K., O'Brian, J., and Lal, H.,** Potentiation of morphine-withdrawal aggression by D-amphetamine, DOPA, or apormorphine, *Pharmacologist,* 13, 280, 1971.
32. **Ary, M., Cox, B., and Lomax, P.,** Dopaminergic mechanism in precipitated withdrawal in morphine-dependent rats, *J. Pharmacol. Exp. Ther.,* 200, 271, 1977.
33. **Miller, R. J., Horn, A. S., and Iversen, L. L.,** The action of neuroleptic drugs on dopamine-stimulated adenosine cyclic 3′, 5′-monophosphate production in rat neostriatum and limbic forebrain, *Mol. Pharmacol.,* 10, 759, 1974.
34. **Carenzi, A., Guidotte, A., Revvelta, A., and Costa, E.,** Molecular mechanisms in the action of morphine and viminol (R2) on rat striatum, *J. Pharmacol. Exp. Ther.,* 194, 311, 1975.
35. **Clouet, D. H. and Iwatsubo, K.,** Dopamine-sensitive adenylate cyclase of the caudate nucleus of rats treated with morphine, *Life Sci.,* 17, 35, 1975.
36. **Iwatsubo, K. and Clouet, D. H.,** Dopamine-sensitive adeylate cyclase of the caudate nucleus of rats treated with morphine or haloperidol, *Biochem. Pharmacol.,* 14, 1499, 1975.
37. **Puri, S. K., Cochin, J., and Volicer, L.,** Effect of morphine sulfate on adenylate cyclase and phosphodiesterase activity in rat corpus striatum, *Life Sci.,* 16, 759, 1975.
38. **Wilkening, D., Mishra, R. K., and Makman, M. H.,** Effects of morphine on dopamine-stimulated adenylate cyclase and on cyclic GMP formation in primate brain amygdaloid nucleus, *Life Sci.,* 19, 1129, 1976.
39. **Lord, J. A. H., Waterfield, A. A., Hughes, J., and Kosterlitz, H. W.,** Multiple opiate receptors, in *Opiates and Endogenous Opioid Peptides,* Kosterlitz, H. W., Ed., Elsevier/North-Holland Biomedical, Amsterdam, 1976, 275.
40. **Simantov, R., and Snyder, S. H.,** Brain-pituitary opiate mechanisms, pituitary opiate receptor binding, radioimmunoassays for enkephalin and leucine enkephalin and ³H enkephalin interactions with the opiate receptor, in *Opiates and Endogenous Opioids Peptides,* Kosterlitz, H. W., Ed., Elsevier/North-Holland Biomedical Press, Amsterdam, 1976, 41.
41. **Pert, C. B. and Taylor, D. B.,** Type 1 and type 2 subclass: scheme based on GTP's differential effect on binding, in *Endogenous and Exogenous Opiate Agonists and Antagonists,* Way, E. L., Ed., Pergamon Press, Elmsford, N.Y., 1979, 87.

INTERACTION OF MONOAMINES WITH OPIOIDS AND RELATED PEPTIDES

Maynard H. Makman

SUMMARY

Significant interactions of opioid and related peptides with dopamine (DA) neurons occur in both invertebrates and vertebrates. These interactions have been studied in nervous tissue of marine molluscs, in rat striatum, and in other brain regions of rat and other mammalian species. In rat striatum, opioid receptors on DA nerve terminals as well as opioid receptors on neurons postsynaptic to DA are regulated by Na^+ and GTP. It is proposed that both populations of opioid receptors interact with N_i and potentially may be linked to and inhibit adenylate cyclase. Also, it is proposed that these interactions occur in both vertebrates and invertebrates. The opioid receptors involved appear to be primarily of the δ subtype. Post-synaptic interactions in some instances involve an adenylate cyclase that is stimulated by DA and inhibited by opioids. The release of DA presynaptically and possibly also the synthesis of DA are inhibited by opioids. These presynaptic effects on DA nerve terminals may be due to inhibition of adenylate cyclase or to regulation of other membrane transduction systems by the opioid receptor-N_i complex. In rat striatum opioid peptides also appear to influence serotonin synthesis and release. In general opioid-DA interactions appear to be more prevalent than opioid-serotonin interactions.

INTRODUCTION

Interactions occur between neurons containing opioid peptides and neurons containing monoamines in a number of vertebrates,[1-10] and recent evidence indicates that these interactions also occur in invertebrates.[11,12] In particular, opioids and dopamine (DA) have been found to influence one another in several invertebrate as well as vertebrate nervous systems. Our studies have concerned (1) the postsynaptic influence of opioids on DA-stimulated adenylate cyclase and on behavioral responses mediated by DA receptors, (2) the influence of DA on neurons containing opioid peptides, and (3) the presynaptic influence of opioids on the synthesis and the release of DA and serotonin. This paper summarizes some of our previous work as well as some recent preliminary findings concerning pre- and postsynaptic regulation of monoamine systems by opioids.

MATERIALS AND METHODS

For measurement of adenylate cyclase activity, tissues were homogenized in Tris malate buffer (2 mM, pH 7.4) and EGTA (0.8 mM) and aliquots incubated at 30°C for 2.5 min in the presence of 80 mM Tris malate buffer (pH 7.4), 5 mM theophylline, 2 mM $MgSO_4$, 0.5 mM ATP, and appropriate test agents. The reaction was terminated and aliquots assayed for cAMP by a protein-binding method previously described.[5,13] For measurement of opioid receptor binding, membrane fractions were incubated with ^{3}H-D-ala^2-Met-enkephalin (^{3}H-DALA) or other opioid radioligand in the presence and absence of dextrorphan, levorphanol, morphine, ketocyclazocine, D-Ala^2-D-Leu^5-enkephalin, or other test substance at appropriate concentrations to assess total specific binding, affinities, and receptor subtype.[5,14-16] Assays were also carried out in the presence and absence of NaCl and GTP as indicated in the text. For studies of opioid receptor-binding and Met-enkephalin levels in rat striatum, unilateral 6-hydroxydopamine lesions of the nigrostriatal pathway and kainic acid lesions of the striatum were carried out. The effectiveness of the lesions was assessed by measurement of DA and its metabolites as well as other markers (e.g., glutamic acid decarboxylase activity) in striatum and also by the rotational responses to DA agonists in vivo.[6]

Conversion of tyrosine to DA in rat striatal synaptosomes was measured as previously described using L-[1-^{14}C]-tyrosine as substrate.[17] In addition, with L-[2,6-^{3}H]-tyrosine as substrate, the amount of newly formed DA was measured following chromatographic isolation on Dowex® 50W × 4 resin. With this latter procedure, newly formed intrasynaptosomal DA and newly formed DA released into the incubation medium could be measured separately. In other studies tyrosine hydroxylase activity of intact synaptosomes was assessed by following the detritiation of L-[3,5-^{3}H]-tyrosine. For measurement of conversion of tryptophan to serotonin, experiments were carried out using L-[1-^{14}C]-tryptophan in a manner analogous to that described for study of DA formation.[18] For studies of DA release from intact tissue slices of *Octopus bimaculatus* brain, tissue was first incubated with ^{3}H-dopamine and then washed and reincubated in a continuous pefusion chamber for study of release of radioactive DA in the presence of opioids and other compounds.[11]

RESULTS AND DISCUSSION

Influence of Opioids on Postsynaptic DA Receptor Systems: Studies of Adenylate Cyclase Activity

Interactions of opioids with adenylate cyclase provides a plausible biochemical basis for at least some of the postsynaptic neuromodulatory action of opioids. In collaboration with Gardner, we have investigated the influence of opioid alkaloids and peptides on DA-stimulated adenylate cyclase activity of rhesus and *Cebus apella* monkey amygdala.[5,13] The amygdala is a brain region rich in both opioid and DA receptors. In the mammalian nervous system DA receptors constitute heterogeneous populations of sites. These include most prominently D_1 receptor sites mediating DA stimulation of adenylate cyclase and D_2 receptor sites either inhibitory or not linked to adenylate cyclase. The D_2 receptors are most readily assessed by binding of antagonist or agonist radioligands to these sites.[19] Opioids were found to profoundly inhibit the DA-stimulated adenylate cyclase of monkey amygdala. The potencies of opioids for this inhibition were similar but not identical to their affinities for ^{3}H-DALA binding sites assessed using the same assay buffer.[5] Thus, IC_{50} values for inhibition of DA-stimulated adenylate cyclase were 0.75, 3.0, and 7000 nM for etorphine, DALA, and morphine, respectively. Corresponding K_i values for ^{3}H-DALA binding sites were 0.35, 3.3, and 770 nM. Naloxone antagonized the inhibitory effects of the opioids and dextrorphan failed to inhibit the DA response. Together these data indicate that the effects were mediated via opiate receptors, most likely of the δ subtype. It should be noted that morphine was clearly less potent in the cyclase system than expected for a δ-receptor interaction. Also, while the relative affinity of morphine in the binding assay was also lower than expected for interaction with δ receptors, this was the case only when the cyclase assay conditions were used for the opioid receptor-binding studies.

Using the same assay conditions that were used for monkey amygdala, we have not observed inhibitory effects of opioids on DA-stimulated adenylate cyclase of rat amygdala or of rat or monkey striatum. However, Gentleman et al. in a study employing different assay conditions reported that certain opioids (including etorphine, dynorphin, β-endorphin, and DALA) inhibit DA-stimulated adenylate cyclase of rat striatum.[20] Recent studies of Stefano et al. indicate that in the pedal ganglia of the marine invertebrate *Mytilus edulis*, opioids interact with specific opioid receptors to inhibit DA-stimulated adenylate cyclase.[21]

On the basis of these results, it appears that opioids may antagonize the action of DA at D_1 receptors in both vertebrates and invertebrates by opioid-receptor-mediated interaction with a common adenylate cyclase system. In many cells, both stimulatory and inhibitory hormones or transmitters regulate activity of the same adenylate cyclase system.[22] Stimulation of the catalytic component of adenylate cyclase by receptor (R_s) component of adenylate cyclase is mediated by a guanyl nucleotide-binding protein, N_s, consisting of subunits a_sB.[22] Inhibition of adenylate cyclase is mediated by another guanyl nucleotide-binding protein, N_i, consisting of subunits a_iB, with the B subunit common to both N_s and N_i.[22] Binding of GTP either to N_s or to N_i permits coupling of the appropriate receptor to adenylate cyclase, transmitting the stimulatory or inhibiting signal, while at the same time decreasing the affinity of the receptor for agonist and causing dissociation of the agonist from receptor. While GTP is required for both inhibitory (R_i) and stimulatory (R_s) receptor interactions, Na^+ has been found to enhance only inhibitory (R_i) interactions with adenylate cyclase.[19,22] With respect to opioids, the adenylate cyclase system studied in most detail is that of NG 108-15 neuroblastoma-glioma hybrid cells. In those cells opioids inhibit adenylate cyclase and stimulate GTP hydrolysis in the presence of Na^+.[23] Also, the effect of opioids on NG-108-15 adenylate cyclase is abolished when N_i is ADP-ribosylated in the presence of *Bordetella pertussis* toxin.[24]

Influence of GTP and Na⁺ on Opioid Receptors Located on DA Nerve Terminals and on Intrinsic Neurons in Rat Striatum

As indicated above, guanyl nucleotides decrease the binding of agonists to receptors that interact with either N_s or N_i, while Na^+ decrease binding to receptors that interact with N_i. In this connection we have found that agonist binding to DA D_2 receptors in striatum and retina is inhibited by both Na^+ and GTP.[19] Agonist binding to opioid receptors is also inhibited by both Na^+ and GTP. In collaboration with Zukin, we carried out studies showing this to be the case for several brain regions including striatum.[25] In lesion studies carried out together with Gardner and Zukin, we found that in rat striatum, about one third of opioid receptors is located on DA nerve terminals, and most or all of the remaining receptors are on neurons with cell bodies intrinsic to the striatum.[26] In more recent studies, carried out by Hirschhorn in our laboratory, we have found that both populations of opioid receptors, whether measured with ³H-DALA, ³H-D-Ala²-D-Leu⁵-enkephalin, or ³H-ketocyclazocine, are primarily of the δ subtype (unpublished studies). Furthermore, as shown in Table 1, both populations of receptors are regulated by Na^+ and by GTP. These results are in accord with the possibility that presynaptic (DA nerve terminal) opioid receptors as well as opioid receptors postsynaptic to DA nerve terminals in striatum are linked to N_i and hence may be coupled to adenylate cyclase. Alternatively, it is possible that receptors may be regulated by Na^+ and GTP without participation of N_i, or it may be that N_i is involved but linked to a system other than adenylate cyclase.

Behavioral Evidence for Pre- and Postsynaptic Regulation of Dopaminergic Systems by Opioids

In collaboration with Hirschhorn et al., we investigated the effects of morphine and naloxone on rotational behavior in the rat.[6] Following chronic unilateral lesion of the nigrostriatal DA neurons, drugs which release DA, such as amphetamine, produce rotations towards the side of the lesion, presumably by releasing DA in the intact striatum. Direct-acting DA drugs such as apomorphine produce rotation away from the lesioned side by a predominant stimulation of "supersensitive" DA receptors in the striatum on the lesioned side. Morphine was found to inhibit both amphetamine- and apomorphine-induced rotation, and naloxone was found to enhance apomorphine-induced rotation. The effects of morphine and naloxone on apomorphine-induced rotation suggest that opiates act at a postsynaptic site in this system. The effect of morphine on amphetamine-induced rotation could also be accounted for by a postsynaptic action, although an additional inhibitory effect on amphetamine-induced release of DA is also possible (see below for further discussion). It is important to note that DA agonist-induced rotational behavior is believed to be mediated primarily if

Table 1

INFLUENCE OF NaCl AND GTP ON BINDING OF 1 n*M* ^{3}H-
DALA TO DOPAMINE NERVE TERMINALS (PRESYNAPTIC
SITES) IN RAT STRIATUM[a]

	fmol ^{3}H-DALA bound/mg protein	
Additions to binding assay	Total	Presynaptic[b]
Control	16.4 ± 0.7	5.5 ± 0.6
50 μ*M* GTP	6.9 ± 0.8	1.5 ± 0.3
100 m*M* NaCl	3.5 ± 0.4	3.3 ± 0.7
100 m*M* NaCl + 50 μ*M* GTP	1.0 ± 0.2	0.4 ± 0.2

[a]   Binding assays were carried out as described in the text with 1 n*M* ^{3}H-DALA (D-
Ala^2-Met-enkephalin) as radioligand.

[b]   Sites retained after intrastriatal kainic acid lesion but lost after 6-OH-dopamine lesion
of the nigrostriatal pathway. Values are means ± S.E.M. for at least 5 separate
experiments for each condition (lesion) studied.

not exclusively by D_2 receptors. Therefore the inhibitory effect of morphine on this behavior
would not readily be accounted for by interaction of opioid and D_1 receptors with a *common*
adenylate cyclase system, nor would one predict opioid- and D_2-receptor interaction with a
common cyclase. The detailed basis for this opioid-DA antagonism remains to be elucidated.

Influence of Lesion of the Nigrostriatal DA Pathway on Striatal Met-Enkephalin Concentration

The influence of naloxone on apomorphine-induced rotational behavior in lesioned rats,
described above, supports a role of endogenous opioids in this dopaminergic system. In
collaboration with Thal et al., we have carried out additional studies indicating that these
may be reciprocal DA opioid interactions in striatum.[7] Following unilateral nigral 6-hydroxy-
dopamine lesion, striatal Met-enkephalin on the lesioned side increased to 245% of that on
the nonlesioned side. This increase was evident only after a lag period of 7 days and the
increase was maintained for at least 2 months after lesion. By contrast, there was no change
in striatal concentration of two other peptides, somatostatin and vasoactive intestinal peptide.
The time course for the increase in striatal Met-enkephalin after lesion followed that for
increase in D_1 and D_2 receptors. Thus, following destruction of nigrostriatal DA neurons,
there gradually develops not only an increase in DA receptors, but also an alteration in the
enkephalinergic system. It is proposed that enkephalins exert a toxic inhibitory influence on
apomorphine activation of DA receptors in the lesioned striatum. Naloxone enhances apomorphine-induced rotation by blocking this effect of enkephalin.

Presynaptic Regulation of Biogenic Amines by Opioids

In studies carried out with Stefano, we have investigated the influence of opioids on the
release of biogenic amines from nervous tissue of the marine invertebrates, *Mytilus edulis*
and *Octopus bimaculatus*. It had been previously found that opioid agonists increase the
concentration of DA in certain ganglia of the molluscs *M. edulis* and *Anodonta cygnea* and
the snail *Helix pomatia*.[27-29] This effect was antagonized by naloxone. In addition, opiate
receptor-binding sites were detected and characterized in *M. edulis*.[30] We found that opioids
inhibited the release of labeled DA from nervous tissue incubated in vitro, in both *M. edulis*
and *O. bimaculatus*.[11] As shown in Table 2, for *O. bimaculatus*, morphine and DALA
suppressed potassium-stimulated release of ^{3}H-dopamine from tissue slices of supraesophageal lobe tissue as well as from subdissected vertical, basal, and frontal lobes of brain. Also,

Table 2
INFLUENCE OF OPIOIDS ON K⁺-EVOKED DOPAMINE RELEASE FROM SUPRAESOPHAGEAL LOBE TISSUE OF *OCTOPUS BIMACULATUS*[a]

Drug additions[b]	Percent of total radioactivity released in 3 min
Supraesophageal lobes (pooled)	
Control (basal)	1.7
KCl	14.5
KCl + morphine	2.7
KCl + morphine + naloxone	20.6
KCl + DALA	1.7
KCl + DALA + naloxone	14.8
Vertical lobe	
Control (basal)	1.7
KCl	15.7
KCl + DALA	1.4
KCl + DALA + naloxone	16.1
Basal lobe	
Control	1.1
KCl	12.0
KCl + DALA	0.9
Frontal lobe	
Control	1.2
KCl	11.3
KCl + DALA (1 μM)	3.5
KCl + DALA	1.1

[a]   Conditions for labeling *O. bimaculatus* brain tissue with ³H-DA and for studying ³H-DA release in vitro were as described in Reference 11. Values represent means of 3 to 5 experiments with S.E.M. less than ±1.5% for all values.

[b]   Unless otherwise indicated, concentrations were 25 μM for morphine, 50 μM for naloxone, 10 μM for DALA, and 50 mM for (elevated) KCl.

the inhibitory effects of the opioids were not seen in the presence of naloxone. In other studies, release of ³H-serotonin was not altered by opioids. Comparable results were obtained with *M. edulis*. The potassium-stimulated release of labeled DA from the pedal, cerebral, and visceral ganglia of *M. edulis* was inhibited by DALA and morphine, and this effect was also prevented by naloxone. In contrast, release of serotonin was not affected and ³H-norepinephrine release was inhibited to only a slight extent by high concentrations of opioids in pedal and visceral ganglia. It should be noted that enkephalin-like immunoreactivity has been detected in proximity to DA-containing structures in the pedal ganglia of *M. edulis*.[12] In addition, lesion studies suggest the presence of opiate-binding sites on DA neurons in the pedal ganglia.[31]

In the mammalian nervous system, the synthesis and the release of catecholamines appear to be regulated by several distinct but interrelated processes; also, multiple regulatory processes are involved in serotonin synthesis and release.[32,33] An important regulatory process is the activation of tyrosine hydroxylase (TH), the rate-determining step in catecholamine biosynthesis, due to phosphorylation of TH by cAMP-dependent protein kinase.[34] In collaboration with Katz et al., we have found that forskolin, an activator of adenylate cyclase in intact cells as well as membrane and soluble preparations, stimulates the conversion of tyrosine to DA in slice and synaptosomal preparations of striatal and other CNS tissue (Table

Table 3
INFLUENCE OF KCl, AMPHETAMINE, COCAINE, AND
PHENCYCLIDINE (PCP) ON DOPAMINE FORMATION AND
RELEASE BY STRIATAL SYNAPTOSOMES[a]

| | pmol DA formed/mg protein/hr | | |
Additions	Synaptosomes	Medium	Total
I			
Basal (control)	12	20	32
Amphetamine (1 μM)	21	52	73 (+128%)
KCl (50 mM)	30	55	85 (+166%)
Forskolin (10 μM	37	39	76 (+138%)
Forskolin + amphetamine	48	103	152 (+375%)
Forskolin + KCl	54	91	145 (+353%)
Forskolin + KCl + amphetamine	54	158	212 (+563%)
II			
Basal (control)	48	40	88
Forskolin (10 μM)	116	56	172 (+95%)
Forskolin + cocaine (1 μM)	136	120	256 (+191%)
Forskolin + cocaine (10 μM)	140	140	280 (+241%)
Forskolin + PCP (1 μM)	136	92	218 (+152%)
Forskolin + PCP (10 μM)	124	132	256 (+191%)

[a]   Synaptosomes were incubated in triplicate in each experiment with 2,6-^{3}H-tyrosine as substrate. Other conditions were as described in the text.

3).[17] This effect of forskolin appeared to be mediated by cAMP. More recently, we have found the forskolin effect to involve a stable activation of TH. Since the activity of forskolin-stimulated cyclase can still be regulated via receptor-mediated processes, particularly those involving N_i, it appeared to us that an effect of forskolin on DA biosynthesis might serve as a specific, though indirect, probe for studying that component of adenylate cyclase that is present in catecholaminergic neurons. We have also now extended this concept to the serotonergic system, since we find that forskolin as well as certain cAMP analogs stimulate serotonin formation in synaptosomal preparations of rat striatum and substantia nigra (Table 4).[18]

Using this approach, we have been able to demonstrate striking inhibitory effects of adenosine A_1 agonists on striatal DA formation (unpublished studies). Also in preliminary studies we have found that opioids inhibit both DA and serotonin formation in striatal synaptosomes. Some of these data are shown in Table 5. Met-enkephalin and Met-enkephalin-Arg-Phe (YGGFMRF) were the most potent peptides tested thus far. Also, etorphine but not morphine was inhibitory at the concentrations tested. These findings suggest the possibility that a particular subtype of presynaptic opioid receptors in striatum may be involved in these effects. A relative insensitivity to morphine was also found in studies of opioid effects on DA-stimulated adenylate cyclase of amygdala as described earlier. Additional studies will be required to elucidate the receptor characteristics as well as the mechanism and possible physiological relevance of the opioid effects on striatal DA and serotonin formation. Evidence has been reported indicating that serotonergic neurons modulate the functional activity of enkephalinergic neurons in rat striatum. Evidence for reciprocal DA-enkephalinergic interactions in striatum has been referred to earlier.

One possible mode of action of the opioids in the striatum is to produce a primary inhibition of monoamine release, resulting in intrasynaptosomal accumulation of DA or serotonin, in turn producing allosteric feedback inhibition of TH or tryptophan hydroxylase. Addition of

Table 4
CONVERSION OF TRYPTOPHAN TO SEROTONIN BY RAT STRIATAL AND SUBSTANTIA NIGRA SYNAPTOSOMES

Additions	pmol/mg Protein/hr[a]	Relative rate[b]
Striatum		
Basal (control)	55 Δ	
8-Thiomethyl-cyclic AMP (1 mM)	86 (31)	1.6
Forskolin (50 μM)	88 (33)	1.6
Forskolin + amphetamine (10 μM)	86 (31)	1.6
Forskolin + KCl (50 mM)	84 (29)	1.5
Forskolin + cocaine (10 μM)	114 (59)	2.1
Forskolin + phencyclidine (10 μM)	101 (46)	1.8
Substantia nigra		
Basal (control)	86 Δ	
Forskolin (50 μM)	130 (44)	1.5

[a]   Values are means for at least 3 separate experiments each with triplicate determination for each condition studied.

[b]   Relative rate is the ratio of serotonin formation from carboxyl-labeled ^{14}C-tryptophan in the presence to that in the absence of the additions as indicated.

Table 5
INFLUENCE OF OPIOIDS ON DOPAMINE AND SEROTONIN FORMATION[a] BY RAT STRIATAL SYNAPTOSOMES IN THE PRESENCE OF 10 μM FORSKOLIN

	Relative rate[b]		
	Dopamine formation		Serotonin formation
Additions (10 μM)	(5 mM KCl)	(50 mM KCl)	(5 mM KCl)
Met-enkephalin	0.38[c]	0.43[c]	0.52[c]
Met-enkephalin-Arg-Phe	0.37[c]	0.45[c]	—
Etorphine	0.70[c]	0.68[c]	0.75[c]
Morphine	0.93	0.91	0.91

[a]   Synaptosomal suspensions were incubated with either carboxyl-labeled ^{14}C-tyrosine (1 μM) or carboxyl-labeled ^{14}C-tryptophan (1 μM) as substrate. For DA formation, incubations were carried out in normal (5 mM KCl) medium or medium with elevated (50 mM) KCl. Other conditions were as described in the text.

[b]   Relative rate is the ratio of DA or serotonin formation in the presence of opioid to that in the absence of opioid.

[c] $p < 0.01$.

DA to the incubation medium indeed leads to inhibition of synaptosomal DA formation and addition of serotonin leads to inhibition of serotonin formation. On the other hand, agents such as amphetamine, KCl, cocaine, and phencyclidine stimulate DA synthesis, entirely or in large part by releasing DA and hence decreasing allosteric feedback inhibition of TH by DA (Table 3). A similar mechanism may be responsible for stimulation of serotonin synthesis by cocaine (Table 4). Another possible mode of action of opioids is to inhibit the presynaptic adenylate cyclase that is stimulated by forskolin. Either mode of action could conceivably involve the GTP regulatory protein N_i. These studies provide a basis for elucidation of the mechanisms involved. It seems highly likely that similar processes for presynaptic regulation of monoamines by opioids will prove to be present in invertebrate as well as in vertebrate species.[35]

ACKNOWLEDGMENTS

These studies were supported by USPHS Grants NS09469 and AG-00374.

REFERENCES

1. **Pickel, V., Joh, T. H., Reis, D. J., Leeman, S. E., and Miller, R. J.,** Localization of substance P and enkephalin in axon terminals related to dendrites of catecholaminergic neurons, *Brain Res,,* 160, 387, 1979.
2. **Pollard, H., Llorens-Cortes, C., and Schwartz, J. C.,** Enkephalin receptors on dopaminergic neurons in rat striatum, *Nature (London),* 268, 165, 1977.
3. **Palmer, M. R., Seiger, A., Hoffer, B. J., and Olson, L.,** Modulatory interactions between enkephalin and catecholamines: anatomical and physiological substrates, *Fed. Proc. Fed. Am. Soc. Exp. Biol.,* 42, 2934, 1983.
4. **Costa, E., Guidotti, A., Hanbauer, I., and Saiani, L.,** Modulation of nicotinic receptor function by opiate recognition sites highly selective for met^5-enkephalin [arg^{6}phe^7], *Fed. Proc. Fed. Am. Soc. Exp. Biol.,* 42, 2946, 1983.
5. **Walczak, S. A., Makman, M. H., and Gardner, E. L.,** Acetylmethadol metabolites influence opiate receptors and adenylate cyclase in amygdala, *Eur. J. Pharmacol.,* 72, 343, 1981.
6. **Hirschhorn, I. D., Hittner, D., Gardner, E. L., Cubells, J., and Makman, M. H.,** Evidence for a role of endogenous opioids in the nigrostriatal system: influence of naloxone and morphine on nigrostriatal dopaminergic supersensitivity, *Brain Res.,* 270, 109, 1983.
7. **Thal, L. J., Sharpless, N. S., Hirschhorn, I. D., Horowitz, S. G., and Makman, M. H.,** Striatal Met-enkephalin concentration increases following nigrostriatal denervation, *Biochem. Pharmacol.,* 32, 3297, 1983.
8. **Dellavedova, L., Parenti, M., Tirone, F., and Groppetti, A.,** Interactions between serotonergic and enkephalinergic neurons in rat striatum and hypothalamus, *Eur. J. Pharmacol.,* 85, 29, 1982.
9. **Edley, S. M. and Herkenham, M.,** Comparative development of striatal opiate receptors and dopamine revealed by autoradiography and histofluorescence, *Brain Res.,* 305, 27, 1984.
10. **Murrin, C., Coyle, J. T., and Kuhar, M. J.,** Striatal opiate receptors: pre- and postsynaptic localization, *Life Sci.,* 21, 1175, 1980.
11. **Stefano, G. B., Hall, B., Makman, M. H., and Dvorkin, B.,** Opioid inhibition of dopamine release from nervous tissue of *Mytilus edulis* and *Octopus bimaculatus, Science,* 213, 928, 1981.
12. **Stefano, G. B. and Martin, R.,** Enkephalin-like immunoreactivity in the pedal ganglion of *Mytilus edulis* (Bivalvia) and its proximity to dopamine containing structure, *Cell Tissue Res.,* 230, 147, 1983.
13. **Walczak, S. A., Wilkening, D., and Makman, M. H.,** Interaction of morphine, etorphine and enkephalins with dopamine-stimulated adenylate cyclase of monkey amygdala, *Brain Res.,* 160, 105, 1979.
14. **Snyder, S. H. and Goodman, R. R.,** Multiple neurotransmitter receptors, *J. Neurochem.,* 35, 5, 1980.
15. **Chang, K. J., Hazum, E., and Cuatrecasas, P.,** Possible role of distinct morphine and enkephalin receptors in mediating actions of benzomorphan drugs (putative K and δ agonists), *Proc. Natl. Acad. Sci. U.S.A.,* 77, 4469, 1980.
16. **Chavkin, C., James, I. F., and Goldstein, A.,** Dynorphin is a specific endogenous ligand of the K opioid receptor, *Science,* 215, 413, 1982.
17. **Katz, I. R., Smith, D., and Makman, M. H.,** Forskolin stimulates the conversion of tyrosine to dopamine in catecholaminergic neural tissue, *Brain Res.,* 264, 173, 1983.
18. **Smith, D. M. and Makman, M. H.,** Serotonin formation in rat striatum and substantia nigra: stimulation by forskolin and 8-thiomethyl cyclic AMP, *Soc. Neurosci. Abstr.,* 10, 219, 1984.
19. **Makman, M. H., Dvorkin, B., and Klein, P. N.,** Sodium ion modulates D_2 receptor characteristics of dopamine agonist and antagonist binding sites in striatum and retina, *Proc. Natl. Acad. Sci. U.S.A.,* 79, 4212, 1982.
20. **Gentleman, S., Parenti, M., Neff, N. H., and Pert, C. B.,** Inhibition of dopamine-activated adenylate cyclase and dopamine binding by opiate receptors in rat striatum, *Cell. Mol. Neurobiol.,* 3, 17, 1983.
21. **Stefano, G. B., Catapane, E. J., and Kream, R. M.,** Characterization of the dopamine stimulated adenylate cyclase in the pedal ganglia of *Mytilus edulis:* interactions with etorphine, β-endorphins, DALA and methionine enkephalin, *Cell. Mol. Neurobiol.,* 1, 57, 1981.
22. **Gilman, A.,** Guanine nucleotide-binding regulatory proteins and dual control of adenylate cyclase, *J. Clin. Invest.,* 73, 1, 1984.

23. **Koski, G. and Klee, W. A.,** Opiates inhibit adenylate cyclase by stimulating GTP hydrolysis, *Proc. Natl. Acad. Sci. U.S.A.,* 78, 4185, 1981.
24. **Hsia, J. A., Moss, J., Hewlett, E. L., and Vaughan, M.,** ADP-ribosylation of adenylate cyclase by pertussis toxin, *J. Biol. Chem.,* 259, 1086, 1984.
25. **Zukin, R. S., Walczak, S., and Makman, M. H.,** GTP modulation of opiate receptors in regions of rat brain and possible mechanism of GTP action, *Brain Res.,* 180, 218, 1980.
26. **Gardner, E. L., Zukin, R. S., and Makman, M. H.,** Modulation of opiate receptor binding in striatum and amygdala by selective mesencephalic lesions, *Brain Res.,* 194, 232, 1980.
27. **Stefano, G. B. and Catapane, E. J.,** Enkephalins increase dopamine levels in the CNS of a marine mollusc, *Life Sci.,* 24, 1917, 1979.
28. **Stefano, G. B. and Hiripi, L.,** Methionine enkephalin and morphine alter monoamine and cyclic nucleotide levels in cerebral ganglia of the freshwater bivalvia *Anodonta cygnea, Life Sci.,* 25, 291, 1979.
29. **Osborne, N. N. and Neuhoff, V.,** Are there opiate receptors in the invertebrates?, *Pharm. Pharmocol.,* 31, 481, 1979.
30. **Kream, R. M., Zukin, R. S., and Stefano, G. B.,** Demonstration of two classes of opiate binding sites in the nervous tissue of the marine mollusc, *Mytilus edulis, J. Biol. Chem.,* 255, 9218, 1980.
31. **Stefano, G. B., Zukin, R. S., and Kream, R. M.,** Evidence of the presynaptic localization of a high affinity opiate binding site on dopamine neurons in the pedal ganglia of *Mytilus edulis, J. Pharmacol. Exp. Ther.,* 222, 759, 1982.
32. **Goldstein, M., Bronaugh, R. L., Ebstein, B., and Roberge, C.,** Stimulation of tyrosine hydroxylase activity by cyclic AMP in synaptosomes and in soluble striatal enzyme preparations, *Brain Res.,* 189, 563, 1976.
33. **Mandell, A. J.,** Redundant mechanisms regulating brain tyrosine and tryptophan hydroxylase, *Ann. Rev. Pharmacol. Toxicol.,* 18, 461, 1978.
34. **Joh, T. H., Park, D. H., and Reis, D. J.,** Direct phosphorylation of brain tyrosine hydroxylase by cyclic AMP-dependent protein kinase: mechanism of enzyme activation, *Proc. Natl. Acad. Sci. U.S.A.,* 75, 4744, 1978.
35. **Makman, M. H. and Stefano, G. B.,** Marine mussels and cephalopods as models for study of neuronal aging, in *Selected Invertebrate Models for Aging Research,* Mitchell, D. and Johnson, T. E., Eds., CRC Press, Boca Raton, Fla., 1984, 165.

Index

INDEX

A

ABRM, see Anterior byssus retractor muscle
Absorption
 liquid-phase, 217
 solid-phase, 107, 110—111
Acetylation, 40
Acetylcholine, 12, 27, 29, 104, 165, 177
Acetylcholinesterase, 177
Acid extract of pedal ganglia of mollusc, 41
Acnidaria, 103
ACTH, see Corticotropin
Address sequences, 93
Adenosine A, 268
Adenylate cyclase, 35, 245, 255, 258, 259, 264—
 266
 assay for, 249
Adipokinetic hormone, 105
Adrenal glands, 4, 16
 bovine, 28
 chromaffin cells in, 16, 46
 denervated, see Denervated adrenal gland
 medulla of, 3, 27
 proenkephalin gene expression in, 3—12
Adrenocorticotropin, see Corticotropin
D-Ala^2-D-Leu^5-enkephalin, 263
^{3}H-D-Ala^2-Met-enkephalin (^{3}H-DALA), 170, 264,
 265, 267
D'-Ala^2-Met^5-enkephalinamide (DAMA), 256, 259
Albumin
 bovine serum, 246
 human serum, 156
N-Allylnormorphine, 249, 251
Altered thermal choices, 140
Amidation, 136
Amines, see specific types
Amino acids, see also specific types, 129
 cyclic, 136
 determination of, 44—45
γ-Aminobutyric acid, 165—177
Aminopeptidase, 34
Amphetamines, 247, 265, 269
Amphimorphic peptides, 96—98
Analgesic peptide, 39
Ancestral peptide, 93
Angiotensin, 105
Animals, see also specific animals, 94
Anodonta cygnea, see Mussel
Anopla, 161
Anterior byssus retractor muscle (ABRM), 98
Anterior vena cava, 50
Anthopleura elegantissima, see Sea anemone
Anthozoa, see also specific types, 103, 104, 106
Antibodies, see also specific types
 characterization of, 214
 FMRFa, 50, 53
 Leu-enkephalin, 96
 synthetic FMRFa, 50

Antibovine pancreatic polypeptide (aPP), 107, 225
Anticalcitonin, 225
Anticholecystokinin (aCCK), 225, 227
Anticorticotropin, 225, 231
Anticorticotropin-releasing factor (aCRF), 225
Antidopamine (aDA), 225, 232, 233
Antidymorphin (aDyn), 225, 237
Anti-β-endorphin, 2225 227
Antienkephalin, 237
AntiFMRFamide, 111, 129, 225
Antigastrin (aGas), 225, 229, 233
Antigenicity, 75
Antigens, see also specific types
 in leech, 175—177
 substance P, 175
Antiglucagon, 225
Antiglucose-dependent insulmotropic peptide
 (aGIP), 225
Antihistamine, 225, 232
Antiinsulin, 225
Anti-Leu-enkephalin (aLE), 225, 232
Antimelanotropin (aMSH), 232, 233, 237
Anti-α-melanotropin, 225
Anti-Met-enkephalin (aME), 129, 225, 232, 237,
 246, 247
Anti-α-neoendorphin (aneoEnd), 225
Antineuropeptide Y, 110
Antineurophysin, 225, 227
Antioctopamine, 225
Antioxytocin, 225
Antipancreatic polypeptide, 107
Antisecretin, 225
Antisera, see also specific types, 156, 213, 225
 crossreactivity of with peptides, 103
 evaluation of, 214
 generation of, 66
Antiserotonin (a5HT), 225, 232, 233
Antisomatostatin, 225
Antisubstance P, 175, 225
Antivasoactive intestinal polypeptide (aVIP), 225
Antivasopressin (aVP), 225
Antivasotocin (aVT), 225, 229, 231, 233
Anti-YGGFMRFamide, 129
Antineurophysin II, 225
Aplysia sp., see Sea hare
Apomorphine, 247, 252, 258, 259, 265
Arg-Leu-enkephalin, 8
Arg-Met-enkephalin, 8
Arg-Met-enkephalin-Arg-Gly-Leu, 8
Arg-Met-enkephalin-Arg-Phe, 8
Arg-Phe-Gly, 136
Artemia salina (brine shrimp), 246
Ascidians, see also specific types, 171
Aspartate, 129
Assays, see also specific types; Bioassays; Radioim-
 munoassay, 133
 adenylate cyclase, 249
 LCED, 247—249